Analysis and Design Principles
of MEMS Devices

Analysis and Design Principles of MEMS Devices

By

Minhang Bao
Department of Electronic Engineering
Fudan University
Shanghai, China

2005

ELSEVIER

Amsterdam • Boston • Heidelberg • London • New York • Oxford • Paris
San Diego • San Francisco • Singapore • Sydney • Tokyo

ELSEVIER B.V.
Radarweg 29
P.O. Box 211, 1000 AE Amsterdam
The Netherlands

ELSEVIER Inc.
525 B Street, Suite 1900
San Diego, CA 92101-4495
USA

ELSEVIER Ltd
The Boulevard, Langford Lane
Kidlington, Oxford OX5 1GB
UK

ELSEVIER Ltd
84 Theobalds Road
London WC1X 8RR
UK

First edition 2005

Library of Congress Cataloging in Publication Data
A catalog record is available from the Library of Congress.

British Library Cataloguing in Publication Data
A catalogue record is available from the British Library.

ISBN: 0 444 51616 6

♾ The paper used in this publication meets the requirements of ANSI/NISO Z39.48-1992 (Permanence of Paper).

Transferred to Digital Printing in 2011

The Author Dedicates This Volume

**To the Centennial of
Fudan University, China
(1905 — 2005)**

献给　复旦大学百年华诞
(1905 — 2005)

Preface

I. MEMS, the acronym of "Micro Electro Mechanical Systems", is generally considered as micro systems consisting of micro mechanical sensors, actuators and micro electronic circuits. As microelectronics is a well-developed technology, the research and development of MEMS is concentrated on the research and development of micro mechanical sensors and actuators, or micro mechanical transducers (Note that the word "transducer" is often used as a synonym of "sensor". However, it is sometimes read as "sensors and actuators".). Or, we can say that micro mechanical sensors and actuators are the basic devices for MEMS. Therefore, this book studies the analysis and design principles of the basic devices used to construct MEMS, the micro mechanical sensors and actuators (or the micro mechanical transducers).

MEMS have been under development for several decades. A variety of MEMS devices have been developed and many of them also mass-produced. However, the analysis and design principles on MEMS devices are still widely scattered in the literature over many different disciplines. This situation is not convenient for the researchers and engineers who are developing MEMS devices or for students who are studying in this interesting area.

As a professor and researcher in this area, I have been interested in providing as much information as possible on the essential theories for the analysis and design of MEMS in a book. My previous book with Elsevier (entitled "Micro Mechanical Transducers — Pressure Sensors, Accelerometers and Gyroscopes", Elsevier 2000, Vol. 8 in Handbook of Sensors and Actuators, edited by Simon Middelhoek) is mostly devoted to this purpose. This book is modified, updated and supplemented with more materials so that readers can study the theories more profoundly and apply them to practical applications more easily.

A new feature is the inclusion of problems at the end of each chapter (except for the first chapter), with answers to the problems provided at the end of the book. Most of the problems are practical examples of the analysis and design of MEMS devices. Therefore, this book is an advanced level textbook especially useful for graduate students who are studying MEMS as well as for researchers and engineers who are developing new MEMS devices.

II. The theories essential for the analysis and design of MEMS devices can be basically divided into three categories: the dynamics of micro mechanical structures (the micro-dynamics), the sensing schemes (micro sensors) and microelectronics theory.

Microdynamics is the main content of this book and is treated more systematically and thoroughly than ever before. The coverage includes the mechanics of beam and diaphragm structures, air damping and its effect on the motion of mechanical structures, the electrostatic driving of mechanical structures and the effects of electrical excitation on capacitive sensing.

The book explains the three most widely used sensing schemes: frequency sensing, capacitive sensing and piezoresistive sensing. The reason for choosing these three schemes is as follows. Frequency sensing is naturally related to mechanical structures and it benefits are its high resolution, high stability and digital compatibility. Capacitive sensing is the sensing scheme most compatible with MEMS devices and the most popular sensing scheme currently used in MEMS. Piezoresistive sensing was the first sensing scheme that led to the development of micro mechanical transducers since the 1960s and is still widely used in MEMS devices today.

An explanation of microelectronics is not included in this book as there are already many other books that cover this topic in detail. Therefore, following an introductory chapter, the content of this book is summarized in the table below.

Microdynamics	Sensing schemes
Basic mechanics (Chap. 2)	Frequency sensing (Chap. 2)
Air damping (Chap. 3)	Capacitive sensing (Chap.5)
Electrostatic driving (Chap. 4)	Piezoresisitve sensing (Chap.6)

III. It is my privilege to dedicate this book to Fudan University, where I have been studying and working since 1956. It would be a great opportunity to express my sincere gratitude to the Shanghai Institute of Microsystem and Information Technologies (SIMIT), where I have been teaching and working as a consultant professor since early 1990s. I would also like to take this opportunity to thank all my colleagues and students in Fudan and SIMIT. The excellent academic environment in the two institutes makes it possible for me to do research and teaching on MEMS, to accumulate all the knowledge needed for the book and to write books in a prolonged time period using the necessary resources.

Finally, I would like to express my sincere gratitude to my wife, Huiran, for her constant encouragement, patience and help with my continuous writing over many years.

Minhang Bao
Fudan University, Shanghai, China
November 26, 2004

Contents

Summary

MEMS have been under development for several decades. A variety of MEMS devices have been developed and many of them also mass-produced. However, the analysis and design principles for MEMS devices are still widely scattered in the literature over many different disciplines. This situation is not convenient for the researchers and engineers who are developing MEMS devices or for students who are studying in this interesting area.

This book studies the analysis and design principles of MEMS devices (the micro mechanical sensors and actuators) more systematically than ever before. The theories essential for the analysis and design of MEMS includes the dynamics of micro mechanical structures (the microdynamics) and the sensing schemes of micro sensors, in addition to microelectronics theory.

Microdynamics is the main content of this book and is treated more systematically and thoroughly than ever before. The microdynamics chapters cover the mechanics of beam and diaphragm structures, air damping and its effect on the motion of mechanical structures, the electrostatic driving of mechanical structures and the effects of electrical excitation on capacitive sensing.

For sensing schemes, the three most widely used methods are considered. These are frequency sensing, capacitive sensing and piezoresistive sensing. Frequency sensing features high resolution, high stability and digital compatibility. Capacitive sensing is the sensing scheme most compatible with MEMS devices and the most popular sensing scheme currently used in MEMS. Piezoresistive sensing was the first sensing scheme that led to the development of micro mechanical transducers since the 1960s and is still widely used in MEMS devices today.

A problem section is included at the end of each chapter (except for chapter 1), with the answers to the problems provided at the end of the book. Most of the problems deal with practical examples of analysis and design of MEMS devices. Therefore, this book is ideally suited as an advanced level textbook, especially useful for graduate students who are studying MEMS, but also a useful reference work for researchers and engineers who are developing new MEMS devices.

Chapter 1

Introduction to MEMS Devices

MEMS, the acronym of "Micro Electro Mechanical Systems", are generally considered as micro systems consisting of micro mechanical sensors, actuators and micro electronic circuits. As microelectronics is a well-developed technology, the research and development of MEMS is concentrated on the research and development of micro mechanical sensors and actuators, or micro mechanical transducers. Or, we can say that micro mechanical sensors and actuators are the basic devices for MEMS (Note that the word "transducer" is often used as a synonym of "sensor". However, it is sometimes read as "sensors and actuators".) Therefore, this book studies the analysis and design principles of the basic devices of MEMS, the micro mechanical sensors and actuators (or the micro mechanical transducers).

Before we can study the fundamental theories of micro mechanical transducers, readers are expected to have some basic knowledge on micro transducers. This chapter is to give the readers some material to get familiar with some important MEMS devices.

Generally, the scope of MEMS devices is very broad. As MEMS devices are the offspring of microelectronics and micro mechanical technologies, the most important MEMS devices are sensors using piezoresistive, capacitive and vibration sensing schemes, and the actuators using electrostatic driving. This introduction and the theories studied in the following chapters will be restricted in these respects.

It is assumed that the readers have had enough knowledge on microelectronics and micro machining technologies so that they can understand the schematic drawings of the device structures and the processing steps for the devices in this chapter.

For those who do not have enough knowledge on microelectronics and micro mechanical technologies, reading of relevant material [1,2,3] prior to this study is advisable. On the contrary, for those who are quite familiar with MEMS devices already, they may skip this chapter and proceed to chapter 2 directly.

§1.1. Piezoresistive Pressure Sensor

§1.1.1. Piezoresistance Effect

(1) Metal Strain Gauge

The metal strain gauge was discovered long before the discovery of the piezoresistance effect in semiconductors and has still been widely used for mechanical transducers in industries. Due to the affinity between metal strain gauge sensors and piezoresistive sensors, the metal strain gauge is first briefly introduced in this section.

Consider a metal filament with a circular cross section. If the radius of the cross section is r, the length of the filament is l and the resistivity of the material is ρ, the resistance of the filament is $R=\rho l/\pi r^2$. If the filament is stretched by an external force F, the stress in the filament is $T=F/\pi r^2$ and the strain (the relative elongation) in the filament is $\varepsilon \equiv \Delta l/l = T/E$, where E is the Young's Modulus of the material. As metal is usually a polycrystalline material with a fine grain structure, its mechanical and electrical properties are isotropic. Thus, the relative change in resistance caused by the force is

$$\frac{\Delta R}{R} = \frac{\Delta l}{l} - \frac{2\Delta r}{r} + \frac{\Delta \rho}{\rho}$$

As well known in mechanics, the longitudinal stretch of a filament is always accompanied with it a lateral contraction, i.e. $\Delta r/r = -\nu(\Delta l/l)$, where ν is the Poisson ratio of the material. For most materials, ν has a value of about 0.3. Thus we have

$$\frac{\Delta R}{R} = (1+2\nu)\varepsilon + \frac{\Delta \rho}{\rho}$$

Usually, the relative change of resistivity, $\Delta \rho / \rho$, is a function of stress/strain and is expressed as $\pi T = \pi E \varepsilon$, where π is the piezoresistive coefficient of the material. Therefore, we have

$$\frac{\Delta R}{R} = (1+2\nu+\pi E)\varepsilon \equiv G\varepsilon$$

where G, the relative change in resistance per unit strain, is referred to as the gauge factor, or, G factor, of the filament.

As π is negligible for metal materials, the gauge factor is just a little larger than unity, i.e., $G \approx 1+2\nu = 1.5 \sim 2.0$. As the maximum strain of the gauge is in the order of 10^{-3}, the relative change of the resistance is also in the order of 10^{-3}.

(2) Strain Gauge Sensors

Strain gauges can be made of metal foil as well as metal wire. Fig. 1.1.1(a) and (b) schematically show a force sensor using four metal foil strain gauges as sensing elements. The strain gauges $R_1 \sim R_4$ are glue-bonded onto the metal beam supported by a metal cylinder. The strain gauges are interconnected into a Wheatstone bridge as shown in Figure 1.1.1(c). As the output of the bridge, ΔV, is proportional to the force F, a force sensor is formed.

As the strain gauges are bonded onto the mechanical structure that is fabricated by conventional machining technique, the strain gauge force sensor is referred to as a conventional mechanical sensor.

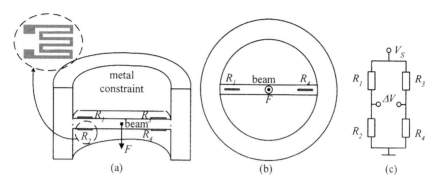

Fig. 1.1.1. A strain gauge force sensor (a) cross-sectional view (insert: the serpentine pattern of a metal foil strain gauge); (b) top-view; (c) Wheatstone bridge

(3) Piezoresistance Effect

S. C. Smith discovered in 1954 that the change in resistance of a strained (or stressed) germanium or silicon filament was much larger than that of a metal strain gauge [4]. He discovered that the change in resistance was mainly caused by the change in resistivity of the material instead of the change of the geometric dimensions.

Therefore, the effect is referred to as the piezoresistance effect. Though piezoresistance effect is quite similar to the strain gauge effect of metal but the difference between them is significant

(a) The effect of metal strain gauge is caused by the geometric deformation of the resistor, whereas piezoresistance effect is caused by the change in resistivity of the material. As a result, the effect of piezoresistance can be two orders of magnitude larger than that of the metal strain gauge effect.

(b) The effect of metal strain gauge is isotropic whereas the effect of piezoresistance is generally anisotropic. This means that, $(\Delta R/R)$, (π) and (T) are tensors and the relation among them, $(\Delta R/R) \cong (\pi)(T)$, is a tensor equation. Further discussion on the piezoresistive tensors of silicon will be given in Chapter 6.

With the discovery of piezoresistance effect, people realized that the large effect of resistance change would have important applications in sensors, especially in mechanical sensors dominated at that time by metal strain gauges. Soon a semiconductor piezoresistive sensing element (a semiconductor strain gauge or a piezoresistor) was developed and found applications in mechanical sensors. Though a semiconductor strain gauge has much higher sensitivity than a metal one, the metal strain gauge matches the metal substrate better and shows better stability than a semiconductor strain gauge. Therefore, semiconductor strain gauge has not been successful in replacing the metal strain gauge sensors.

§1.1.2. Piezoresistive Pressure Transducer

(1) Silicon as a Mechanical Material

With the rapid development of silicon technology in the 1960s, the excellent mechanical properties of silicon material were understood in addition to its versatile electrical and thermal properties. Therefore, efforts were made to make use of silicon as a mechanical material. First, piezoresistors were made by selective diffusion into a silicon wafer by planar process so that the silicon wafer could be used as a mechanical diaphragm with integrated piezoresistors on it. When the diaphragm was bonded to a glass or metal constraint by epoxy as schematically shown in Fig. 1.1.2, a pressure transducer was formed [5]. For the first time, silicon was used as both the mechanical and the sensing material in a sensor.

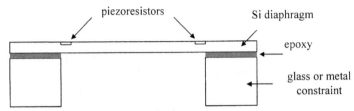

Fig. 1.1.2. A pressure transducer using a silicon diaphragm

Significant progress was made around 1970 when the silicon substrate with sensing elements on it was shaped by mechanical drilling to form an integrated diaphragm-constraint complex [6]. A pressure transducer formed by this technique is schematically shown in Fig. 1.1.3. As the whole structure is made out of bulk silicon material, the mechanical performance of the device is greatly improved.

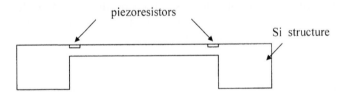

Fig. 1.1.3. A silicon piezoresistive pressure transducer made out of a bulk silicon material

(2) Micro Mechanical Pressure Sensor

The processing technology for the silicon structure shown in Fig. 1.1.3 was further improved in the mid-1970s when anisotropic etching technology was used for silicon pressure transducers. By using anisotropic etching, silicon pressure transducers could be batch-fabricated with the planar process steps, such as oxidation, diffusion, photolithography, etc., originally developed for silicon transistors and integrated circuits [7,8]. Thanks to the advanced micro fabrication technology, the dimensions of pressure sensors are reduced significantly. The silicon "chip" of the pressure sensor is schematically shown in Fig. 1.1.4.

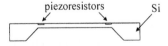

Fig. 1.1.4. A micro mechanical silicon piezoresistive pressure transducer

As the dimensions of the mechanical structures processed can be controlled to an accuracy in the order of microns, the technologies are often referred to as micro machining technologies. Some basic features of the pressure transducer shown in Fig. 1.1.5 are:

(i) The silicon material can be used for the mechanical structure as well as sensing elements and electronic components;

(ii) The mechanical structure of silicon can be batch-fabricated by micromachining technologies.

Numerous improvements and innovations have been made for silicon pressure transducers in the years followed and the production volume of silicon pressure transducers has been growing steadily since then, but the basic features remain unchanged even today, i.e., the structure of the silicon pressure transducer shown in Fig. 1.1.4 is basically a typical structure of a present-day silicon pressure transducer. For further understanding the working principles of the device, more detailed illustrations are given in Fig. 1.1.5.

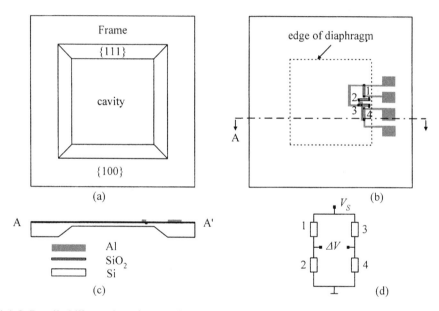

Fig. 1.1.5. Detailed illustrations for a typical silicon pressure transducer
(a) bottom view; (b) top view; (c) cross-sectional view; (d) Wheatstone bridge

Fig. 1.1.5(a) is a bottom view of the sensor chip. With etching masks (SiO_2 or Si_3N_4) on the frame region, the cavity is etched using an anisotropic etchant (typically, aqueous KOH). As the wafer is (100)-oriented and the edges of the etching windows are along the <110> directions of the silicon crystal, the sidewalls of the cavity are {111} planes. As the angle between the {111} sidewalls and the (100) bottom is 54.74°, the bottom of the cavity (the diaphragm) is smaller than the etching window by $\sqrt{2}d$, where d is the depth of the cavity. Therefore, the size of the diaphragm can be controlled by the size of the etching window and the etching depth.

Fig. 1.1.5(b) shows the front side of the chip (for clarity, the area of the diaphragm is delineated by dotted lines). Schematically shown on the right-hand side of the diaphragm are four piezoresistors formed by boron diffusion or ion implantation on an n-type silicon diaphragm. The cross section along the AA' line is shown in Fig. 1.1.5(c). The four piezoresistors are interconnected using Al lines to form a Wheatstone bridge as schematically shown in Fig. 1.1.5(d). The four bonding pads on frame are used to connect the Wheatstone bridge to outside by wire bonding for power supply and signal output.

When a pressure is applied on top of the sensor chip, i.e., the pressure on the front is larger than that on the back of the diaphragm by Δp, the silicon diaphragm will bend downwards. This causes stress in the diaphragm. The stress, in turn, causes a change in resistance of the resistors. For a typical layout as shown in Fig. 1.1.5(b), the resistance of R_2 and R_3 goes up and that of R_1 and R_4 goes down by ΔR. Thus, the output of the Wheatstone bridge is

$$\Delta V = \frac{\Delta R}{R} V_S$$

As ΔR and, in turn, ΔV is in proportion to the pressure difference Δp, the device is a pressure transducer.

Generally, the output of the bridge can be more than 100 mV with good linearity for a 5V power supply. This usually determines the nominal maximum operation range of silicon pressure sensors. The operation range of a pressure transducer is usually from 1 kPa to 50 MPa determined basically by the size and the thickness of the diaphragm.

(3) Gauge-, Absolute- and Differential- Pressure Transducers
Before the sensor chips can be put into practical applications, they must be encapsulated. An example of the encapsulated pressure transducer is shown in Fig. 1.1.6. The silicon sensor chip is first electrostatically bonded to a glass plate with a hole at center. The chip-glass combination is then mounted onto the base of a package (also with a hole at center). Then, pads on the chip are electrically connected to the leads of the package by wire-bonding. A cap with an input port is then hermetically sealed to the base of the package so that the pressure to be measured can be applied through the input port of the cap.

To meet different application needs, pressure transducers can be packaged to form three types of devices. They are gauge pressure transducers (GP), absolute pressure transducers (AP) and differential pressure transducers (DP). The pressure transducer shown in Fig. 1.1.6 is a gauge pressure transducer. This kind of pressure transducer measures a pressure measurand with reference to the environmental pressure around the device.

An absolute pressure transducer measures a pressure measurand with reference to an absolute reference pressure. The reference pressure is usually a vacuum so that it is not temperature dependent.

A differential pressure transducer measures the difference between two pressure measurands. Therefore, a differential pressure transducer has two input ports for the two pressures. Generally, the sensor chips for the three types of pressure transducers are similar, but the packaging techniques are different.

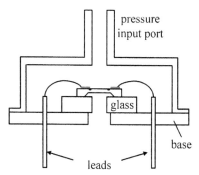

Fig. 1.1.6. An encapsulated silicon pressure transducer (a gauge pressure transducer)

According to the brief description given above, the analysis and design of a piezoresistive pressure transducer are based on much theory, including the stress distribution in a diaphragm caused by pressure and the piezoresistive effect of silicon. The stress distribution in a diaphragm will be discussed in Chapter 2 and the piezoresistance effect in silicon will be discussed in Chapter 6. Readers interested in the detail designs of pressure transducer are referred to reference [9].

§1.2. Piezoresistive Accelerometer

The success of silicon piezoresistive pressure transducers stimulated research for a silicon piezoresistive accelerometer. However, since the mechanical structure of an accelerometer is more difficult to fabricate than that of a piezoresistive pressure transducer, silicon accelerometers were not successfully developed until the late 1970s. The first silicon accelerometer prototype was developed in Stanford University in 1979 [10]. The silicon sensor chip was a cantilever beam-mass structure made out of a single crystalline silicon wafer. Fig. 1.2.1(a) is the top view of the sensor chip and Fig. 1.2.1(b) shows the cross-sectional view of the chip. As shown in the figures, the accelerometer sensor chip consists of a frame, a seismic mass and a thin beam. Two resistors are formed by selective diffusion on the chip, one on the beam and the other on the frame, and they are connected by metal tracks to form a half bridge.

The basic working principle of the sensor is rather simple. When the device (the frame of the structure) is moving with an acceleration normal to the chip plane, the inertial force on the mass forces the beam to bend and causes stress in the beam. The stress, in turn, causes a change in resistance of the resistor on the beam. Therefore, the output of the half bridge is directly proportional to the acceleration.

Though the structure of the accelerometer has been improved since its introduction for better performances, such as higher sensitivity, lower cross-axis sensitivity, etc., the basic principle of the piezoresistive accelerometer remains the same even today.

However, the silicon piezoresistive accelerometer was not successful in mass production and industrial applications until the late 1980s. The main reason is that the dynamic characteristics of the piezoresistive accelerometer were not well controlled to meet application requirements. The early devices were plagued with two main problems:

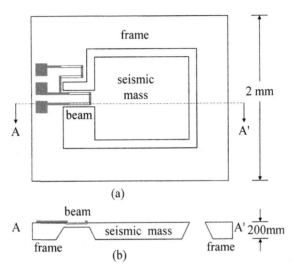

Fig. 1.2.1. Schematic structure of the first silicon piezoresistive accelerometer
(a) top view; (b) cross-sectional view of the chip

(a) Difficulty in Over-range Protection

For high sensitivity, the beam should be narrow and thin and the mass should be large. In addition to the difficulty in processing such a fragile structure, the beam can easily break due to a drop on to the ground or an inadvertent crash of the device onto a hard surface. For example, the operation range of an accelerometer could be only a few gravitational accelerations (a few *g*'s), but the drop of a packaged device from one meter high onto a cement floor could cause an acceleration over 1000 *g*. As the critical displacement, which may cause the beam to break, is only a few microns, the stop movement mechanism for over-range protection used in conventional mechanical accelerometers can hardly be used for a silicon accelerometer.

(b) Difficulty in Damping Control

An acceleration signal cannot be constant over a long period; it is either a short-lived fast varying signal or an alternating signal. Therefore, one of the most important characteristics of an accelerometer is its dynamic performance. An ideal accelerometer should have a uniform sensitivity for a large frequency bandwidth.

As the beam-mass structure is basically a spring-mass system (a second order system), it has a natural vibration frequency of $\omega_o = \sqrt{k/m}$, where k is the spring constant of the beam and m is the effective mass of the seismic mass. Therefore, the amplitude-frequency relation of the system has a peak near ω_o, as shown by curve *a* in Fig. 1.2.2. The ratio of the peak amplitude (when it exists) to the amplitude at low frequency is referred to as "Quality factor" (also "*Q* factor" or simply "*Q*") of the mechanical system.

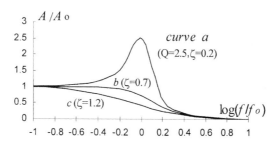

Fig. 1.2.2. The amplitude-frequency relations for three damping conditions: (curve *a*) slight damping; (curve *b*) optimum damping; (curve *c*) over-damping

The Q factor of a system is determined by the energy dissipation mechanism (the damping effect) of the system. The larger the damping effect, the smaller the Q factor. For slight damping (an under-damping state) Q is large and the peak is high as shown by curve *a* in Fig. 1.2.2. For very large damping (an over-damping state) the amplitude drops significantly at frequencies well below ω_o, as shown by curve *c*. For an optimum damping, the amplitude-frequency relationship has a maximum bandwidth as shown by curve *b* in the figure. Note that in the conditions of optimum damping and over-damping, the damping ratio ζ is used instead of the Q factor as there is no resonant peak in the curves. Readers are referred to Chapter 2 for relations between Q and ζ.

For an isolated micromachined silicon beam-mass structure, the Q factor in air can be as high as a few hundred. For example, the Q factor for the accelerometer by Roylance [10] is reported to be 109. When the device is immersed in liquids with viscosity higher than that of air (such as acetone, methanol, deionized water, and isopropyl), the Q factors are reduced but still not small enough. Though the method of viscous damping in oil has been used for conventional mechanical accelerometers, it is not an easy technique to control for micromechanical silicon accelerometers.

The above mentioned difficulties were solved using micromechanical technologies in the 1980s. Fig. 1.2.3 shows a typical micro structure for a practical silicon piezoresistive accelerometer [11].

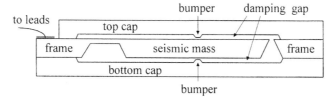

Fig. 1.2.3. Micro accelerometer with squeeze-film air damping and over-range stop mechanisms

The sensor chip with a beam-mass structure is sandwiched between a top cap and a bottom cap. Both top and bottom caps are made of silicon with etched cavities. The air in the cavities provides squeeze-film air damping [12]. As the damping force is inversely proportional to the cube of the cavity depth, a depth of 20-40 μm can usually provide a optimum damping for the accelerometer. Squeeze-film air damping shows much better performance than viscous damping by a liquid. The damping force is easy to control and the temperature coefficient is very small.

The bumpers in the cavities have a height that is a few microns smaller than the cavity depth so that there are small gaps between the seismic mass and the bumper tips. The bumpers restrict the displacement of the seismic mass, thus the beam will not be damaged by acceleration much larger than the nominal operation range of the accelerometer [13]. The typical distance of the gaps for over-range protection is 5-10 μm.

Due to the success in damping control and over-range protection, piezoresistive silicon accelerometers have been mass-produced since the late 1980s.

According to the brief description above, the analysis and design of a piezoresistive accelerometer has a diverse theoretical basis, including the piezoresistive effect of silicon, the stress distribution in a beam by the inertial force on a seismic mass, the forced vibration of a mechanical system and squeeze-film air damping. The stress distribution in a beam and the forced vibration of a beam-mass system will be analyzed in Chapter 2, the squeeze-film air damping will be studied in Chapter 3, the piezoresistive effect of silicon will be discussed in Chapter 6. Readers who are interested in the detailed design of piezoresistive accelerometers are referred to Chapter 7 of reference [9].

§1.3. Capacitive Pressure Sensor, Accelerometer and Microphone

Piezoresistive pressure transducers and accelerometers have been well developed and widely used in industry and consumer applications due to their high sensitivity, high linearity, ease of signal processing and so on. However, piezoresistive sensors do have their drawbacks:

(i) As piezoresistive sensors are stress sensitive, their performances are closely related to the packaging technologies. Any mechanical stress or thermal mismatch between the sensor chip and the packaging material may cause an offset voltage and a temperature drift of the output.

(ii) The temperature coefficient of piezoresistance is of the order of 10^{-3}, which is quite large. For many applications, the piezoresistive transducer must be carefully calibrated and the temperature coefficient of the device must be compensated for. That is usually very time consuming and costly.

(iii) Piezoresistance is susceptible to junction leakage and surface contamination. These factors may cause significant stability problems.

Due to these problems, efforts have been made in the development of micromechanical sensors using capacitive sensing schemes. As capacitive sensors are related to the mechanical properties of the material, they are more stable than piezoresistive sensors. However, capacitive sensors are inherently nonlinear and the measurement of small capacitance is very difficult due to the parasitic and stray effects, and the electromagnetic interference from the environment. Therefore, progress in this development was not significant until the late 1980s.

With the advance of micromachining technologies and the integration of micromechanical structures with integrated circuits for signal processing, progress on capacitive accelerometers has been significant since the early 1990s.

§1.3.1. Capacitive Pressure Transducer

A basic capacitive pressure transducer is schematically shown in Fig. 1.3.1 [14,15,16]. A silicon chip with a thin diaphragm is hermetically bonded to a glass plate face-to-face by electrostatic bonding. The silicon diaphragm serves as one electrode of the capacitor and the other electrode is a metal film on the glass under the cavity. The distance between the two electrodes is a few microns, which is controlled by the shallow cavity etch on the front surface (the surface being bonded to the glass) of the silicon wafer. The diaphragm is formed by a deep etch from the back of the silicon wafer. The chamber formed by the silicon chip and the glass is usually evacuated so that an absolute pressure transducer is formed.

The operation principle of the device is very straightforward. As the displacement of the diaphragm is dependent on the applied pressure P, the capacitance is a function of the pressure. However, there are many difficulties associated with a capacitive pressure transducer:

(i) It is difficult to interconnect the metal electrode out of the vacuum chamber without jeopardizing the hermetic seal;

(ii) The vacuum in the cavity is not easy to maintain due to the degassing of the material in the process of electro-static bonding and after encapsulation.

(iii) The measurement of capacitance and the linearization of the output signal are difficult and it is further complicated by the parasitic and stray capacitance, especially when the signal conditioning circuit is not monolithically integrated with the sensor chip.

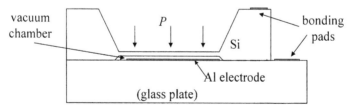

Fig. 1.3.1. A schematic view of a basic capacitive pressure transducer

It is easy to understand that these difficulties are somewhat inherent to a capacitive pressure transducer. Though there have been many improvements in the designs and fabrication technologies, few capacitive pressure transducers have been accepted by the market for wide applications.

§1.3.2. Capacitive Accelerometer

The situation of capacitive accelerometers is quite contrary to that of capacitive pressure transducers, though the development of capacitive accelerometers were not quite successful until the 1990s. The capacitive accelerometers based on bulk micromachining technologies feature high resolution and have found some specific applications. The capacitive accelerometers based on surface micromachining allow low cost mass production and have been widely used in applications such as air-bag control systems in automobiles.

Fig. 1.3.2 shows a capacitive accelerometer made by bulk micromachining technology [17]. A silicon cantilever beam-mass structure is sandwiched between two Pyrex glass plates by electrostatic bonding. The seismic mass acts as a movable electrode and two fixed electrodes are on the glass plates. Movement of the seismic mass due to an acceleration of the device changes the capacitances between the mass and the two fixed electrodes. The capacitance difference between the two capacitors (ΔC) can be used as a measure of the acceleration. The symmetric design and differential sensing reduces the effect of thermal mismatch to a minimum and linearizes the ΔC~acceleration relationship.

To reduce the effects of parasitic and stray capacitance and to achieve high sensitivity, the capacitance should be large. This can be achieved by making the gap between the movable electrode and the fixed electrodes small (2~5 μm). However, a small gap results in large squeeze-film air damping. This may reduce the bandwidth of the accelerometer significantly. To reduce the air damping effect, through-holes have to be opened in the mass. The air damping effect can also be reduced by a force-balanced measurement scheme.

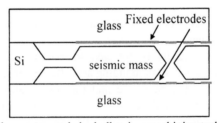

Fig. 1.3.2. A capacitive accelerometer made by bulk micromachining technology

The capacitive accelerometer described above is fabricated using bulk micro machining technology, so it is relatively bulky and can hardly be integrated monolithically with integrated circuit for signal processing. However, if a capacitive accelerometer is fabricated using surface micro machining technology, the device will be much smaller in size and can easily be integrated with a signal processing circuit monolithically. For a typical surface micromachined capacitive accelerometer integrated with signal processing circuits, readers are referred to §1.7.1 some more details.

§1.3.3. Capacitive Microphone

A microphone is a sensor for detecting sound. It is a device capable of transforming a pressure signal of sound into an electric voltage signal. Therefore, a microphone is in essence a very sensitive pressure sensor for an alternative pressure in sound frequency (several tens of Hz to several tens of kHz).

There have been many conventional types of microphone in the market, including electromagnetic, piezoelectric, capacitive and electret types. Among them, the most widely used ones are electromagnetic microphone (featuring high fidelity) and electret microphones (featuring low cost and miniaturization).

With the wide applications of portable audio-video systems and mobile phones, miniaturized microphone with the capability of being integrated into a printed circuit board (PCB) or even with an IC are highly demanded. These demands stimulate the research and development efforts for micro mechanical microphones. Almost all types of microphone have been investigated using micro machining technologies and prototypes have been demonstrated. However, among them, the micro machined capacitive microphones are the most successful ones. Some of them are being marketed [18].

The structure of a micro-mechanical capacitive microphone is schematically shown in Fig, 1.3.3 [19]. Basically, the microphone is a capacitor consisting of two electrodes, one on the thin diaphragm and the other on the back-plate as shown in the figure. The diaphragm is very thin so that it is very compliant, but, the back plate is usually much thicker than the diaphragm so that it is quite stiff.

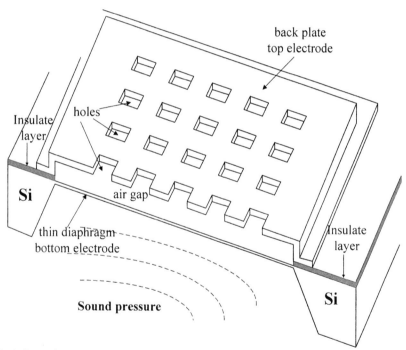

Fig. 1.3.3. Schematic structure of a micro-machined capacitive microphone

With the air pressure of sound on the diaphragm, the diaphragm vibrates with the sound pressure and the capacitance between the two electrodes changes. If the change in capacitance is picked up by a circuitry sensitive to the capacitance, the sound signal is detected.

As a sound often contains a band of frequencies (normally, 20 Hz to 20 kHz for music and 50Hz to 5kHz for voice), a microphone should be able to work in this bandwidth with a uniform sensitivity. The bandwidth of a microphone is determined by the stiffness of the diaphragm and the squeeze-film air damping effect between the two electrodes.

For large bandwidth, the diaphragm should be stiff enough. This will reduce the sensitivity of the microphone. On the other hand, for high sensitivity, the gap between the two electrodes should be small. However, when the gap goes small the squeeze-film air damping effect between the two electrodes becomes large. The bandwidth of the microphone can be reduced by the damping effect significantly, as air damping reduces the sensitivity at high frequency more significantly. For both high sensitivity and low damping effect, back plate is often perforated, as shown in Fig. 1.3.4. The holes on the back plate allow air to flow through them when the diaphragm is vibrating so that the air damping effect can be reduced.

According to the brief description above, the analysis and design of capacitive accelerometers has a diverse theoretical basis, including the displacement of seismic mass under an inertial force, the forced vibration of a mechanical system, air damping, capacitive sensing and the electrostatic force related to capacitive sensing. The displacement of a mass and the forced vibration of a beam-mass system will be studied in Chapter 2, air damping will be discussed in Chapter 3 and capacitive sensing will be discussed in Chapter 5.

§1.4. Resonant Sensor and Vibratory Gyroscope

§1.4.1. Resonant Sensor

Similar to the vibration frequency of a violin string, the vibration frequency of a mechanical beam (or a diaphragm) changes with the extent it is stretched. This phenomenon has been used in conventional mechanical sensors since the 1970s. These sensors are classified as resonant sensors, which are noted for their high stability and high resolution. The stability is determined only by the mechanical properties of the resonator material, which can be very stable. In addition, the frequency output is highly insensitive to possible environmental interference and is easy to interface with computer systems.

Resonant sensors based on microstructures have been developed in recent years [20]. They are getting increasingly interesting due to the advantages of high resolution, long term stability, immunity to environmental interference and ease of interfacing to computers.

A micro resonant pressure transducer is schematically shown in Fig. 1.4.1 [21]. The sensor chip consists of a frame (400 μm thick), a thin diaphragm (100 μm thick and 1200 μm by 1200 μm in size) and a double-clamped micro beam (600 μm long, 40 μm wide and 6 μm thick). The micro beam resides over a small cavity located at the center of the silicon diaphragm. A driving electrode is formed by diffusion beneath the micro beam on the bottom of the cavity. The chip is bonded to a glass and encapsulated as shown in Fig. 1.4.2.

The micro beam is driven into vibration by an alternating force on the beam by an alternating electrical signal on the driving electrode. The operating voltage level is typically a few mV due to the high Q factor of the micro beam structure in a vacuum. The vibration of the beam can be sensed by the piezoresistive sensing elements on the top surface near one end of the beam. The signal output from the sensing elements is sent to a circuit consisting of amplifiers, detectors, automatic gain control, etc. The output of the circuit is fed back to the driver to control the driving signal level for maintaining a close-loop electro-mechanical oscillation with constant amplitude at the mechanical resonant frequency of the beam.

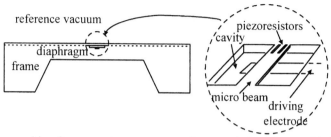

Fig. 1.4.1. The sensor chip of a resonant pressure transducer made by micromachining technology

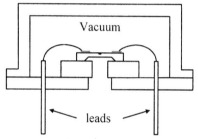

Fig. 1.4.2. The packaged resonant pressure transducer

With an applied pressure on the back of the diaphragm, the silicon diaphragm deforms in direct proportion to the pressure. The deformation, in turn, stretches the micro beam and changes its resonant frequency. Therefore, the oscillation frequency of the micro electro mechanical system can be used as a measure of the pressure.

For the device developed, the resonant frequency of the beam at zero pressure is about 110 kHz. Typical variation of resonant frequency for the full operation pressure range (100 kPa) is about 15 kHz. In practice, resonant sensors can attain measurement accuracy more than ten times better than piezoresistive or capacitive sensors. However, it is important to maintain a good vacuum environment for high Q factor.

§1.4.2. Vibratory Gyroscope

A gyroscope is a sensor for measuring angular displacement. It is important for the attitude control of a moving object. The critical part of a conventional gyroscope is a wheel spinning at a high speed. Therefore, conventional gyroscopes are accurate but bulky and very expensive. Their main applications are in the navigation systems of large vehicles, such as ships, airplanes, spacecrafts, etc.

As recognized by many researchers, gyroscopes would have many more applications if the cost of the devices could be drastically reduced [22]. Low cost micro gyroscopes could find wide applications in automobiles (such as short range navigation as a supplement to Global Positioning System, i.e. GPS, traction control systems and ride-stabilization systems), consumer electronics (such as compensation systems for the movement of video camera and model aircraft control), computer applications (such as inertial mouse) and, of course, military applications (such as tactical weapon control). Therefore, micro-gyroscopes have been under vigorous development efforts and many prototypes have been developed in research institutes worldwide.

Micro mechanical gyroscopes developed so far are exclusively vibratory gyroscopes. A simplified model for a vibratory gyroscope is shown in Fig. 1.4.3. The system is a two-dimensional vibration system with two orthogonal vibration modes. One vibration mode corresponds to the vibration of the mass in the x-direction. The vibration frequency of the vibration mode is ω_x. The other vibration mode corresponds to the vibration of the mass in the y-direction with a frequency of ω_y. The values of ω_x and ω_y are usually quite close to each other.

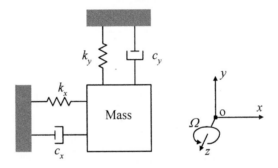

Fig. 1.4.3. A simplified model for a vibratory gyroscope

For operation, the mass is driven into vibration in the x-direction with a frequency ω_d (the driving frequency), which is close to ω_y. Then, if the system rotates around the z-axis (normal to the paper plane) with an angular rate of Ω, an alternating force in the y-direction is induced by the Coriolis force. Thus, the system is driven into vibration in the y-direction with a frequency of ω_d. The vibration amplitude in the y-direction can be detected by some sensing schemes and used as a measure of angular rate Ω.

As a Coriolis force is usually extremely weak, it is important to make full use of the amplification effect of mechanical resonance and to keep the noise level low in the signal bandwidth. Therefore, the driving frequency, ω_d, and the two resonant frequencies, ω_x and ω_y, have to be designed carefully, and sophisticated electronic circuits have to be used.

Fig. 1.4.4 shows a schematic diagram of a comb-drive resonator micro-gyroscope [23]. The plate (mass) at the center has two orthogonal vibration directions (in the x- and z-directions). The mass is driven into vibration in the x-direction by an electrostatic force generated by an alternating voltage applied on the comb-actuators on both sides of the plate. The Coriolis force produced by an angular rate Ω around the y-axis causes the resonator to vibrate in the z-direction (normal to the plate). This vibration can be detected by the change in the capacitance between the plate and the substrate.

As the signal of the gyroscope is very weak, it is important to reduce the effect of interference from the environment. For this purpose, two identical comb drive resonators are used to constitute a gyroscope as shown in Fig. 1.4.5. The resonators are driven into vibration in the *x*-direction with the same amplitude but out of phase. Thus, the vibrations in the *z*-direction of the two comb-drive resonators induced by the angular rate Ω will have the same amplitude but out of phase.

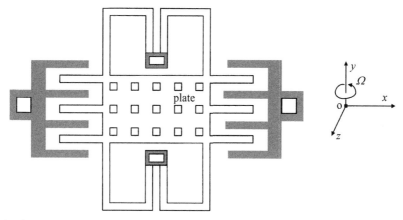

Fig. 1.4.4. A schematic drawing of a comb-driver micro-gyroscope

The differential capacitance signal of the two comb-drive resonators is now used as a measure of the angular rate. In this way, the gyroscope will have much higher immunity to the interference from environment, as the interference are usually common mode signals. As the two masses are oscillating out of phase, the dual resonator structure is often referred to as a tuning fork structure.

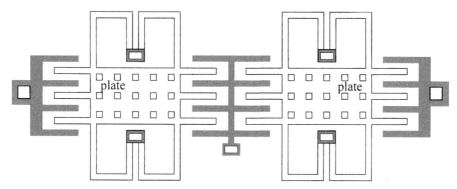

Fig. 1.4.5. A schematic drawing of a tuning fork micro-gyroscope

For a practical gyroscope device, the design, fabrication, packaging and signal conditioning circuitry are quite complicated. The design is related to driving schemes, damping control and sensing schemes. The fabrication is difficult as the requirement for the resonant frequencies of the structure is strict and, quite often, the structure has to be encapsulated in a vacuum. The signal detection is difficult because of the extremely weak signal and the phase differences among the electrical driving signals, driving vibration and the detection vibration caused by mechanical reasons and the air damping effect.

The brief description of micro resonant sensors and micro gyroscopes has shown that the analysis and design of these devices are based on much background theory, including the vibration frequency of a beam-mass structure, the forced vibration of a micro mechanical structure, air damping and its effect on vibration, electrostatic driving, and piezoresistive or capacitive sensing. The vibration frequency of a mechanical structure and the forced vibration will be discussed in Chapter 2, air damping and its effects will be discussed in Chapter 3, the electrostatic driving will be studied in Chapter 4, capacitive sensing and piezoresistive sensing will be discussed in and Chapter 5 and Chapter 6, respectively.

§1.5. Micro Mechanical Electric and Optical Switches

§1.5.1. Micro Mechanical Electric Switch

Electric switches (or relays) are very important devices in communication and control systems. For example, electric switches are widely used in microwave and millimeter wave integrated circuits (MMICs) for telecommunication applications including signal routing, impedance matching networks, adjustable gain amplifiers and so forth. The state-of-the-art technology uses solid-state switches such as GaAs MESFETs and PIN diodes. However, these solid-state switches have a large insertion loss in the "on" state, and a poor electrical isolation in the "off" state when the signal frequency is in GHz level.

As micro mechanical switches make use of mechanical contact and separation for the "on" and "off" states, its performance will be much better than solid-state switches. Therefore, micro mechanical switches are considered as promising alternatives for solid-state switches in telecommunication systems, especially for integrated systems. Fig. 1.5.1 schematically shows a surface micro machined miniature switch for RF applications [24].

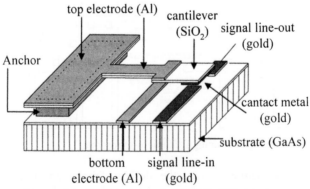

Fig. 1.5.1. Schematic picture of a micro mechanical switch for RF signals

The switch uses a suspended silicon dioxide cantilever beam as a moveable arm. When there is no voltage applied between the top and bottom electrodes, the signal line-in and the signal line-out are not connected electrically. The switch is called in an "off" state. When an electric voltage is applied between the top and bottom electrodes a, the electrostatic force applied on the top electrode pulls the cantilever down so that the contact metal bridges the signal line-in and the signal line-out electrically. Thus, the switch is turned to an "on" state.

Obviously, the "on" state resistance of a micro mechanical switch is usually much lower than that of a solid-state switch. For an "off" state, electrical isolation of the switch depends on the capacitive coupling between the two signal lines. As the coupling between the two signal lines is mainly through the substrate, semi-insulating GaAs or ceramic substrate is often chosen for RF switch instead of silicon substrate. It is easy to understand that, in this case, the capacitance between the two signal lines is lower than that of a solid-state switch for an "off" state.

§1.5.2. Micro Mechanical Optical Switch

With the rapid development of optical communication, miniaturized micro optical switches are widely demanded. In recent years, a variety of micro mechanical optical switches have been developed. Two typical designs are briefly introduced in this section.

(1) Optical Switch With Torsion Bar Actuator

A torsion bar optical switch is schematically shown in Fig. 1.5.2 [25]. The incident light beam from an optic fiber is reflected towards position 1 by the metal-coated refractive plate supported by two torsion bars when the plate is at its original, horizontal position. This corresponds to an "off" state.

When a driving voltage V is applied between the refractive plate and the metal electrode on the substrate under one side of the plate, the plate will turn and hit the substrate at its one end as shown in Fig. 1.5.2(b) by the dotted lines. In this case, the incident light beam will be conducted towards position 2, corresponding to an "on" state of the optical switch.

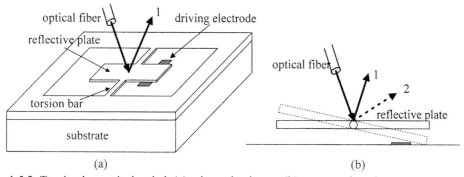

Fig, 1.5.2. Torsion bar optical switch (a) schematic picture; (b) cross section view

(2) Optical Switch With Comb Drive Actuator

For optical communication applications, optical switches are often arranged in arrays. For the switches shown in Fig. 1.5.2, light beams travel out of the surface plane of the device. This operation mode facilitates the formation of a multi-in and multi-out (N×M) switch array. However, the alignment of the light beams to specific directions is difficult.

Fig. 1.5.3 shows an optical switch featuring in-plane traveling of light beams [26]. In an original state (or, an "off" state), the incident light beam is reflected towards position 1 by the vertical micro mirror attached to the movable electrode. When a driving voltage is applied between the stationary electrode and the movable electrode, the movable electrode and the mirror move away from the light beam by the operation of the come drive actuator, and the light beam goes to the position 2. The switch is thus turned "on".

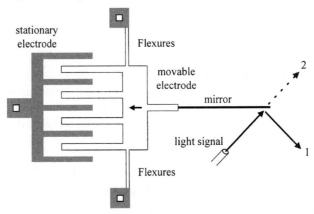

Fig. 1.5.3. Schematic picture of a comb drive actuator optical switch (top view)

As the size of the mirror has to be large enough to completely cut off the beam light in the "off" state and the travel distance of the mirror large enough to let the light beam pass in an "on" state, the structure is usually processed using one of the high aspect ratio technologies, such as the deep reactive ion etching technology.

According to the brief description above, the analysis and design of micro switches are closely related to the displacement of mechanical structures caused by an electrostatic force. For some applications the switching speed is important. In this case, the natural vibration frequency of the structure, the damping effect and the magnitude of the electrostatic force play important roles. The displacement of a mass and vibration frequency of a mechanical structure will be studied in Chapter 2, air damping will be discussed in Chapter 3 and electrostatic driving will be discussed in Chapter 4.

§1.6. Micro Mechanical Motors

As an important actuator and a symbol of MEMS devices, micro motors are interesting to researchers from the early development stage of micro mechanical technology. In 1988, L. S. Fan, Y. Ch. Tai and R. Muller demonstrated the first micro motor [27]. As the rotor of the motor is driven step and step by the voltage applied on selected stators, the motor is called an electrostatic step motor or an electrostatic synchronous motor.

In the years followed, a variety of micro motors were demonstrated, including the electrostatic comb drive motors, electromagnetic micro motors, ultrasonic micro motors, motors using scratch force, etc. However, most of the devices are not suitable for practical applications. Some of them are too weak in power, some are too complicated in fabrication process, some are not rugged enough, etc. As a result, the applications of micro motors have not been as wide as expected except for the comb driver micro motors, which are considered as one of the most suitable designs for micromachining technology.

For the situation mentioned above, we will only introduce two types of micro motors in this section: the electrostatic step motor and the electrostatic comb drive motor.

§1.6.1. Electrostatic Step Motor

A schematic top view of electrostatic step motor is shown in Fig. 1.6.1. The rotor of the motor has 4 vanes and measures about 100μm in diameter. The stator around the rotor consists of 12 stationary electrodes. The cross-sectional view of the motor is shown in Fig. 1.6.2. It shows that the motor is composed of mechanical parts in several polysilicon levels using a surface micro machining technology. Each polysilicon layer is about 2 μm thick and the gaps between the polysilicon layers are 1-2μm. The gaps are controlled by the thickness of the sacrificial layers (e.g., phospho-silicate glass — PSG) temporarily filling the gaps in the process and later removed by selective etch when the rotor is to be released.

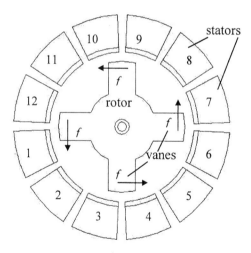

Fig, 1.6.1. Schematic top view of electrostatic step motor

The rotor of the motor is driven into rotational movement by the electric force between the stationary electrodes and the vanes. To do so, the rotor is grounded and the electric voltage is applied on some selected stationary electrodes. For an original rotor position as shown in Fig. 1.6.1, the voltage is applied on electrodes 1, 4, 7 and 10 so that the electrostatic forces on the vanes unanimously drive the rotor to rotate counter-clockwisely.

Once the vanes of the rotor have moved to the position in alignment with the electrodes 1,4,7 and 10, the voltage is switched to the next electrodes, i.e. the electrodes 2, 5, 8 and 11. Thus, the electrostatic forces combined with the momentum of rotation drive the rotor to move on. The motor keeps running if the above step is repeated continuously.

Though the first micro step motor worked in 1988, so far the applications of step motor are rarely seen. The main difficulties are the small torque and relatively large friction and stiction. The torque T on the rotor can be expressed as $T = nfr$, where n is the vane number of the rotor ($n=4$ in Fig. 1.6.1), r the radius of the rotor and f the tangential force on each vane. When the width of the vane is much larger than the gap distance d between the vanes and the stators, the tangential force on each vane is

$$f = \frac{\varepsilon_o t}{d} V^2$$

where t is the thickness of the rotor as shown in Fig. 1.6.2, ε_o the permitivity of free space and V the driving voltage applied between the rotor and the stators. For a specific fabrication process and a given driving voltage, the only way to increase the torque is to increase the vane number of the rotor. As the vane number n can hardly be larger than 10 for conventional designs, there is little room for increasing the torque of the electrostatic step motor.

Stiction is the phenomenon that two micro mechanical parts tend to stick together once they are brought into contact and can only be separated by a relatively large force. With stiction, extra force is needed to make the motor to start. As the friction coefficient for micro mechanical parts is larger than that for conventional mechanical parts [28] and it is difficult to apply lubricant in between two moving parts to reduce friction, a voltage larger than expected has to be applied to keep the motor running.

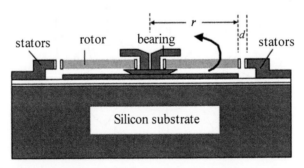

Fig, 1.6.2. Schematic cross-sectional view of electrostatic step motor

§1.6.2. Comb Drive Vibratory Motor

Commonly, a motor is considered as a rotation machine. As a matter of fact, a motor is traditionally described as "a machine that supplies power or cause motion" by dictionaries (say, The Advanced Learner's Dictionary of Current English, 2nd edition, Oxford University Press, Amen House, London). Therefore, a motor is not necessarily a rotational one. It can also be a machine that provides other forms of power or movement. For example, there are vibratory motors, linear motors, etc.

As mentioned above, micro step motors suffer from the difficulties in small torque, relatively large friction and stiction. Therefore, a micro motor with larger driving force and free from the problems of friction and stiction will be more promising in applications.

As we have seen in section §1.4.2 and in §1.5.2, the comb drive structures can provide back-and-forth movement to turn the switches "on" and "off". As each side of a movable finger of the comb drive structure provides a tangential force as a vane does in a rotational step motor and the finger number in a comb drive structure can be increased without limit in principle, the electrostatic driving force of a comb drive motor can be increased greatly.

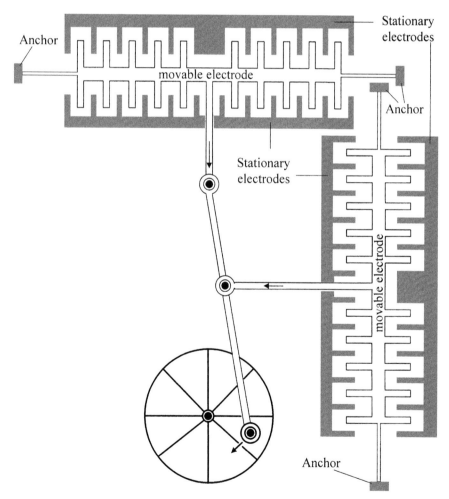

Fig. 1.6.3. Transformation from translation movement to rotation movement (not to scale)

On the other hand, the movable parts are suspended over the substrate without a direct contact with the stationary fingers or substrate. The problems of stiction and friction can be alleviated greatly though the resistance to the movement from surrounding air (the air damping) still exists. Therefore, the comb drive micro motor is a more reasonable design than a rotational micro motor for micro mechanical world. As the comb structure can provide "back-and-forth" movement, it is a vibratory motor. So far a variety of comb drive motors have been developed and applied in MEMS.

It has been demonstrated that the "back-and-forth" movement can also be transformed into a rotational motion using a transformation mechanism with cranks, pin joints, wheels and gears [28], as shown in Fig. 1.6.3. Two comb drive micro motors are used to provide two orthogonal forces: one for driving in the x-direction and the other for the y-direction. With some control on the phase difference between the two motors, the driving forces in the two directions can be incorporated to drive the wheel into a rotational movement. In this case, stiction and friction appear again, but the problem is now alleviated by the increased driving force by using a large number of fingers.

According to the brief description above, the analysis and design of electrostatic micro motors are related to the mechanics of beam-mass structures, the electrostatic forces between two plates, the air damping effect on the movement of mechanical parts and the stiction and friction of micro parts. The mechanical performance of beam-mass structures will be discussed in Chapter 2, the air damping will be studied in Chapter 3 and the electrostatic force between mechanical plates will be discussed in Chapter 4. As step motors are not practically used and the problems on stiction and friction are much related to microscopic theories, stiction and friction will not be discussed in this book.

§1.7. Micro Electro Mechanical Systems

Based on the micro mechanical and microelectronics technology, micro mechanical sensors, actuators and electronic circuits can be integrated into a system or subsystem on a single chip. The devices are often referred to as Micro Electro Mechanical System (MEMS).

Needless to say, MEMS are very attractive for varies applications. However, the development of MEMS is often very challenging in technology. It requires large investment and needs long development time period. Therefore the development of MEMS can only be started in some specific applications where mass production is most likely. Fortunately, some successes have so far been achieved. In this section, two notable examples will be given as follows.

§1.7.1. MEMS Accelerometer

Surface micro machining technology has been used for the development of silicon micro accelerometers with on-chip signal detection circuit since the late 1980s. A few micro accelerometers have been developed and some of them have been commercialized. In this section, the operation principle of a MEMS accelerometer is introduced based on the well-known device ADXL50 developed by Analog Devices and Siemens [29].

Basically, the device can be divided into two portions: the mechanical capacitive sensor and the signal detection circuit. The mechanical sensor formed using surface micro-machining technology is schematically shown in Fig. 1.7.1.

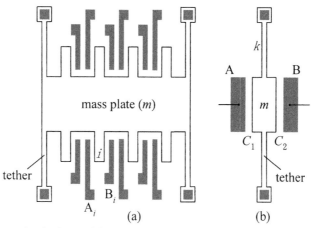

Fig. 1.7.1. A micro mechanical capacitive sensor (a) Schematic structure; (b) lumped model

The mechanical structure of the sensor is made of a layer of polysilicon material about 2μm thick. A central mass plate with fingers on both sides is suspended over the substrate by four thin flexures (tethers). The central mass plate (working as the seismic mass of accelerometer) can be displaced by an inertial force in the surface plane and in perpendicular to the beam flexures.

For a finger of the central plate (say, the i'th finger), there are two fixed strips, A_i and B_i, on both sides at a distance of 1.3μm. All the fixed strips, A_i, are electrically connected together to form a stationary electrode, A. Also, all fixed strips, B_i, are electrically connected together to form another stationary electrode, B. All the fingers of the central plate form the movable electrode. The movable plate and the two stationary electrodes constitute two mechanical capacitors C_1 and C_2. With a lateral acceleration, the plate is displaced and the capacitances C_1 and C_2 changes in a push-pull way. The differential capacitance C_1-C_2 can be used as a measure of the acceleration.

Typically, the original capacitance is about 0.1pF. The variation of the capacitance is about 3fF for a full operation range of 50g. To detect such a small change in capacitance, a carefully designed signal detection circuit is monolithically integrated with the mechanical sensor. The equivalent circuitry is shown in Fig. 1.7.2, where the mechanical capacitive sensor is schematically shown in the dotted block.

The signal detection circuit consists of a sensing circuit excited by two opposite excitation signals, a buffer amplifier, a high pass filter (HPF), a synchronous demodulator (sync. demo.), an electro-mechanical feedback loop with a feedback amplifier (FB Amp.) and a feedback resistance of 3MΩ. Thus, the accelerometer works in a force-balanced mode. As a result, the sensitivity at the output is about 20 mV/g for an operation range of 0 to 50g.

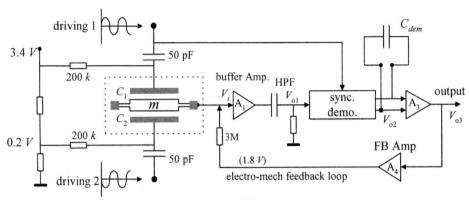

Fig. 1.7.2. The equivalent circuit of the integrated MEMS accelerometer

§1.7.2. Digital Micro Mirror Device (DMD)

Display devices are widely used in information technologies and consumer electronics. For quite a long time, cathode ray tube (CRT) dominated the display market. Although the display quality of CRTs has been very good and is improving continuously, they are too bulky, too heavy and too large in power consumption for many modern information systems, especially for mobile communication systems. Therefore, a variety of new display technologies appeared in recent years to challenge the CRT display. The new devices having had mass applications include the liquid crystal display (LCD), the plasma display panel (PDP), and the digital micro mirror devices (DMD), just to name a few. Among these new display devices, DMD of Taxas Instruments (TI) is the only device developed using sophisticated micro machining and microelectronics technology [30]. Due to its high performance, DMD has been widely accepted in market for high-resolution, large screen display applications.

In short, DMD is a device with a two-storage structure. The first storage is an array of Static Random Access Memory (SRAM) cells on silicon substrate. The size of the memory determines the resolution of the display. For the first DMD devices, the memory is a 768 bit by 576 bit CMOS SRAM with 96 16-bit shift registers in the peripheral for data input. The schematic configuration of the SRAM is shown in Fig. 1.7.3. At present, the memory can be larger than 1 million bits so that they can display for high-definition television (HDTV). Each memory cell represents a pixel in a picture — say, a "1" state of the cell for a bright pixel and a "0" state of the cell for a dark pixel of the picture.

The second storage of the DMD device is an array of micro mechanical mirrors, with each mirror measuring 17μm×17μm and located on top of a memory cell as schematically shown in Fig. 1.7.4.

The mirror can be turned to one of the two stable states, a -10° state and a $+10^\circ$ state, according to the state of the memory cell beneath the mirror. For example, if the memory cell is in a "1" state, the mirror flips to a $+10^\circ$ state and if the memory cell is in a "0" state, the mirror flips to a -10° state. When the device is used for a display, the data for a frame of picture are loaded into the memory by a serial-parallel input operation through the shift-registers. For a memory in the state of "1", the incident light on the mirror is reflected onto the screen to present a light spot. On the contrary, if the state of a memory is in a "0" state, the incident light on the mirror is reflected out of the screen, leaving a dark spot on the screen as shown in Fig. 1.7.5. In this way, all the data in the memory are translated to a frame of picture on the screen.

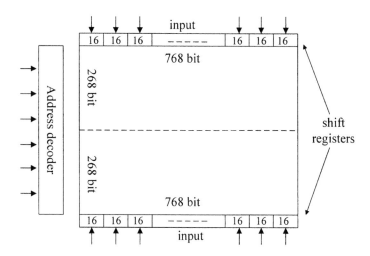

Fig. 1.7.3. The schematic configuration of the SRAM

As the mirror is very small, it can be flip-flopped in high frequency and the rate of bright display on a pixel represents the gray scale of the pixel. For the display of color pictures, three DMD devices, each for green, blue and red images, have to be used, or, a single DMD device is used with a color wheel to split the time for the images of the three colors.

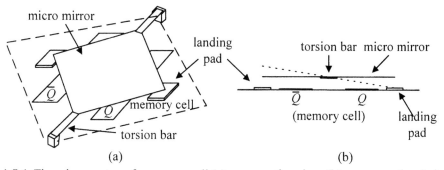

Fig. 1.7.4. The mirror on top of a memory cell (a) a perspective view; (b) a cross sectional view

At present, projectors using DMD have been widely used at conferences, demonstration sites and classrooms. As light signals are processed digitally in these projectors, they are known as digital light processing (DLP) projectors.

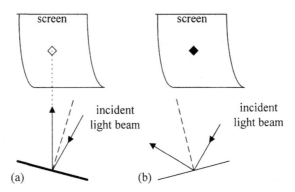

Fig. 1.7.5. The display of a pixel (a) a bright spot for a "1" state; (b) a dark spot for a "0" state

§1.8. Analysis and Design Principles of MEMS Devices

A typical micro electro mechanical system is a monolithically integrated system or subsystem consisting of micro sensors, micro actuators and microelectronic circuits, as schematically shown in Fig. 1.8.1. The bold arrows in the figure indicate the mechanical interaction between the device and the outside world, and the thin arrows indicate the electrical or other non-mechanical signals among the components of the device, and between the device and the outside world. Note that not every MEMS device makes use of all the components and operations indicated in the figure, but each involves a micro mechanical structure of some kind.

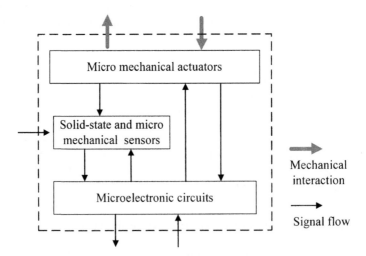

Fig. 1.8.1. Basic configuration of a MEMS device

The movement of mechanical structures is detected by mechanical sensors. The other parameters (say, temperature, pressure, acceleration, and so on) can be sensed either by mechanical sensors or by solid-state sensors. Obviously, the sensors for the state of mechanical structure are more essential than those for other parameters.

The micro mechanical actuators are driven and controlled by a variety of means. However, electrostatic driving is the most commonly used one nowadays. Electronic circuits are widely used for signal detection, signal processing, the control of the system and providing electrostatic driving force for mechanical actuators.

Based on the description above, the analysis and design principles of the MEMS device fall into three basic categories: the dynamics of micro mechanical structure (or, the microdynamics), the sensing schemes and microelectronics, as illustrated in Fig. 1.8.2.

Micro mechanical technology has been under development for several decades. So far a variety of MEMS devices have been developed and many of them have been mass-produced. However, the analysis and design principles on MEMS devices are still widely scattered in literatures of different disciplines, such as mechanics, electrostatics, hydrodynamics, solid-state physics, microelectronics, etc. This situation is not convenient for the researchers and engineers who are developing MEMS devices and for students who are studying in this interesting area.

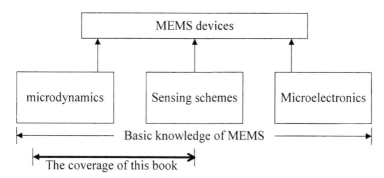

Fig. 1.8.2. Basic analysis and design principles of MEMS and the coverage of this book

To continue the effort of this author to provide as much as possible the essential theories for the analysis and design of MEMS in a book, the previous book of this author [9] is modified, updated and supplemented with more materials so that the readers can understand the theories and put them into practical applications more easily. Besides, problems will be given at the end of each chapter (except for the first chapter) and the answers to the problems are given at the end of the book. Therefore, this book will be especially useful for graduate students who are studying MEMS as well as for researchers and engineers who are developing new MEMS devices.

Microdynamics will be the main content of this book and be treated more thoroughly than ever before. It includes the basic principles of mechanics of beam and diaphragm structures, air damping and its effect on the motion of mechanical structures, and the electrostatic driving of mechanical structures.

For sensing schemes, we will restrict ourselves to frequency sensing, capacitive sensing and piezoresistive sensing under the following reasons. Frequency sensing is naturally related to mechanical structures and promising for its high resolution, high stability and digital compatibility. Capacitive sensing is the sensing scheme most compatible with MEMS devices and the most popular sensing scheme currently used in MEMS. Piezoresistive sensing is the first sensing scheme that led to the development of micro mechanical transducers since the 1960s and is now still widely used in MEMS.

The knowledge on microelectronics will not be included in this book as there are already too many books around. Therefore, the content of this book is indicated by the bold arrow in Figure 1.8.2 and summarized in Table 1.8.1

Table 1.8.1. The main content of this book

Microdynamics	Sensing schemes
Basic mechanics (Chap. 2)	Frequency sensing (Chap. 2)
Air damping (Chap. 3)	Capacitive sensing (Chap.5)
Electrostatic driving (Chap. 4)	Piezoresisitve sensing (Chap.6)

References

[1] S. M. Sze (editor), VLSI Technology, 2nd edition, McGraw-Hill, New York, 1988

[2] S. M. Sze (editor), Semiconductor Sensors, John Wiley & Sons, Inc., 1994

[3] J. Brysek, K. Petersen, J. K. Mallon, L. Christel, F. Pourahmadi, Silicon Sensors and Micro Structures, NOVA Sensor, Fremont, 1990

[4] C.S. Smith, Piezoresistance in germanium and silicon, Physics Review, Vol. 94 (1954) 42-49

[5] O. N. Tufte, P.W. Chapman, D. Long, Silicon diffused-element piezoresistive diaphragms, Journal of Applied Physics, Vol. 33 (1962) 3322-3327

[6] A. C. M. Gieles, G.H.J. Somers, Miniature pressure transducers with a silicon diaphragm, Philips Tech. Rev., Vol. 33, No.1 (1973) 14-20

[7] Samaun, K. D. Wise, J. B. Angell, An IC piezoresistive pressure sensor for biomedical instrumentation, IEEE Trans. on Biomedical Engineering, Vol. BME-20 (1973) 101-109

[8] W. Ko, J. Hynecek, S. F. Boettcher, Development of a miniature pressure transducer for biomedical applications, IEEE Trans. on Electron Devices, Vol. ED-26 (1979) 1896-1905

[9] M-H Bao, "Micro Mechanical Transducers — Pressure Sensors, Accelerometers and Gyroscopes", Elsevier 2000, Vol. 8 in Handbook of Sensors and Actuators, edited by Simon Middelhoek

[10] L. M. Roylance, J. B. Angell, A batch-fabricated silicon accelerometer, IEEE Trans. on Electron Devices, Vol. ED-26 (1979) 1911-1917

[11] P. W. Barth, F. Pourahmadi, R. Mayer, J. Poydock, K. E. Petersen, A monolithic accelerometer with integrated air damping and overrange protection, Proc. of IEEE Solid-State Sensor and Actuator Workshop, Hilton Head Island, SC, USA, June 6-9, 1988, 35-38

[12] J. Starr, Squeeze-film damping in solid-state accelerometers, Proc. of IEEE Solid-State Sensor and Actuator Workshop, Hilton Head Island, SC, USA, 1990, 44-47

[13] H. Chen, S. Shen, M. Bao, Over-range capacity of a piezoresistive micro-accelerometer, Sensors and Actuators A58 (1997) 197-201

[14] S. K. Clark, K. D. Wise, Pressure sensitivity in anisotropically etched thin-diaphragm pressure sensors, IEEE Trans. on Electron Devices, Vol. ED-26 (1979) 1887-1896

[15] W. Ko, M. Bao, Y. Hong, A high sensitivity integrated circuit capacitive pressure transducer, IEEE Trans. on Electron Devices, Vol. ED-29 (1982) 48-56

[16] W. Ko, Q. Wang, Q. Wu, Long term stability capacitive pressure sensor for medical implant, Digest of Technical Papers, The 7th International Conference on Solid-State Sensors and Actuators, Yokohama, Japan, June7–10, 1993 (Transducers'93) 592-595

[17] F. Rudolf, A micromechanical capacitive accelerometer with a two-point inertial-mass suspension, Sensors and Actuators, Vol. 4 (1983) 191-198

[18] See, for example, www. Knowles.com

[19] M. Fuldner, A. Dehe, R. Aigner, T. Bever, R. Lerch, Silicon microphones with low stress membranes, Digest of Technical Papers, The 11th Intl. Conf. on Solid-State Sensors and Actuators, Munich, Germany, June 10-14, 2001 (Transducers'2001) 126-129

[20] J. D. Zook, D. W. Burns, H. Guckel, J. J. Sniegowski, R. L. Engelstad, Z. Feng, Resonant microbeam strain transducers, Digest of Technical Papers, The 6th Intl. Conf. on Solid-State Sensors and Actuators, San Francisco, CA, USA, June 24-27, 1991 (Transducers'91) 529-532

[21] K. Petersen, F. Pourahmadi, J. Brown, P. Parson, M. Skinner, J. Tudor, Resonant beam pressure sensor fabricated with silicon fusion bonding, Digest of Technical Papers, The 6th International Conference on Solid-State Sensors and Actuators, San Francisco, CA, USA, June 24-27, 1991 (Tansducers'91) 664-667

[22] J. Soderkvist, Micromachined gyroscopes, Digest of Technical Papers, The 7th Intl. Conf. on Solid-State Sensors and Actuators, Yokohama, Japan, June 7-10, 1993 (Transducers'93) 638-641

[23] K. Tanaka, Y. Mochida, M. Sugimoto, K. Moriya, T. Hasegawa, K. Atsuchi, K. Ohwada, A micromachined vibrating gyroscope, Sensors and Actuators A50 (1995) 111-115

[24] J. J. Yao, M. F. Chang, A surface micromachined miniature switch for telecommunications applications with signal frequencies from DC up to 4 GHz, Digest of Technical Paper, The 8th Intl. Conf. on Solid-State Sensors and Actuators, Stockholm, Sweden, June 25-29, 1995 (Transducers'95) 384-387

[25] W. Dotzel, T. Gensser, R. Hahn, Ch. Kaufmann, K. Kehr, S. Kurth, J. Mehner, Silicon mirrors and micromirror array for spatial laser beam modulation, Digest of Technical Papers, The 9th International Conference on Solid-State Sensors and Actuators, Chicago, IL, USA, June 16-19, 1997 (Transducers'97) 81-84

[26] W. Juan, S. Pang, Batch-micromachined, high aspect ratio Si mirror arrays for optical switching applications, Digest of Technical Papers, The 9th International Conference on Solid-State Sensors and Actuators, Chicago, IL, USA, June 16-19, 1997 (Transducers'97) 93-96

[27] L. S. Fan, Y. C. Tai, R. Muller, IC-processed electrostatic micromotor, Sensors and Actuators, Vol. 20 (1989) 49-55

[28] S. Miller, J. Sniegowski, O LaVigne, P. McWhorter, Friction in surface micromachined microengines, Proc. SPIE, Vol. 2722, Smart Electronics and MEMS, San Diego, Feb. 28-29, 1996, 197-204

[29] Kuehnel, S. Sherman, A surface micromachined silicon accelerometer with on-chip detection circuitry, Sensors and Actuators A 45 (1994) 7-16

[30] J. B. Sampsell, The digital micro mirror device and its application to project displays, Digest of Technical Papers, The 7th International Conference on Solid-State Sensors and Actuators, Yokohama, Japan, June 7-10, 1993 (Transducers'93) 24-27

Chapter 2

Mechanics of Beam and Diaphragm Structures

Crystalline silicon is an excellent mechanical material as well as an excellent electronic material. The mechanical properties of bulk silicon are quite ideally governed by the theory of elasticity in a large temperature range. It has been speculated that the mechanical properties of silicon may change when the geometries of the mechanical structure are scaled down. Fortunately, however, no significant changes in the mechanical properties have been observed so far for silicon mechanical structures in micrometer scale. Therefore, it will be assumed throughout this book that the mechanical properties of a silicon micro structure are ideally elastic. This assumption implies that, if the deformation produced by external forces does not exceed a certain limit, it disappears once the forces are removed (i.e., Hooke's law, see §2.1.3).

As a crystalline material is anisotropic, the mechanical properties of silicon are orientation dependent and the relations among mechanical parameters are tensor equations, which are quite complicated as shown in §2.1. Thus, for the simplicity of analysis, homogeneous assumption is used in most part of this book. The homogeneous assumption simplifies analytical analysis greatly without causing significant errors in the results.

Even with these assumptions, approximations have to be made for analytical analysis of most practical problems. However, the results are generally accurate enough for design optimization, as in most cases the performance of mechanical sensors is more significantly affected by the variations of geometric parameters determined by process control rather than by the assumptions and approximations. If a more precise result is required, a numerical analysis has to be made using a computer aided design (CAD) tool in addition to the analytical analyses. This will be beyond the scope of this book.

As silicon mechanical structures with beams and diaphragms are the most important parts for MEMS devices, mechanics of beam and diaphragm structures will be studied in this chapter according to the theory of elasticity for homogeneous material [1,2].

§2.1. Stress and Strain

§2.1.1. Stress

According to the theory of elasticity, external forces acting on a solid-state body produce internal forces between the portions of the body and cause deformation. If the external forces do not exceed a certain limit, the deformation disappears once the forces are removed. To describe the internal forces, the stress tensor is introduced. Mathematically, stress is a second rank tensor, which has nine components as shown by the matrix

$$T = \begin{pmatrix} T_{XX} & T_{XY} & T_{XZ} \\ T_{YX} & T_{YY} & T_{YZ} \\ T_{ZX} & T_{ZY} & T_{ZZ} \end{pmatrix} \tag{2.1.1}$$

where the three diagonal components are referred to as normal stresses and the six off-diagonal components are called shearing stresses.

To illustrate the definition of the components of stress tensor, let us examine an elemental cube inside the body as shown in Fig. 2.1.1. The six faces of the cube are denoted as x, $\bar{x}$, y, $\bar{y}$, z, $\bar{z}$, according to the normal of the faces. (Note: a bar over a letter indicates a negative sign for the letter.)

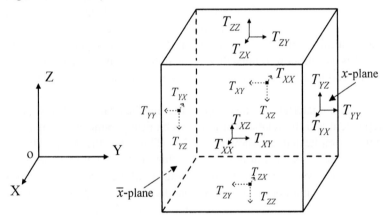

Fig. 2.1.1. Components of the stress tensor

In the figure, T_{ij}, the component of a stress tensor, is defined as the force in a specific direction j (the second subscript) on a unit area on a specific surface i (the first subscript) of the elemental cube. For examples, T_{XX} in Fig. 2.1.1 is the normal force per unit area of the x-plane, T_{XY} is the force in the y-direction applied on a unit area of the x-plane and T_{XZ} is the force in the z-direction per unit area of x-plane, and so forth.

The signs of the tensor components are determined by the direction of the force relative to the normal of the plane. For example, for the x-plane, the normal stress component caused by a force in the x-direction is defined as positive but that caused by a force in the $\bar{x}$-direction is defined as negative. The stress component caused by the tangential forces on the x-plane and in the y- and z-directions are defined as positive while those by tangential forces in the $\bar{y}$- and $\bar{z}$-directions are defined as negative. Furthermore, for the $\bar{y}$ plane, the stress component caused by a force in the $\bar{y}$-direction is positive but that caused by a force in the y-direction is negative. Similarly, the stress components caused by the tangential forces in the $\bar{x}$ - and $\bar{z}$ -directions on the $\bar{y}$ -plane are defined as positive and those by the forces in the x- and z- directions on the $\bar{y}$ -plane are defined as negative.

According to the condition of force balance, the T_{XX} in two opposite parallel planes (x- and $\bar{x}$ -planes) should be equal in quantity and sign, and the same is true for the T_{YY} and the T_{ZZ}. Also from the condition of torque balance, we have

$$T_{XY} = T_{YX,}\ T_{YZ} = T_{ZY},\ T_{XZ} = T_{ZX} \tag{2.1.2}$$

This means that the matrix of stress tensor T is symmetric and has only six independent components. Therefore, they are often denoted by a simplified notation system

$$T_1 = T_{XX,}\ T_2 = T_{YY,}\ T_3 = T_{ZZ,}\ T_4 = T_{YZ,}\ T_5 = T_{XZ},\ T_6 = T_{XY} \tag{2.1.3}$$

Therefore, equation (2.1.1) can be written as

$$(T) = \begin{pmatrix} T_1 & T_6 & T_5 \\ T_6 & T_2 & T_4 \\ T_5 & T_4 & T_3 \end{pmatrix}$$

§2.1.2. Strain

With stresses, deformation is produced inside the material. For easy to understand, let us first look at the deformation of a one-dimensional material as shown in fig.2.1.2. If the displacement of an original position x is $u(x)$ and the displacement of an original position $x+\Delta x$ is $u(x+\Delta x)$, the strain (the relative elongation of the material) in the one-dimensional material has only one component,

$$e = \lim_{\Delta x \to 0} \frac{u(x + \Delta x) - u(x)}{\Delta x} = \lim_{\Delta x \to 0} \frac{\Delta u}{\Delta x} = \frac{\partial u}{\partial x}$$

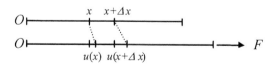

Fig. 2.1.2. Deformation of a one-dimensional material

For a three dimensional material, the deformation of the material is schematically shown in Fig. 2.1.3. The displacement components for the point r (x,y,z) are $u(x,y,z)$, $v(x,y,z)$ and $w(x,y,z)$ in the x-, y- and z-directions, respectively, and the displacement components for the point r ' $(x+\Delta x,\ y+\Delta y,\ z+\Delta z)$ are $u+\Delta u$, $v+\Delta v$ and $w+\Delta w$, respectively. The three components are all functions of x, y and z.

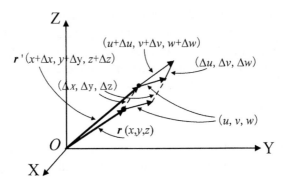

Fig. 2.1.3. Deformation in a three-dimensional material

Therefore, the incremental displacement between point r (x, y, z) and point r ' $(x+\Delta x, y+\Delta y, z+\Delta z)$ can be expressed as

$$
\begin{pmatrix} \Delta u \\ \Delta v \\ \Delta w \end{pmatrix} = \begin{pmatrix} \dfrac{\partial u}{\partial x} & \dfrac{\partial u}{\partial y} & \dfrac{\partial u}{\partial z} \\ \dfrac{\partial v}{\partial x} & \dfrac{\partial v}{\partial y} & \dfrac{\partial v}{\partial z} \\ \dfrac{\partial w}{\partial x} & \dfrac{\partial w}{\partial y} & \dfrac{\partial w}{\partial z} \end{pmatrix} \begin{pmatrix} \Delta x \\ \Delta y \\ \Delta z \end{pmatrix}
$$

$$
= \begin{pmatrix} \dfrac{\partial u}{\partial x} & 0 & 0 \\ 0 & \dfrac{\partial v}{\partial y} & 0 \\ 0 & 0 & \dfrac{\partial w}{\partial z} \end{pmatrix} \begin{pmatrix} \Delta x \\ \Delta y \\ \Delta z \end{pmatrix} + \begin{pmatrix} 0 & \dfrac{1}{2}(\dfrac{\partial u}{\partial y}+\dfrac{\partial v}{\partial x}) & \dfrac{1}{2}(\dfrac{\partial u}{\partial z}+\dfrac{\partial w}{\partial x}) \\ \dfrac{1}{2}(\dfrac{\partial u}{\partial y}+\dfrac{\partial v}{\partial x}) & 0 & \dfrac{1}{2}(\dfrac{\partial v}{\partial z}+\dfrac{\partial w}{\partial y}) \\ \dfrac{1}{2}(\dfrac{\partial u}{\partial z}+\dfrac{\partial w}{\partial x}) & \dfrac{1}{2}(\dfrac{\partial v}{\partial z}+\dfrac{\partial w}{\partial y}) & 0 \end{pmatrix} \begin{pmatrix} \Delta x \\ \Delta y \\ \Delta z \end{pmatrix}
$$

$$
+ \begin{pmatrix} 0 & \dfrac{1}{2}(\dfrac{\partial u}{\partial y}-\dfrac{\partial v}{\partial x}) & \dfrac{1}{2}(\dfrac{\partial u}{\partial z}--\dfrac{\partial w}{\partial x}) \\ \dfrac{1}{2}(\dfrac{\partial v}{\partial x}-\dfrac{\partial u}{\partial y}) & 0 & \dfrac{1}{2}(\dfrac{\partial v}{\partial z}-\dfrac{\partial w}{\partial y}) \\ \dfrac{1}{2}(\dfrac{\partial w}{\partial x}-\dfrac{\partial u}{\partial z}) & \dfrac{1}{2}(\dfrac{\partial w}{\partial y}-\dfrac{\partial v}{\partial z}) & 0 \end{pmatrix} \begin{pmatrix} \Delta x \\ \Delta y \\ \Delta z \end{pmatrix} \qquad (2.1.4)
$$

For a solid-state body rotating around the origin O of a coordinate system with an angular velocity $\vec{\omega}$, the velocity of the end of the vector $\vec{r}(x, y, z)$ is $\vec{v} = \vec{\omega} \times \vec{r}$. According to equations $\nabla \times \vec{v} = \nabla \times \vec{\omega} \times \vec{r} = (\nabla \cdot \vec{r}) \cdot \vec{\omega} - (\vec{\omega} \cdot \nabla)\vec{r} = 3\vec{\omega} - \vec{\omega} = 2\vec{\omega}$, we have $2\vec{\omega} = \nabla \times \vec{v}$. Therefore, an angular displacement $\vec{\Phi} = \vec{\omega} \Delta t$ can be expressed as

$$\vec{\Phi} = \vec{i} \frac{1}{2}(\frac{\partial w}{\partial y} - \frac{\partial v}{\partial z}) + \vec{j} \frac{1}{2}(\frac{\partial u}{\partial z} - \frac{\partial w}{\partial x}) + \vec{k} \frac{1}{2}(\frac{\partial v}{\partial x} - \frac{\partial u}{\partial y})$$

$$\equiv \vec{i} \Phi_X + \vec{j} \Phi_Y + \vec{k} \Phi_Z$$

(2.1.5)

where

$$\Phi_X = \frac{1}{2}(\frac{\partial w}{\partial y} - \frac{\partial v}{\partial z}), \quad \Phi_Y = \frac{1}{2}(\frac{\partial u}{\partial z} - \frac{\partial w}{\partial x}), \quad \Phi_Z = \frac{1}{2}(\frac{\partial v}{\partial x} - \frac{\partial u}{\partial y})$$

Therefore, the last term on the right side of Eq. (2.1.4) is

$$\begin{pmatrix} 0 & -\Phi_Z & \Phi_Y \\ \Phi_Z & 0 & -\Phi_X \\ -\Phi_Y & \Phi_X & 0 \end{pmatrix} \begin{pmatrix} \Delta x \\ \Delta y \\ \Delta z \end{pmatrix}$$

If no rotational movement for the solid-state material is allowed, the last term on the right side of Eq. (2.1.4) is zero and the equation can be written as

$$\begin{pmatrix} \Delta u \\ \Delta v \\ \Delta w \end{pmatrix} = \begin{pmatrix} \frac{\partial u}{\partial x} & \frac{1}{2}(\frac{\partial u}{\partial y} + \frac{\partial v}{\partial x}) & \frac{1}{2}(\frac{\partial u}{\partial z} + \frac{\partial w}{\partial x}) \\ \frac{1}{2}(\frac{\partial u}{\partial y} + \frac{\partial v}{\partial x}) & \frac{\partial v}{\partial y} & \frac{1}{2}(\frac{\partial v}{\partial z} + \frac{\partial w}{\partial y}) \\ \frac{1}{2}(\frac{\partial u}{\partial z} + \frac{\partial w}{\partial x}) & \frac{1}{2}(\frac{\partial v}{\partial z} + \frac{\partial w}{\partial y}) & \frac{\partial w}{\partial z} \end{pmatrix} \begin{pmatrix} \Delta x \\ \Delta y \\ \Delta z \end{pmatrix}$$

(2.1.6)

The 3 by 3 matrix in the equation is referred to as a strain tensor in solid-state physics. The three diagonal components in the matrix are called the normal strain components of the strain tensor

$$e_{xx} = \frac{\partial u}{\partial x}, \quad e_{yy} = \frac{\partial v}{\partial y}, \quad e_{zz} = \frac{\partial w}{\partial z}$$

(2.1.7)

It is quite clear that the three normal components of strain tensor shown in Eq. (2.1.7) are the relative elongations along the three coordinate axes. The six off-diagonal components are referred to as the shearing strain components of the strain tensor

$$e_{xy} = \frac{1}{2}\left(\frac{\partial u}{\partial y} + \frac{\partial v}{\partial x}\right) = e_{yx}, e_{yz} = \frac{1}{2}\left(\frac{\partial v}{\partial z} + \frac{\partial w}{\partial y}\right) = e_{zy}, e_{zx} = \frac{1}{2}\left(\frac{\partial u}{\partial z} + \frac{\partial w}{\partial x}\right) = e_{xz}$$

(2.1.8)

Therefore, Equation (2.1.6) is written as

$$\begin{pmatrix} \Delta u \\ \Delta v \\ \Delta w \end{pmatrix} = \begin{pmatrix} e_{xx} & e_{xy} & e_{xz} \\ e_{yx} & e_{yy} & e_{yz} \\ e_{zx} & e_{zy} & e_{zz} \end{pmatrix} \begin{pmatrix} \Delta x \\ \Delta y \\ \Delta z \end{pmatrix} = (e) \begin{pmatrix} \Delta x \\ \Delta y \\ \Delta z \end{pmatrix} \qquad (2.1.9)$$

where (e) represents the strain tensor, a tensor of the second rank. As the strain tensor (e) is a symmetrical tensor with only six independent components, simplified notations can be used

$e_1 = e_{xx}, e_2 = e_{yy}, e_3 = e_{zz}\ e_4 = e_{yz} = e_{zy}, e_5 = e_{zx} = e_{xz}, e_6 = e_{xy} = e_{yx}$

Thus, we have

$$(e) = \begin{pmatrix} e_1 & e_6 & e_5 \\ e_6 & e_2 & e_4 \\ e_5 & e_4 & e_3 \end{pmatrix}$$

It has been found that the three shearing strain components are related to the angular distortion of the material. To verified these results, we consider the distortion of angle $\angle APB$, a right angle included by the elemental sections of PA=dx and PB=dy in the XY plane as shown in Fig. 2.1.4. Due to a deformation, the original positions A, P and B move to A', P', and B', respectively. If $u(x, y)$ and $v(x, y)$ are the displacements in the x- and y-directions for point P(x,y), respectively, the displacement of point A in the y-direction is

$$v(x + dx, y) = v(x, y) + (\frac{\partial v}{\partial x})dx$$

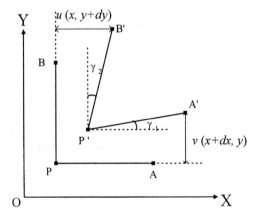

Fig. 2.1.4. Angular deformation by shearing stress

And the displacement of point B in the *x*-direction is

$$u(x, y+dy) = u(x, y) + (\frac{\partial u}{\partial y})dy$$

The line segment P'A' is now inclined against its initial direction, PA, by a small angle $\gamma_1 = \dfrac{v(x+dx, y) - v(x, y)}{dx} = \dfrac{\partial v}{\partial x}$. Similarly, the line segment P'B' is inclined against PB by a small angle $\gamma_2 = \dfrac{u(x, y+dy) - u(x, y)}{dx} = \dfrac{\partial u}{\partial y}$. Therefore, the initial right angle $\angle$APB is now deformed to $\angle$A'P'B' by a shearing angle

$$\gamma_1 + \gamma_2 = (\frac{\partial v}{\partial x} + \frac{\partial u}{\partial y})$$

By referring to Equation (2.1.8), we have $e_6 = (\gamma_1 + \gamma_2)/2$. This means that the shearing strain $e_6 = e_{xy}$ is a half of the angular distortion. The components e_4 and e_5 have the similar geometric meaning in the YZ-plane and the XZ-plane, respectively.

It may be mentioned here that, in engineering, strain is often defined as

$$\varepsilon = \begin{pmatrix} \dfrac{\partial u}{\partial x} & \dfrac{\partial u}{\partial y} + \dfrac{\partial v}{\partial x} & \dfrac{\partial u}{\partial z} + \dfrac{\partial w}{\partial x} \\ \dfrac{\partial u}{\partial y} + \dfrac{\partial v}{\partial x} & \dfrac{\partial v}{\partial y} & \dfrac{\partial v}{\partial z} + \dfrac{\partial w}{\partial y} \\ \dfrac{\partial u}{\partial z} + \dfrac{\partial w}{\partial x} & \dfrac{\partial v}{\partial z} + \dfrac{\partial w}{\partial y} & \dfrac{\partial w}{\partial z} \end{pmatrix} = \begin{pmatrix} \varepsilon_1 & \varepsilon_6 & \varepsilon_5 \\ \varepsilon_6 & \varepsilon_2 & \varepsilon_4 \\ \varepsilon_5 & \varepsilon_4 & \varepsilon_3 \end{pmatrix} \quad (2.1.10)$$

Thus we have $\varepsilon_1 = e_1, \varepsilon_2 = e_2, \varepsilon_3 = e_3, \varepsilon_4 = 2e_4, \varepsilon_5 = 2e_5$ and $\varepsilon_6 = 2e_6$. Now, the shearing component, $\varepsilon_6 = (\gamma_1 + \gamma_2)$, is exactly the angular distortion between the ZX plane and the ZY plane in the material instead of a half of the angular distortion, and so on.

According to either of the definitions described above, strain is a dimensionless quantity. However, a unit called *microstrain* is introduced for convenience. A *microstrain* (1$\mu\varepsilon$) is defined as a strain of 10^{-6}. Thus, a strain of 10^{-4} is often called 100 $\mu\varepsilon$.

§2.1.3. Hooke's Law

Linear relations between stress and strain are known as Hooke's Law. For homogeneous materials, we consider an elemental rectangular parallelepiped with its edges parallel to the coordinate axes as shown in Fig. 2.1.5. If the opposite sides of the parallelepiped perpendicular to *x*-axis are subjected to the action of a normal stress T_{XX}, the relative elongation of the element is given by

$$e_{xx} = \frac{T_{XX}}{E} \quad (2.1.11)$$

where E is a material-related constant called Young's modulus. E is usually very large in comparison with allowable stresses so that the elongation ε_{ii} is usually smaller than 0.01. Therefore, superposition can be used for deformations for most applications without causing significant errors.

Fig. 2.1.5. An elemental rectangular parallelepiped with a normal stress T_{XX},

The extension in the x-direction is always accompanied by contraction in lateral directions (the y- and z- directions)

$$e_{yy} = -\nu e_{xx}, e_{zz} = -\nu e_{xx} \tag{2.1.12}$$

where ν is another material-related constant called Poisson's ratio.

If the element is subjected to the action of normal stresses T_{XX}, T_{YY} and T_{ZZ}, the resultant strain components is obtained from Eqs. (2.1.11) and (2.1.12) by superposition

$$e_{xx} = [T_{XX} - \nu(T_{YY} + T_{ZZ})] / E$$
$$e_{yy} = [T_{YY} - \nu(T_{XX} + T_{ZZ})] / E \tag{2.1.13}$$
$$e_{zz} = [T_{ZZ} - \nu(T_{XX} + T_{YY})] / E$$

In Eq. (2.1.13), the relations between normal strain components and normal stress components are completely defined by two material-related constants: Young's modules, E, and Poisson ratio, ν.

For homogeneous materials, the relations for three shearing strain components and three shearing stress components are

$$e_{xy} = \frac{T_{XY}}{2G}, e_{yz} = \frac{T_{YZ}}{2G}, e_{zx} = \frac{T_{ZX}}{2G} \tag{2.1.14}$$

or

$$\varepsilon_{xy} = \frac{T_{XY}}{G}, \varepsilon_{yz} = \frac{T_{YZ}}{G}, \varepsilon_{zx} = \frac{T_{ZX}}{G} \tag{2.1.15}$$

where G is called the Shear modulus. Eqs. (2.1.14) and (2.1.15) indicate that a shearing strain component depends only on the corresponding shearing stress component.

It can be proved by coordinate transformation relations (See §6.3.3) that for a homogeneous material, G is not an independent parameter but one related to E and ν

$$G = \frac{E}{2(1+\nu)} \tag{2.1.16}$$

§2.1.4. General Relations Between Stress and Strain

Generally speaking, a crystal is anisotropic in its mechanical properties. Therefore, the general relation between stress (T) and strain tensors (e) is described by a matrix equation

$$(e) = (\Sigma)(T)$$

A component of the strain tensor (e) can be expressed as

$$e_{ij} = \sum_{k,l} \Sigma_{ijkl} T_{kl} \quad (i,j,k,l=1,2,3)$$

As (e) and (T) are both second rank tensors with 3×3 components so (Σ) is a forth rank tensor with 9×9 components. Tensor (Σ) is often referred to as a compliance coefficient tensor. As (T) and (e) are both symmetrical tensors, they have only six independent components each. Using simplified notation, we have

$$e_i = \sum_j \Sigma_{ij} T_j \quad (i,j=1,2,..6) \tag{2.1.17}$$

Thus, (Σ) is now a matrix of 6×6 components. As silicon material has a diamond lattice structure, (Σ) can be reduced further due to the cubic symmetry of crystal. It can be verified that in a crystallographic coordinate system, there are only three independent non-zero components, Σ_{11}, Σ_{12} and Σ_{44}, in the compliance coefficient tensor

$$(\Sigma) = \begin{pmatrix} \Sigma_{11} & \Sigma_{12} & \Sigma_{12} & 0 & 0 & 0 \\ \Sigma_{12} & \Sigma_{11} & \Sigma_{12} & 0 & 0 & 0 \\ \Sigma_{12} & \Sigma_{12} & \Sigma_{11} & 0 & 0 & 0 \\ 0 & 0 & 0 & \Sigma_{44} & 0 & 0 \\ 0 & 0 & 0 & 0 & \Sigma_{44} & 0 \\ 0 & 0 & 0 & 0 & 0 & \Sigma_{44} \end{pmatrix} \tag{2.1.18}$$

The three non-zero components are related to the three material constants by

$$\Sigma_{11} = \frac{1}{E}, \Sigma_{12} = -\frac{\nu}{E}, \Sigma_{44} = \frac{1}{2G} \tag{2.1.19}$$

Note that G is now an independent parameter as the material is not homogenous.

If the engineering definition for strain (ε) is used, we have

$$(\varepsilon) = (S)(T) \tag{2.1.20}$$

where (S) is a matrix of 6×6 components and is referred to as a coefficient of compliance. For silicon in its crystallographic coordinate system, we have

$$(S) = \begin{pmatrix} S_{11} & S_{12} & S_{12} & 0 & 0 & 0 \\ S_{12} & S_{11} & S_{12} & 0 & 0 & 0 \\ S_{12} & S_{12} & S_{11} & 0 & 0 & 0 \\ 0 & 0 & 0 & S_{44} & 0 & 0 \\ 0 & 0 & 0 & 0 & S_{44} & 0 \\ 0 & 0 & 0 & 0 & 0 & S_{44} \end{pmatrix}$$

(2.1.21)

where $S_{11}=\Sigma_{11}$, $S_{12}=\Sigma_{12}$ and $S_{44}=2\Sigma_{44}$.

Sometimes, Eq. (2.1.20) is expressed in its inverse form

$$(T) = (C)(\varepsilon)$$

(2.1.22)

where (C) is a matrix of 6×6 components and is often referred to as the coefficient of elasticity. For a cubic crystal such as silicon in its crystallographic coordinates, there are only three independent non-zero components

$$(C) = \begin{pmatrix} C_{11} & C_{12} & C_{12} & 0 & 0 & 0 \\ C_{12} & C_{11} & C_{12} & 0 & 0 & 0 \\ C_{12} & C_{12} & C_{11} & 0 & 0 & 0 \\ 0 & 0 & 0 & C_{44} & 0 & 0 \\ 0 & 0 & 0 & 0 & C_{44} & 0 \\ 0 & 0 & 0 & 0 & 0 & C_{44} \end{pmatrix}$$

(2.1.23)

The relations among E, G, ν, the components of (C), (S) and (Σ) are

$$\frac{1}{E} = S_{11} = \Sigma_{11} = \frac{C_{11}+C_{12}}{(C_{11}-C_{12})\cdot(C_{11}+2C_{12})}$$

$$\nu = -\frac{S_{12}}{S_{11}} = -\frac{\Sigma_{12}}{\Sigma_{11}} = \frac{C_{12}}{C_{11}+C_{12}}$$

(2.1.24)

$$\frac{1}{G} = S_{44} = 2\Sigma_{44} = \frac{1}{C_{44}}$$

From reference [3], the generally accepted data for silicon in a crystallographic coordinate system are: $C_{11}=1.674\times10^{11}$ Pa, $C_{12}=0.652\times10^{11}$ Pa, and $C_{44}=0.796\times10^{-11}$Pa. The matrix for the coefficient of elasticity is

$$(C) = \begin{pmatrix} 1.674 & 0.652 & 0.652 & 0 & 0 & 0 \\ 0.652 & 1.674 & 0.652 & 0 & 0 & 0 \\ 0.652 & 0.652 & 1.674 & 0 & 0 & 0 \\ 0 & 0 & 0 & 0.796 & 0 & 0 \\ 0 & 0 & 0 & 0 & 0.796 & 0 \\ 0 & 0 & 0 & 0 & 0 & 0.796 \end{pmatrix} \times 10^{11}\,\text{Pa}$$

By using equation (2.1.24), we can find the three independent components for (Σ). The compliance coefficient tensor (Σ) is

$$(\Sigma) = \begin{pmatrix} 0.764 & -0.214 & -0.214 & 0 & 0 & 0 \\ -0.214 & 0.764 & -0.214 & 0 & 0 & 0 \\ -0.214 & -0.214 & 0.764 & 0 & 0 & 0 \\ 0 & 0 & 0 & 0.628 & 0 & 0 \\ 0 & 0 & 0 & 0 & 0.628 & 0 \\ 0 & 0 & 0 & 0 & 0 & 0.628 \end{pmatrix} \times 10^{-11}\,\text{/Pa}$$

And the three mechanical parameters are found to be $E = 1.31 \times 10^{11}$ Pa, $\nu = 0.28$ and $G = 0.796 \times 10^{11}$ Pa

However, in an arbitrary coordinate system (x'-y'-z') instead of a crystallographic coordinate system (X-Y-Z), the mechanical parameters are orientation dependent. As an example, let us consider a coordinate system with its x'-axis in [110] direction, y'-axis in [$\bar{1}$10] direction and z'-axis in [001] direction. The compliance coefficient tensor (Σ') in the coordinate system (x'-y'-z') can be found from (Σ) in the crystallographic coordinate system through coordinate transformation of the tensor of the forth rank. (Readers interested in the coordinate transformation of tensors are referred to Chapter 6).

The compliance coefficient tensor in the coordinate system (x'-y'-z') is

$$(\Sigma') = \begin{pmatrix} 0.589 & -0.039 & -0.214 & 0 & 0 & 0 \\ -0.039 & 0.589 & -0.214 & 0 & 0 & 0 \\ -0.214 & -0.214 & 0.764 & 0 & 0 & 0 \\ 0 & 0 & 0 & 0.628 & 0 & 0 \\ 0 & 0 & 0 & 0 & 0.628 & 0 \\ 0 & 0 & 0 & 0 & 0 & 0.978 \end{pmatrix} \times 10^{-11}\,\text{/Pa} \qquad (2.1.25)$$

From Eq. (2.1.25), we can find that the mechanical properties are anisotropic. For example, from the value of Σ'_{11} and Σ'_{22}, we obtain the Young's modulus in the x'- and y'-direction $E_{x'x'} = E_{y'y'} = 1.7 \times 10^{11}$ Pa. However, from the value of Σ'_{33}, we have the Young's modulus in the z'-direction $E_{z'z'} = 1.31 \times 10^{11}$ Pa. Similar anisotropic conditions appear for shearing modulus G and Poisson ratio v too. For shearing modulus, we have $G= 0.796 \times 10^{11}$ for $T_{y'z'}$ and $T_{x'z'}$, but $G= 0.511 \times 10^{11}$ Pa for $T_{x'y'}$. For a normal stress $T_{x'x'}$, Poisson ratio is $v=0.066$ in the y'-direction but $v=0.36$ for the z'-direction.

As anisotropic characteristics of the mechanical property will not be considered in the rest part of this book (except for some of the problems in §2.7.), some approximate values for mechanical parameters will be used in this book for the simplicity of calculation. These values are: $E=1.7 \times 10^{11}$ Pa, $v=0.30$ and $G= 0.65 \times 10^{11}$ Pa.

§2.2. Stress and Strain of Beam Structures

§2.2.1. Stress, Strain in a Bent Beam

Consider a thin straight beam with a rectangular cross section of width b and thickness h. The length of the beam L is much larger than b and h. A coordinate system is taken with its origin at the center of a cross section and the x-axis along the beam. The z-axis is in the thickness direction of the beam and downward, as shown in Fig. 2.2.1. If the beam is bent in the x-z plane in an upward direction on both ends due to the bending moment, M, as shown in Fig. 2.2.2, the top (the concave side) of the beam is compressed and the bottom (the convex side) of the beam is stretched. There is a neutral plane somewhere in between which is neither stretched nor compressed by the bending. It will be clear later that the neutral plane is right at the middle of the rectangular beam for pure bending (by "pure bending" we mean that there is no net axial force over the cross section of the beam.)

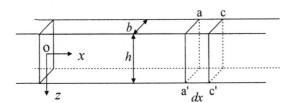

Fig. 2.2.1. An elemental section of a beam

Consider an elemental section, dx, of the beam between two vertical planes aa' and cc'. Generally, the displacement of the beam in the z-direction, w, is a function of x, i.e., $w=w(x)$, and the function is referred to as the displacement function of the beam. If the radius of curvature of the elemental section dx is OO'$=r$, as shown in Fig. 2.2.2, for a section of horizontal layer at the central plane ($z=0$), we have $dx = rd\theta$.

However, for a layer of beam away from the central plane ($z\neq0$), the material is stretched or compressed in the x-direction due to the bending deformation. The elongation of the material in the x-direction for the layer at z is $\Delta(dx)=(r+z)d\theta-rd\theta=zd\theta$. The strain of the layer is the relative elongation of the material, i.e.,

$$\varepsilon(z) = \frac{\Delta(dx)}{dx} = \frac{zd\theta}{rd\theta} = \frac{z}{r} \qquad (2.2.1)$$

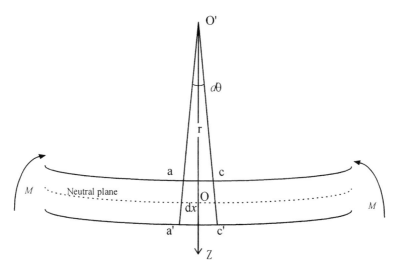

Fig. 2.2.2 Bending of a beam under a bending moment

According to Hooke's Law, the stress in the layer at z is

$$T_{XX}(z) = \frac{Ez}{r} \qquad (2.2.2)$$

The sign of the strain or stress is positive for stretch deformation and negative for compression. Note that the other two normal stresses and the three shearing stresses are not considered in the discussion here, as the beam is assumed to be thin so that the problem is one-dimensional.

When a beam is bent, the displacement of its central plane is a function of position. This means that $w=w(x)$, which is often referred to as a displacement function. In mathematics, the reciprocal of the radius of curvature of a curve $w(x)$ is the absolute value of the second derivative of $w(x)$, i.e.,

$$\frac{1}{r} = |w''(x)| \qquad (2.2.3)$$

From Eqs. (2.2.2) and (2.2.3), we have

$$|T(x,z)| = Ez|w''(x)| \qquad (2.2.4)$$

For the bending condition as shown in Fig. 2.2.2 (bending up), we have $w''(x) < 0$. For a layer with $z>0$ in the element section (below the neutral plane), the layer is stretched, i.e., $T>0$, and vice versa. Therefore, we can establish the algebraic relation for Eq. (2.2.4)

$$T(x,z) = -Ezw''(x) \qquad (2.2.5)$$

§2.2.2. Bending Moment and the Moment of Inertia

(1) Axial Force of a Beam

The axial force of a beam is the integral force on its cross section. From Eq. (2.2.5), the axial forxce of a rectangular beam caused by pure bending is

$$F = \int T(z)bdz = -Ebw''(x) \int_{-h/2}^{h/2} zdz = 0 \qquad (2.2.6)$$

As a matter of fact, the zero axial force condition is often used to find the position of the neutral plane for pure bending.

(2) Bending Moment and the Moment of Inertia

Inside a bent beam there exist internal forces across its cross sections. The bending moment is the integral force moment of the forces (against neutral plane) across a cross section of the beam

$$M(x) = \int zdF = \int zT(z)dA = -\int Ez^2 w''(x)bhdx \qquad (2.2.7)$$

Note that the origin of the z-axis is on the neutral plane for the integration. If the origin is not at the neutral plane and the position of the neutral plane is at $z=a$, then $(z-a)$ must be used to replace the z in Eq. (2.2.7). In addition, if the width of the cross section is not uniform, i.e., $b = b(z)$, we have $dA=b(z)dz$ and

$$M(x) = -Ew''(x) \int z^2 b(z)dz \qquad (2.2.8)$$

The integral is referred to as the moment of inertia of the beam and designated as I

$$I = \int_{-h/2}^{h/2} z^2 b(z)dz \qquad (2.2.9)$$

For a beam with a rectangular cross section of width b and thickness h, the moment of inertia of the beam is $I = bh^3/12$. With the moment of inertia of the beam, Eq. (2.2.8) can be written as

$$-EIw''(x) = M(x) \qquad (2.2.10)$$

and from Eqs. (2.2.5) and (2.2.10), we have

$$T(x,z) = \frac{zM(x)}{I}$$ (2.2.11)

Eqs. (2.2.10) and (2.2.11) are useful in finding the displacement function, $w(x)$, and the stress in the beam once $M(x)$ is known.

According to the definition, the bending moment is positive in sign if the beam is bending up on both sides of the element section because dF is positive for $z>0$ and negative for $z<0$, as shown in Fig. 2.2.3. For the external forces that create the bending moments, the rule for signs is: if we look at the left side of the element section dx, the moment caused by a clockwise force is positive. However, if we look at the right side of the element section dx, the moment caused by a counter clockwise force is positive. On the contrary, the bending moment is negative if the beam is bent down on both ends.

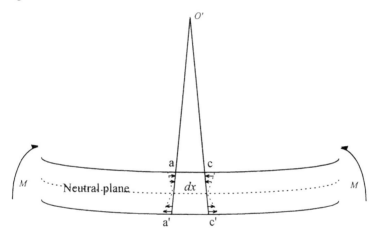

Fig. 2.2.3. The forces causing a bending moment

Quite often, the beams are made of bulk silicon by anisotropic etching. In this case, the cross section of the beam is actually a trapezoid instead of a rectangle. As a useful example, we calculate the moment of inertia for a trapezoid cross section as shown in Fig. 2.2.4.

First of all, the neutral plane has to be found for the beam. As the bottom angle is $\alpha = \tan^{-1}\sqrt{2}$ (i.e., $\alpha=54.74°$), if the origin of the z-axis is taken on the top, we have

$$b(z) = b_1 + z\sqrt{2}$$ (2.2.12)

According to the zero axial force condition for pure bending, we have

$$\int_0^h b(z)T(z)dz = 0$$ (2.2.13)

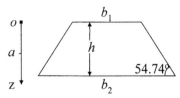

Fig. 2.2.4. The trapezoid cross section of a beam

If the position of the neutral plane is $z = a$, we have $T(z) \sim (z-a)$. Position a can be found using the condition of

$$\int_0^h b(z)(z-a)dz = 0 \tag{2.2.14}$$

From Eq. (2.2.14), a is found to be

$$a = \frac{(b_1 + 2b_2)h}{3(b_1 + b_2)} \tag{2.2.15}$$

If b_1, $b_2 \gg h$, we have $a \cong h/2$, i.e., it can be approximated as a rectangular beam. On the other extreme, if $b_1 = 0$ (i.e., a triangle beam), we have $a = 2h/3$. The result is now quite different from the rectangular beam.

Once the position of the neutral plane is found, the moment of inertia of the beam can be calculated. According to the definition, we have

$$I = \int_0^h b(z)(z-a)^2 dz \tag{2.2.16}$$

From Eqs. (2.2.15) and (2.2.16), the moment of inertia of the trapezoid beam is

$$I = \frac{h^3(b_1^2 + 4b_1b_2 + b_2^2)}{36(b_1 + b_2)} \tag{2.2.17}$$

§2.2.3. Displacement of Beam Structures Under Weight

If the geometrical dimensions, the material parameters of the beam, the external forces and the boundary conditions are known, the displacement function $w(x)$ of the beam can be found by Eq. (2.2.10) and the boundary conditions. Once $w(x)$ is found, the stresses in the beam can be calculated using Eq. (2.2.5). Some commonly used beam structures will be discussed in the following examples.

(1) Cantilever Beam With a Concentrated End Loading

Consider a rectangular cantilever beam with a concentrated end loading, as shown in Fig. 2.2.5. The width, thickness and the length of the beam are b, h, and L, respectively. Let

F be the loading force caused by a mass, M, attached to the free end of the beam and the gravitational acceleration, i.e., $F=Mg$. Also assume that the end mass, M, is much larger than the beam mass, so that the gravitational force of the beam can be neglected. To balance the loading force, F, there must be a supporting force, F_o, acting on the beam at the clamped end. The force balance in the z-direction requires that $F_o=F$. In the meantime, there must be a restrictive bending moment, m_o, at the clamped end of the beam to balance the bending moment caused by the loading force, F.

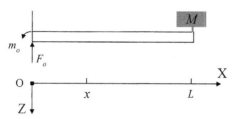

Fig. 2.2.5. Cantilever beam with a concentrated end loading

The force moment produced by the loading force against the clamped end is FL, in a clockwise direction. For the balance of bending moments, the restrictive bending moment, m_0, must be equal to FL, in a counter-clockwise direction. Generally, m_0 cannot be found as easily as in the simple discussion above. It might have to be found using equation solving.

Looking at position x as shown in Fig. 2.2.5, the bending moment on the left is $(-m_o+F_ox)$ and the bending moment on the right side is $-F \cdot (L-x)$. As a matter of fact, the bending moment on the left is equal to that on the right, i.e., $M(x)=-F \cdot (L-x)$. Therefore, the differential equation for $w(z)$ is

$$- EIw''(x) = -F \cdot (L-x) \tag{2.2.18}$$

The boundary conditions for the problem are

$$w(0) = 0, w'(0) = 0 \text{ and } w''(L) = 0 \tag{2.2.19}$$

The solution to Eqs. (2.2.18) and (2.2.19) is

$$w(x) = \frac{F(3L-x)x^2}{6EI} = \frac{2Mg(3L-x)x^2}{Ebh^3} \tag{2.2.20}$$

The maximum displacement at the free end (i.e., at $x=L$) is

$$w_{max} = w(L) = \frac{FL^3}{3EI} = \frac{4MgL^3}{Ebh^3} \tag{2.2.21}$$

If the beam is considered as a spring-board, its spring constant is

$$k = \frac{Ebh^3}{4L^3} \qquad (2.2.22)$$

From Eq. (2.2.5), the stress on the top surface of the beam (at $z = -h/2$) is

$$T(x) = -E \cdot \left(-\frac{h}{2}\right)w''(x) = \frac{Fh(L-x)}{2I} = \frac{6Mg(L-x)}{bh^2} \qquad (2.2.23)$$

The stress has a maximum at $x = 0$ and decreases linearly to zero at $x = L$. From Eq. (2.2.23), the maximum stress at $x = 0$ is

$$T_{max} = \frac{FhL}{2I} = \frac{6MgL}{bh^2} \qquad (2.2.24)$$

(2) Bending of Cantilever Beam Under Weight

If there is no concentrated mass at the free end of the cantilever beam and the loading is just the weight of the beam, the loading is uniformly distributed. By using the same notations for the geometries and the mechanical parameters as in the previous example and the coordinate system as shown in Fig. 2.2.6, the differential equation for the beam is

$$-EIw''(x) = M_b gx - m_o - \int_0^x \frac{M_b g}{L}(x-s)ds \qquad (2.2.25)$$

where M_b is the total mass of the beam, i.e., $M_b = \rho bhL$, g is the gravitational acceleration and the last term on the right is the bending movement caused by the distributed weight of the beam between 0 and x. After integration, we have the equation

$$-EIw''(x) = M_b gx - m_o - \frac{M_b gx^2}{2L} \qquad (2.2.26)$$

The boundary conditions for the beam are

$$w(0) = 0, \ w'(0) = 0 \ \text{and} \ w''(L) = 0 \qquad (2.2.27)$$

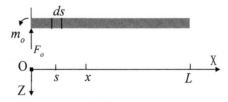

Fig. 2.2.6. Cantilever beam with a distributed loading

By solving equations (2.2.26) and (2.2.27), the restrictive bending moment is found to be $m_o = M_b gL/2 = \rho bhgL^2/2$ and the displacement of the beam is

$$w(x) = \frac{M_b gx^2(x^2 - 4Lx + 6L^2)}{24ELI} = \frac{M_b gx^2(x^2 - 4Lx + 6L^2)}{2EbLh^3} \tag{2.2.28}$$

The maximum displacement at the end of the beam is

$$w_{max} = \frac{M_b gL^3}{8EI} = \frac{3M_b gL^3}{2Ebh^3} \tag{2.2.29}$$

or,

$$w_{max} = \frac{3\rho gL^4}{2Eh^2} \tag{2.2.30}$$

The stress on the top of the beam is

$$T(x) = -E \cdot \left(-\frac{h}{2}\right) w''(x) = \frac{M_b g(L-x)^2}{4LI} = \frac{3M_b g(L-x)^2}{bh^2 L} \tag{2.2.31}$$

The maximum stress at $x = 0$ is

$$T_{max} = \frac{M_b ghL}{4I} = \frac{3M_b gL}{bh^2} = \frac{3\rho gL^2}{h} \tag{2.2.32}$$

(3) Double-clamped Beam (Bridge)

For a beam with both ends clamped (often called a bridge) as shown in Fig. 2.2.7, if the loading of the beam is again its own weight, the equation for the displacement of the beam is similar to Eqs. (2.2.25) or (2.2.26), but, as the beam is supported on both ends, the two supporting forces are $F_o = M_b g/2$ for both ends. Therefore, the equation for the displacement function is

$$-EIw''(x) = \frac{M_b gx}{2} - m_o - \frac{M_b gx^2}{2L} \tag{2.2.33}$$

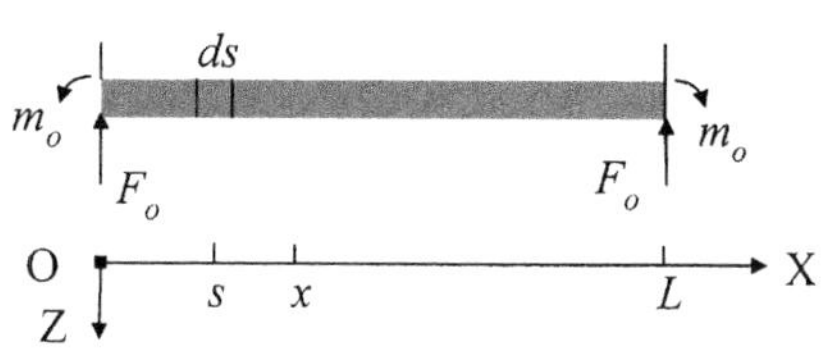

Fig. 2.2.7. Double-clamped beam (bridge) and its coordinates

The boundary conditions are

$$w(0) = 0, \ w'(0) = 0, \ w(L) = 0 \text{ and } w'(L) = 0 \tag{2.2.34}$$

From Eqs. (2.2.33) and (2.2.34), m_o is found to be

$$m_o = \frac{M_b g L}{12} = \frac{\rho g b h L^2}{12} \tag{2.2.35}$$

and the displacement function of the beam is

$$w(x) = \frac{M_b g}{24 E L I} x^2 (L-x)^2 = \frac{\rho g}{2 E h^2} x^2 (L-x)^2 \tag{2.2.36}$$

The maximum displacement at the center of the beam (at $x=L/2$) is

$$w_{max} = \frac{\rho g L^4}{32 E h^2} \tag{2.2.37}$$

The stress on the top of the beam ($z = -h/2$) is

$$T(x) = -E \cdot \left(-\frac{h}{2} \right) w''(x) = \frac{M_b g h}{24 I L} (L^2 - 6xL + 6x^2) \tag{2.2.38}$$

The maximum stress at $x = 0$ (also at $x=L$) is

$$T_{max} = \frac{M_b g h L^2}{24 I L} = \frac{\rho g L^2}{2h} \tag{2.2.39}$$

When Eq. (2.2.39) is compared with Eq. (2.2.32), we can find that the maximum stress in a bridge is smaller than that in a cantilever beam by a factor of 6 for the same conditions. Also, if Eq. (2.2.37) is compared with Eq. (2.2.30), we can find that the maximum displacement of a bridge is smaller than that of a cantilever beam by a factor of 48.

(4) Double-clamped Beam With a Central Mass

The structure of two clamped beams supporting a central mass is often used for inertial sensors (such as accelerometers and gyroscopes), resonators and mechanical actuators (such as switches). A schematic drawing of the structure is shown in Fig. 2.2.8 and, as a typical condition, the central mass is much wider and/or thicker than the beams so that the bending of the mass can be negligible. If the loading of the beam-mass structure is its own weight in the z-direction, the displacement of the mass is piston-like due to the symmetry of the structure in the x- and y-direction. Therefore, only a half of the structure has to be considered. For the left beam, the displacement is similar to Eq. (2.2.33)

$$-EIw''(x) = \frac{Mg}{2}x - m_o - \frac{\rho bhgx^2}{2} \qquad (2.2.40)$$

where $I = bh^3/12$, m_o is the restrictive bending moment to be determined and M the total mass of the beam-mass structure

$$M = 2(L - a_1)BH\rho + 2bha_1\rho$$

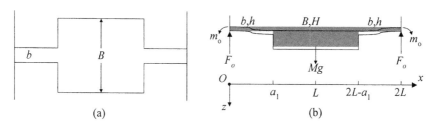

(a)   (b)

Fig. 2.2.8. The schematic drawing of a double-supported beam-mass structure (a) top view; (b) cross sectional view

Under the condition that the mass of the beams is much smaller than that of the central mass, the equation can be simplified as

$$-EIw''(x) = \frac{Mg}{2}x - m_o \qquad (2.2.41)$$

The boundary conditions for the equation are

$$w(0) = 0, \; w'(0) = 0, \; w'(a_1) = 0 \text{ and } w''(\frac{1}{2}a_1) = 0 \qquad (2.2.42)$$

From equations (2.2.41) and (2.2.42), we have

$$m_o = \frac{1}{4}Mga_1$$

Therefore, the stress on the top of the beam is

$$T(x) = \frac{3Mg}{bh^2}\left(\frac{1}{2}a_1 - x\right) \qquad (2.2.43)$$

The distribution of the stress is shown in Fig. 2.2.9. The stresses on the beam surface vary linearly from a positive maximum at one end to a negative maximum at the other end. The values for both the positive and negative maximums are the same

$$T_{max} = \frac{3a_1}{2bh^2} Mg$$

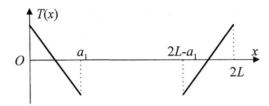

Fig. 2.2.9. The stress distribution on the beam surface

The displacement of the left beam is

$$w(x) = \frac{Mg}{Ebh^3} x^2 (\frac{3}{2}a_1 - x)$$
(2.2.44)

The displacement of the central mass is

$$w(a_1) = \frac{Mg}{2Ebh^3} a_1^3$$

If the beams are considered as a spring, the spring constant is

$$k = \frac{2Ebh^3}{a_1^3}.$$
(2.2.45)

§2.2.4. Bending of Layered Composite Beam by Residual Strain

For practical applications, silicon beams are often covered with layers of films such as silicon dioxide, silicon nitride, resist, aluminum or other metals for insulation, passivation, actuation and many other purposes. For example, the most widely used beam structure is the silicon beam with silicon dioxide film on top. As the silicon dioxide film is grown or deposited at an elevated temperature and the coefficient of thermal expansion of the film is different from that of silicon, internal strain or stress develops when the SiO_2/Si composite beam is cooled down to a room temperature. The strain/stress is often referred to as a thermal strain/stress or a residual strain/stress.

The residual strain will cause the beam to bend or curl and the normal operation of the beam might be jeopardized. Therefore, the bending effect of the residual strain should be considered for MEMS devices. First, the general expressions for neutral plane and the curvature of a double-layer beam are derived following the line of [4] with some modification. With these results, the thermal stress/strain in the beam can be found. As an application of the equations, Stoney equation is derived, which relates the stress/strain in a thin film to the curvature of the beam. Finally, the respective expressions for multi-layer beam structures are given.

(1) Double-layer Composite Beam

For simplicity, let us first consider the bending of a SiO_2/Si composite cantilever beam as shown in Fig. 2.2.10, where t_{ox} is the thickness of silicon dioxide and t_{si} is the thickness of silicon, and the thickness of the beam is supposed to be much smaller than the width and the length of the beam. Suppose that the silicon dioxide is grown at an elevated temperature T' and the length of the composite beam is L' and no internal strain (i.e., no intrinsic strain) appears in the beam at the temperature.

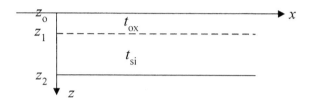

Fig. 2.2.10. The double-layer SiO_2/Si composite beam

If the silicon dioxide and the silicon beams were detached and cooled down to room temperature T, the length of the dioxide film would become

$$L_{ox} = L'(1 - \alpha_{ox}\Delta T) \tag{2.2.46}$$

where $\Delta T = T' - T$ and α_{ox} is the coefficient of thermal expansion of silicon dioxide. The length of the silicon beam would become

$$L_{si} = L'(1 - \alpha_{si}\Delta T) \tag{2.2.47}$$

where α_{si} is the coefficient of thermal expansion of silicon. The commonly used data for the coefficients of thermal expansion are $\alpha_{ox}=0.5\times10^{-6}$/K and $\alpha_{si}=2.6\times10^{-6}$/K [5].

As α_{ox} is smaller than α_{si}, the silicon dioxide is compressed and the silicon is stretched in the SiO_2/Si composite beam at temperature T. Therefore, the resulting length of the composite beam, L, will be somewhere in between L_{si} and L_{ox}, i.e., $L_{ox}>L>L_{si}$. As $\alpha_{ox}\Delta T$ and $\alpha_{si}\Delta T$ are much smaller than unity, the strain in the silicon dioxide film is

$$\varepsilon_{ox} = \frac{L - L_{ox}}{L} \doteq \ln\frac{L}{L_{ox}} \tag{2.2.48}$$

Similarly, the strain in the silicon layer is

$$\varepsilon_{si} = \frac{L - L_{si}}{L} \doteq \ln \frac{L}{L_{si}} \tag{2.2.49}$$

As one end of the beam is free, the total end force is zero. Therefore, we have

$$E_{si} t_{si} \ln \frac{L}{L_{si}} + E_{ox} t_{ox} \ln \frac{L}{L_{ox}} = 0 \tag{2.2.50}$$

By substituting equations (2.2.46) and (2.2.47) into equation (2.2.50), the beam length at temperature T is found to be

$$L = L'(1 - \alpha_{eff} \Delta T) \tag{2.2.51}$$

where α_{eff} is the effective coefficient of thermal expansion of the composite beam

$$\alpha_{eff} = \frac{E_{si} t_{si} \alpha_{si} + E_{ox} t_{ox} \alpha_{ox}}{E_{si} t_{si} + E_{ox} t_{ox}} \tag{2.2.52}$$

Although the total end force acting on the composite beam is null, this strain state does not necessarily satisfy the condition of moment equilibrium. As a general result, the composite beam bends. The strain caused by the bending is added to satisfy the equilibrium of moment. If the displacement function of the beam is designated as $w(x)$, the strain caused by the bending is $-(z - z_o) w''(x)$, where z_o is the position of the neutral plane and $w''(x)$ the curvature of the beam. Generally, the bending effect is much more concerned than the effect of change in length of the beam.

When the strain caused by the bending is considered, the stress in silicon dioxide is

$$T_{ox} = E_{ox} \ln \frac{L}{L_{ox}} - E_{ox}(z - z_o) w''(x)$$

and the stress in silicon is

$$T_{si} = E_{si} \ln \frac{L}{L_{si}} - E_{si}(z - z_o) w''(x)$$

The position of the neutral plane is found to be

$$z_o = \frac{E_{ox} t_{ox}(0 + z_1) + E_{si} t_{si}(z_1 + z_2)}{2(E_{ox} t_{ox} + E_{si} t_{si})}$$

Or,

$$z_o = \frac{E_{ox}t_{ox}(0+t_{ox}) + E_{si}t_{si}(2t_{ox}+t_{si})}{2(E_{ox}t_{ox}+E_{si}t_{si})} \quad (2.2.53)$$

The curvature of the composite beam can be found by using the balance condition of force moment against the neutral plane

$$\int_0^{z_1} E_{ox}\ln\frac{L}{L_{ox}}(z-z_o)dz - E_{ox}w''\int_0^{z_1}(z-z_o)^2dz + \int_{z_1}^{z_2}E_{si}\ln\frac{L}{L_{si}}(z-z_o)dz - E_{si}w''\int_{z_1}^{z_2}(z-z_o)^2dz = 0$$

The expression for the curvature of the composite beam is found to be

$$w''(x) = \frac{\int_0^{z_1}E_{ox}\ln\frac{L}{L_{ox}}(z-z_o)dz + \int_{z_1}^{z_2}E_{si}\ln\frac{L}{L_{si}}(z-z_o)dz}{E_{ox}\int_0^{z_1}(z-z_o)^2dz + E_{si}\int_{z_1}^{z_2}(z-z_o)^2dz} \quad (2.2.54)$$

From equations (2.2.46), (2.2.47), and (2.2.51), we have $\ln(L/L_{ox}) \doteq (\alpha_{ox}-\alpha_{eff})\Delta T = \varepsilon_{ox}$ and $\ln L/L_{si} \doteq (\alpha_{si}-\alpha_{eff})\Delta T = \varepsilon_{si}$. Therefore, the curvature of the beam is

$$w''(x) = 3\frac{E_{ox}(\alpha_{ox}-\alpha_{eff})t_{ox}(\frac{0+z_1}{2}-z_o) + E_{si}(\alpha_{si}-\alpha_{eff})t_{si}(\frac{z_1+z_2}{2}-z_o)}{E_{ox}((z_1-z_o)^3-(0-z_o)^3) + E_{si}((z_2-z_o)^3-(z_1-z_o)^3)}\Delta T \quad (2.2.55)$$

With Eq. (2.2.53) for z_o and Eq. (2.2.55) for $w''(x)$, the stress and strain in the beam can be found.

Now let us consider a commonly encountered situation that the silicon dioxide film is much thinner than the silicon substrate (i.e., $t_{si} \gg t_{ox}$). In this case, the position of neutral plane can be approximated as

$$z_o = \frac{1}{2}t_{si}$$

By substituting this value into Eq. (2.2.55), we have

$$w''(x) = 6\frac{E_{ox}\varepsilon_{ox}t_{ox}}{E_{si}t_{si}^2}$$

Therefore, the residual strain in oxide is

$$\varepsilon_{ox} = \frac{E_{si}t_{si}^2}{6E_{ox}t_{ox}}w''(x)$$

Or, the residual stress in oxide is

$$T_{ox} = \frac{E_{si} t_{si}^{2}}{6 t_{ox}} w''(x) \tag{2.2.56}$$

This is the well-known Stoney equation, which is widely used to relate the bending of the beam to the stress or strain in the thin film.

(2) Multi-layer Composite Beam

By the discussion similatr to that for double layer composite beam, the resultant beam length, the position of the neutral plane and the curvature of the beam can be found for a composite beam consisting of n layers as shown in Fig. 2.2.11. The length of the beam at temperature T is

$$L = L'(1 - \alpha_{eff} \Delta T)$$

where

$$\alpha_{eff} = \frac{E_1 t_1 \alpha_1 + E_2 t_2 \alpha_2 + ... + E_n t_n \alpha_n}{E_1 t_1 + E_2 t_2 + ... + E_n t_n}$$

where E_i, t_i and α_i are Young's modulus, the thickness and the coefficient of thermal expansion of the i'th layer, respectively.

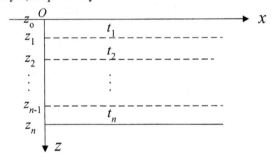

Fig. 2.2.11. Multi-layer composite beam with n layers

The position of the neutral plane is

$$z_o = \frac{E_1 t_1 (0 + z_1) + E_2 t_2 (z_1 + z_2) + ... + E_n t_n (z_{n-1} + z_n)}{2(E_1 t_1 + E_2 t_2 + ... + E_n t_n)} \tag{2.2.57}$$

where $z_i = \sum_{j=1}^{i} t_j$ (i =1,2,3 ... n), i.e., the position of the bottom of the i'th layer. The curvature of the beam is

$$w'' = \frac{N}{D} \tag{2.2.58}$$

where

$$N = \int_0^{z_1} E_1 \ln\left(\frac{L}{L_1}\right)(z - z_o)dz + \int_{z_1}^{z_2} E_2 \ln\left(\frac{L}{L_2}\right)(z - z_o)dz + ... + E \int_{z_{n-1}}^{z_n} e_n \ln\left(\frac{L}{L_n}\right)(z - z_o)dz$$

$$D = E_1 \int_0^{z_1} (z - z_o)^2 dz + E_2 \int_{z_1}^{z_2} (z - z_o)^2 dz + ... + E_n \int_{z_{n-1}}^{z_n} (z - z_o)^2 dz$$

With the similar transformation as made for SiO_2/Si beam, we have

$$N = \Delta T \left\{ \begin{array}{l} E_1(\alpha_1 - \alpha_{eff})t_1\left[\frac{1}{2}(0 + z_1) - z_o\right] + E_2(\alpha_2 - \alpha_{eff})t_2\left[\frac{1}{2}(z_1 + z_2) - z_o\right] + ... \\ + E_n(\alpha_n - \alpha_{eff})t_n\left[\frac{1}{2}(z_n + z_{n-1}) - z_o\right] \end{array} \right\}$$

$$D = \frac{1}{3}\left\{ \begin{array}{l} \left[(z_1 - z_o)^3 - (0 - z_o)^3\right] + \frac{1}{3}\left[(z_2 - z_o)^3 - (z_1 - z_o)^3\right] + ... \\ + \left[(z_n - z_o)^3 - (z_{n-1} - z_o)^3\right] \end{array} \right\} \tag{2.2.59}$$

In practical conditions, however, different layers might be grown or deposited at different temperatures, or there might be intrinsic strain in the as-deposited films at the deposition temperatures. In these cases, the situation will be much more complicated. That will not be considered here.

§2.2.5. Angular Displacement of Torsion Bar Structures

When a torque, T, is applied on the free end of a cantilever beam, the beam is twisted (angular displaced) until the restoring torque of the beam balances the applied torque. Within the elastic limitation, the relation between the twist angle, ϕ, and the torque is

$$T = k_\phi \phi \tag{2.2.60}$$

where k_ϕ is referred to as the torsion constant of the beam. The torsion constant equals the restoring torque per unit angular displacement. The beam for torsion movement is often called a torsion bar.

(1) Circular Torsion Bar

If the torsion bar has a circular cross section, i.e., the torsion bar is a solid cylinder with a radius a and a length L as shown in Fig. 2.2.12, the torsion constant can be found by the following analysis.

Consider an elemental cylinder between radii r and $r+dr$. The effective area dA over which a tangential force, dF, is applied is half the cross section area of the cylinder, i.e., $dA=\pi r dr$. According to the definition of shear modulus G, we have

$$G = \frac{dF}{\gamma dA}$$

where γ is the shearing angle as shown in Fig. 2.2.12. As $\gamma=\phi r/L$, we have

$$dF = \frac{\pi r^2 G\phi}{L} dr$$

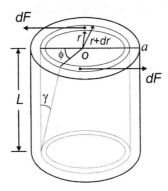

Fig. 2.2.12. Circular torsion bar

The torque on the top surface of the solid cylinder is

$$T = 2\int_0^a r dF = \frac{\pi G a^4}{2L}\phi \equiv k_\phi \phi \qquad (2.2.61)$$

Therefore, the torsion constant of the solid cylinder is

$$k_\phi = \frac{\pi G a^4}{2L} \qquad (2.2.62)$$

The shearing stress on the surface of the cylinder is [1]

$$\tau = \frac{Ga}{L}\phi \qquad (2.2.63)$$

(2) Rectangular Torsion Bar

For a torsion bar with a rectangular cross section, the analysis of its torsion constant is quite complicated. According to Eq. (2.2.81), the theoretical expression [1] for k_ϕ is

$$k_\phi = \frac{1}{3L} Ga^3 b \left(1 - \frac{192}{\pi^5} \frac{a}{b} \sum_{n=1,3,5,}^{\infty} \frac{1}{n^5} \tanh \frac{n\pi b}{2a} \right) \equiv \frac{k_1 Ga^3 b}{L} \qquad (2.2.64)$$

where a and b are the length of the shorter and longer sides of the rectangular cross section (i.e., $b>a$) as shown in Fig. 2.2.13, respectively, and the coefficient k_1 is

$$k_1 = \frac{1}{3} \left(1 - \frac{192}{\pi^5} \frac{a}{b} \sum_{n=1,3,5,}^{\infty} \frac{1}{n^5} \tanh \frac{n\pi b}{2a} \right) \qquad (2.2.65)$$

For $a=b$, we have $k_\phi=0.141$. Several values of k_1 are given in Table 2.2.1.
    In case of a narrow rectangular cross section (i.e., $b>>a$), we have

$$k_1 = \frac{1}{3} \left(1 - 0.63 \frac{a}{b} \right) \qquad (2.2.66)$$

    The maximum shearing stresses on the bar are at the middle points of the long sides of the rectangle as shown in Fig. 2.2.13. The maximum shearing stress is

$$\tau_{max} = \frac{G\phi a}{L} - \frac{8G\phi a}{\pi^2 L} \sum_{n=1,3,5,}^{\infty} \frac{1}{n^2 \cosh(n\pi b/2a)} \equiv k \frac{G\phi a}{L} \qquad (2.2.67)$$

where

$$k = 1 - \frac{8}{\pi^2} \sum_{n=1,3,5,}^{\infty} \frac{1}{n^2 \cosh(n\pi b/2a)} \qquad (2.2.68)$$

For a very narrow rectangular cross section, i.e. $b>>a$, the sum of the infinite series can be neglected so that $k=1$. In case of a square cross section, i.e., $a = b$, we have $k=0.675$.

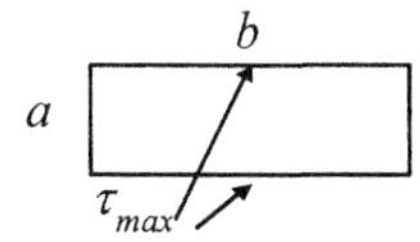

Fig. 2.2.13. The cross section of the rectangular torsion bar

    From Eq. (2.2.67) and Eq. (2.2.61) the maximum shearing stress is

$$\tau_{max} = \frac{T}{k_2 a^2 b} \qquad (2.2.69)$$

where $k_2=k_1/k$. Several values of k and k_2 can also be found in Table 2.2.1.

Table 2.2.1. Factors for torsion relations

b/a	1.0	1.2	1.5	2.0	2.5	3.0	4.0	5.0	10	∞
k	0.675	0.759	0.848	0.930	0.968	0.985	0.997	0.999	1.000	1.000
k_1	0.141	0.166	0.196	0.229	0.249	0.263	0.281	0.291	0.312	0.333
k_2	0.208	0.219	0.231	0.246	0.258	0.267	0.282	0.291	0.312	0.333

The application of torsion movement is getting more and more popular in micro mechanical sensors and actuators in recent years. A typical structure for torsion movement is a micro mechanical pendulum as shown in Fig. 2.2.14, where a plate is suspended on a pair of micro mechanical torsion bars.

Thus, a torque will act on the torsion bars when the plate is subjected to an acceleration (or gravity) in its normal direction and the torsion bars are twisted by the torque. As the twist angle ϕ is proportional to the torque which is in turn proportional to the acceleration, the twist angle, ϕ, can be used as a measure of the acceleration. Accelerometers have been developed using this principle.

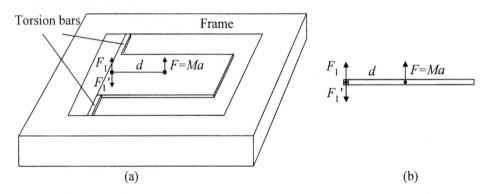

Fig. 2.2.14. A micro Torsion bar structure (a) perspective view; (b) cross-sectional view

(3) Shear Center

For the torsion bar structure as shown in Fig. 2.2.14, if the mass of the plate is M and the mass center of the plate is off the axis of the torsion bars by a distance d, the inertial force $F=Ma$ at the mass center of the plate can be dissolved into two components: a force $F_1=Ma$ acting at the center of the bar axis and a torque $T_1 = Mad$ formed by the forces F_1' and F. Therefore, the plate should have a displacement in the normal direction (though small) due to the bending of the bars caused by F_1 in addition to the torsion movement caused by T_1.

In the more general case when the cross section of the bar is not rectangular so that the axis of the bars is not obviously known, the applied force has to be dissolved referring to an axis called the shear center of the torsion bar. The shear center is a point at the beam cross section where a force perpendicular to the beam at that point does not cause a twist of the beam. Obviously, for rectangular beams, the shear center coincides with the center of the cross section of the beam.

§2.3. Vibration Frequency by Energy Method

§2.3.1. Spring-mass System

(1) Vibration Frequency of a Spring-mass System
A vibratory mechanical structure is often simplified as a mass-spring model as shown in Fig. 2.3.1.

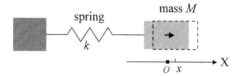

Fig. 2.3.1. A mass-spring model

The origin of the x-axis is at the balance position of the mass center. A displacement of the mass from its balance position, x, causes a recovery force acting on the mass by the spring, $F = -kx$, where k is the elastic constant of the spring. The negative sign implies that the force is pointing back to the origin of the x-axis. If there is no other force acting on the mass, the differential equation for the mass movement is

$$M\ddot{x} = -kx \qquad (2.3.1)$$

Let $\omega^2 = k/M$, we have

$$\ddot{x} + \omega^2 x = 0 \qquad (2.3.2)$$

The general solution to Eq. (2.3.2) is a sinusoidal vibration of the mass

$$x = A\sin(\omega t + \alpha) \qquad (2.3.3)$$

where ω is the radial frequency of the vibration, A the amplitude of the vibration and α a phase lag angle. The radial frequency ω is only decided by the structure parameters k and M

$$\omega = \sqrt{\frac{k}{M}} \qquad (2.3.4)$$

The amplitude A and the phase lag α can be decided by the initial conditions of the system. For example, for $x = x_o$ and $\dot{x} = v_o$ at $t = 0$, we have

$$A = \sqrt{x_o^2 + \left(\frac{v_o}{\omega}\right)^2}$$

and

$$\alpha = \arctan\left(\frac{x_o \omega}{v_o}\right)$$

The vibration frequency, ω, can also be found by the general principle of energy conservation of the system. As the vibration of the spring-mass system can be described by Eq. (2.3.3), the potential energy of the system is

$$E_P = \frac{1}{2}kx^2 = \frac{1}{2}A^2 k \sin^2(\omega t + \alpha)$$

and the kinetic energy of the system is

$$E_K = \frac{1}{2}M\dot{x}^2 = \frac{1}{2}A^2\omega^2 M \cos^2(\omega t + \alpha)$$

The total energy, E, of the system is

$$E = E_P + E_K$$

When $\omega t + \alpha = 0$, we have $E = E_{k,max} = MA^2\omega^2/2$ and when $\omega t + \alpha = \pi/2$, we have $E = E_{P,max} = kA^2/2$. According to the principle of energy conservation, $E_{k,max} = E_{P,max} = E$, we have

$$\frac{1}{2}MA^2\omega^2 = \frac{1}{2}kA^2$$

This leads to the same result as that given by Eq. (2.3.4).

(2) Vibration Frequency of a Cantilever Beam With an End Mass

Now let us consider the vibration frequency of a cantilever beam structure with a loading mass, M, at its free end. Assume that the mass at the end is much larger than the mass of the beam so that the mass of the beam can be neglected. This structure can be considered as a spring-mass system. To find the elastic constant of the spring we check the force-displacement relation of the mass. According to Eq. (2.2.22), the elastic constant of the beam is

$$k = \frac{Ebh^3}{4L^3} \tag{2.3.5}$$

By substituting the k into the Eq. (2.3.4), the radial frequency of the beam-mass structure is

$$\omega = \sqrt{\frac{Ebh^3}{4L^3 M}} \tag{2.3.6}$$

Therefore, the vibration frequency of the beam-mass system is

$$f = \frac{1}{4\pi} \sqrt{\frac{Ebh^3}{L^3 M}} \tag{2.3.7}$$

Note that the vibration frequency is only a function of structural parameters.

(3) Vibration Frequency of a Double Supported Beam-mass Structure

For a double-supported beam-mass structure as shown in Fig. 2.2.8, if the mass of the beams and the bending of the central mass is negligible, the structure can be considered as a spring-mass system and the spring constant is given by Eq. (2.2.45), $k = 2Ebh^3 / a_1^3$. Therefore, the vibration frequency of structure determined by Eq. (2.3.4) is

$$\omega = \sqrt{\frac{2Ebh^3}{Ma_1^3}} \tag{2.3.8}$$

where $M = 2\rho BH(L - a_1)$ is the mass of the central plate.

(4) Vibration Frequency of a Torsion Structure

The same method can be used for torsion vibrations. Consider a torsion bar with a mass at its end. When the bar is twisted by an angle φ at the end, the restoring torque acting on the mass by the torsion bar is

$$T = -k_\varphi \varphi$$

where k_φ is the elastic constant of the torsion bar. If the moment of inertia for rotation of the mass is I_φ, by the Newton's second law on rotation, the differential equation for an angular vibration is

$$I_\varphi \ddot{\varphi} + k_\varphi \varphi = 0 \tag{2.3.9}$$

The solution to Eq. (2.3.9) is an angular oscillation $\varphi = \Phi \sin(\omega t + \alpha)$. The radial frequency of the angular vibration is

$$\omega = \sqrt{\frac{k_\varphi}{I_\varphi}} \tag{2.3.10}$$

Now let us consider an example. Suppose that we have a torsion structure as shown in Fig. 2.3.2. The width and length of the torsion bars are b=2μm and l=20μm, respectively. The width and length of the plate are B=400μm and L=600μm, respectively, and the torsion bars and the plate have the same thickness of h=2μm. The moment of inertia for the rotation of the plate is

$$I_\varphi = \int_{-L/2}^{L/2} \rho Bhx^2 dx = \frac{1}{12}\rho BhL^3$$

The resultant elastic constant of the two torsion bars is

$$k_\varphi = \frac{2Gk_1bh^3}{l}$$

If the structure is made of silicon, we use G=6.5×10^{10} Pa and ρ=2330 Kg/m^3. From Table 2.1.1, we find k_1=0.141. According to Eq. (2.3.10), the radial frequency of the angular oscillation is 20900/sec, or, the frequency is 3.33 kHz.

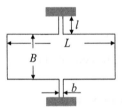

Fig. 2.3.2. A torsion bar structure for the frequency analysis

§2.3.2. Rayleigh-Ritz Method

According to the discussion in the previous section, the vibration frequency of the system with a concentrated mass and a spring can be calculated easily. However, the simple relations cannot be used to calculate the vibration frequency for a system with a distributed mass. Based on the general principle of energy conservation, a more general method called the Rayleigh-Ritz method has been developed for calculating the vibration frequency of systems with distributed masses. Here we will establish the Rayleigh-Ritz method for a uniform beam with a rectangular cross section. The result is applicable to many micro structures with some minor modifications.

First, we consider the energy stored in an elemental section, dx, at position x on the beam as shown in Fig. 2.3.3. The coordinate system is taken as before with the x-axis at the centroidal line of the beam and the z-axis in the thickness direction and downward. The displacement of the beam in the z-direction is a function of x and t, $w(x,t)$.

(1) The Potential Energy

As discussed in §2.2, for a pure bending in the *x-z* plane, the stress and strain in the beam are: $\varepsilon = -zw''(x,t)$ and $T = -Ezw''(x,t)$, respectively.

For a thin layer of thickness *dz* in the section, the areas on its front and rear ends are *bdz* (where *b* is the width of the beam) and the normal forces on both ends are

$$df = Tbdz = -Ezw''(x,t)bdz$$

The elongation of the layer in the *x*-direction due to the forces is

$$\Delta(dx) = \varepsilon dx = -zw''(x,t)dx \qquad (2.3.11)$$

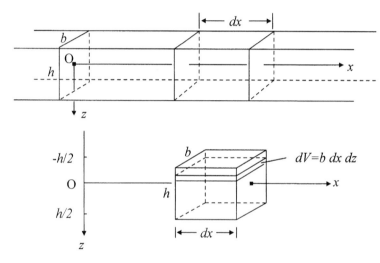

Fig. 2.3.3. Analysis of potential energy in a beam

The potential energy stored in the layer of *dz* caused by the bending deformation is

$$\Delta(dE_P) = \frac{1}{2} df \cdot \Delta(dx)$$

From Eq. (2.3.11), the potential energy in the layer of *dz* is

$$\Delta(dE_P) = \frac{1}{2} Ez^2 w''^2(x,t)bdxdz$$

The potential energy stored in the section, *dx*, of the beam is

$$dE_P = \frac{1}{2} Ew''^2(x,t) \left(\int_{-h/2}^{h/2} bz^2 dz \right) dx = \frac{1}{2} Ew''^2(x,t)Idx$$

For the whole beam with a length L, the total potential energy caused by the bending is

$$E_P = \int_0^L \frac{1}{2} EI{w''}^2(x,t)dx \qquad (2.3.12)$$

(2) The Kinetic Energy

The kinetic energy of the elemental section dx is

$$dE_K = \frac{1}{2}\rho bh dx \left(\frac{dw}{dt}\right)^2$$

And, the kinetic energy of the whole beam is

$$E_K = \int_0^L \frac{1}{2}\rho bh\left(\frac{dw}{dt}\right)^2 dx \qquad (2.3.13)$$

(3) Approximate Shape Functions of Vibration Modes

In general, the vibration of a beam structure can be described by

$$w(x,t) = \sum_{n=0}^{\infty} c_n W_n(x,t)$$

where $W_n(x,t)$ is the wave function of the n'th vibration mode with a sinusoidal vibration of radial frequency ω_n and a shape function, $W_n(x)$

$$W_n(x,t) = W_n(x)\sin(\omega_n t + \alpha_n)$$

Mathematically, $W_n(x,t)$ is an eigen-function of the differential equation for the vibration system to be given in §2.4.

(4) Rayleigh's Quotient

For the vibration mode designated by subscript n, the potential energy is

$$E_P = \frac{1}{2}c_n^2 \int_0^L EI[W_n''(x)]^2 \sin^2(\omega_n t + \alpha_n)dx$$

and the kinetic energy of the mode is

$$E_K = \frac{1}{2}c_n^2 \int_0^L \rho bh W_n^2(x)\omega_n^2 \cos^2(\omega_n t + \alpha_n)dx$$

By the principle of energy conservation, i.e., $E_{Pmax}=E_{Kmax}=E$, the vibration frequency of the mode can be found by the equation below.

$$\omega_n^2 = \frac{\int_0^L EIW_n''^2(x)dx}{\int_0^L \rho bh W_n^2(x)dx}$$ (2.3.14)

Eq. (2.3.14) can be used for a beam with non-uniform cross section if the area of the cross section $A(x)$ is used to substitute for $b \cdot h$ in the denominator.

$$\omega_n^2 = \frac{\int_0^L EIW_n''^2(x)dx}{\int_0^L \rho A(x)W_n^2(x)dx}$$ (2.3.15)

This is the well-known Rayleigh's quotient and the method for the vibration frequency is referred to as Rayleigh-Ritz method.

(5) The Nature of the Rayleigh-Ritz Method

With Rayleigh's quotient, the vibration frequency of a specific vibration mode can be found if the shape function of the vibration mode is known. The problem is that it is generally quite difficult to find the shape functions for a structure. However, the vibration frequency for the fundamental vibration mode (with the lowest vibration frequency) can be found by Rayleigh's quotient with high accuracy when a static displacement function $w(x)$ is used as an approximation for the shape function. In most cases, the basic vibration mode plays an important role and the effect of the higher vibration modes can be neglected as their frequencies are much higher than that of the basic vibration mode.

For some higher vibration modes, if a proper approximation of shape function $W_n(x)$ can be found (based on the information on boundary conditions, nodes, etc.), the vibration frequencies can also be found by Rayleigh's quotient with a reasonably high accuracy.

As the shape function $W_n(x)$ used for the vibration frequency calculation by Rayleigh's quotient is an approximation of the real eigenfunction, the result is always higher than the actual eigenvalue of the corresponding eigenfunction. However, the approximation is usually very good for the fundamental vibration frequency. The results for the first and second higher harmonics can also be quite good approximations if reasonable shape functions are assumed.

§2.3.3. Vibration Frequencies of Beam Structures

Now we will use the Rayleigh-Ritz method to find the vibration frequencies for some typical beam structures. The structures considered are quite basic ones widely used in micro sensors and actuators.

(1) Cantilever Beam

The static displacement function of a cantilever beam given in Eq. (2.2.28) is written as

$$w(x) = \frac{\rho g}{2Eh^2}x^2(x^2 - 4Lx + 6L^2) = cx^2(x^2 - 4Lx + 6L^2)$$ (2.3.16)

This function can be used as an approximate shape function for the calculation of the fundamental vibration frequency of the beam. By substituting Eq. (2.3.16) into Eq. (2.3.14), we find

$$\omega_1 = 3.53\sqrt{\frac{EI}{m_b L^3}} = 1.019\frac{h}{L^2}\sqrt{\frac{E}{\rho}} \tag{2.3.17}$$

where $m_b = bhL\rho$ is the mass of the beam.

(2) Cantilever Beam With an End Mass

For a beam with an end mass, M, that is much larger than the mass of the beam, the static displacement function is

$$w(x) = c(3L - x)x^2 \tag{2.3.18}$$

By substituting Eq. (2.3.18) into Eq. (2.3.14) for $W_n(x)$, we find

$$\omega_1^2 = \frac{\int_0^L EI36(L-x)^2 dx}{4ML^6}$$

or

$$\omega_1 = \sqrt{\frac{3EI}{ML^3}} = \sqrt{\frac{Ebh^3}{4ML^3}} \tag{2.3.19}$$

If the beam mass, m_b, is not negligible, the vibration frequency of the beam-mass structure can still be calculated using the Rayleigh-Ritz method with the same shape function as for a beam with negligible mass. By using Eq. (2.3.18) as the shape function, we find

$$\omega_1^2 = \frac{\int_0^L EI36(L-x)^2 dx}{\int_0^L \rho bh(3Lx^2 - x^3)^2 dx + 4ML^6} = \frac{Ebh^3}{4(M + \frac{33}{140}m_b)L^3}$$

or

$$\omega_1 = \sqrt{\frac{Ebh^3}{4(M + \frac{33}{140}m_b)L^3}} \tag{2.3.20}$$

When Eq. (2.3.20) and Eq. (2.3.19) are compared, we find that the distributed beam mass, m_b, is equivalent to a mass of "$33m_b/140$" at the free end of the beam.

This approximation may be questioned for its accuracy. Here we make a comparison to justify the approximation. As an extreme, we use the approximation of Eq. (2.3.20) for the cantilever beam without an end mass. By letting $M=0$, we have

$$\omega_1 = 1.0299 \sqrt{\frac{Ebh^3}{m_b L^3}} = 1.0299 \frac{h}{L^2} \sqrt{\frac{E}{\rho}} \tag{2.3.21}$$

By comparing Eq. (2.3.21) with Eq. (2.3.17), we find that the two results are extremely close. This reminds us that the Rayleigh-Ritz method is not very sensitive to the details of the assumed shape functions.

If both the mass of the beam and the end mass are considered, it can be shown that the accurate expression of basic vibration frequency is

$$\omega_1 = \sqrt{\frac{Ebh^3}{4ML^3}\left(1 + \frac{3}{4}m' + \frac{3}{20}m'^3\right) \Big/ \left(1 + \frac{552}{560}m' + \frac{26}{80}m'^2 + \frac{26}{720}m'^3\right)} \tag{2.3.22}$$

where $m'=\rho bhL/M$. If Eq. (2.3.22) is approximated to the first power of m', the result is exactly the same as that of Eq. (2.2.20).

(3) Double-clamped Beam

According to Eq. (2.2.36), the displacement function of a double-clamped beam is

$$w(x) = cx^2(L-x)^2 \tag{2.3.23}$$

By substituting this equation into Eq. (2.3.14) for $W_n(x)$, we find

$$\omega_1 = \frac{22.45}{L^2}\sqrt{\frac{IE}{\rho bh}} = \frac{4.738^2}{L^2}\sqrt{\frac{IE}{\rho bh}} = 6.48\frac{h}{L^2}\sqrt{\frac{E}{\rho}} \tag{2.3.24}$$

When Eq. (2.3.24) is compared with Eq. (2.3.17), the basic vibration frequency for the double-clamped beam is higher than that of the cantilever beam by a factor of about 6.4.

Now one more example to justify the argument that the vibration frequency found by the Rayleigh-Ritz method is not very sensitive to the assumed shape function if the assumed shape function meets the basic features of the vibration mode. Suppose that the shape function for the double-clamped beam is

$$w(x) = c(1 - \cos 2\pi \frac{x}{L}) \tag{2.2.25}$$

This function looks quite different from the one shown in Eq. (2.3.23), but the general shape of Eq. (2.3.25) and that of Eq. (2.3.23) are similar as they both meet the same boundary conditions

$$w(0) = 0, \ w'(0) = 0, \ w(L) = 0, \ w'(L) = 0$$

By substituting Eq. (2.3.25) into Eq. (2.3.14), we find

$$\omega_1 = 6.58 \frac{h}{L^2} \sqrt{\frac{E}{\rho}} \tag{2.3.26}$$

This is quite close to, but a little bit larger than, the previous result shown in Eq. (2.3.24). as the exact constant is 6.4585 (instead of 6.48 or 6.58 as will be shown in §2.4). Eq. (2.3.23) is a better approximation for the shape function than Eq. (2.3.25).

(4) Frequencies of Higher Harmonics

It has been widely believed that, the shape function of the first higher harmonic for a double-clamped beam is anti-symmetric with a node at the center point of the beam. Therefore, the shape function has the following boundary conditions

$$w(0) = w(L) = 0; w'(0) = w'(L) = 0; w(L/2) = 0, w''(L/2) = 0 \tag{2.3.27}$$

As the shape function is anti-symmetric and Rayleigh's quotient is an even function of x, the vibration frequency can be found for the beam length between $x=0$ and $x = L/2$.

If the shape function that satisfies the boundary condition takes a polynomial form

$$w(x) = x^4 + ax^3 + bx^2 \ \ (0 \le x \le L/2)$$

By satisfying the boundary conditions as shown in Eq. (2.3.27), we find $a = -5/4L$ and $b = 3L^2/8$. Thus, the approximate shape function is

$$w(x) = c(x^4 - \frac{5}{4}Lx^3 + \frac{3}{8}L^2x^2) \ \ (0 \le x \le \frac{L}{2}) \tag{2.3.28}$$

The shape of the function is shown in Fig. 2.3.4.

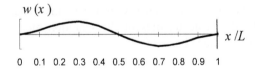

Fig. 2.3.4. An approximate shape function for the first higher vibration mode

By using Eq. (2.3.14) and Eq. (2.3.28) for the region 0~$L/2$, the radial frequency for the first higher harmonic is found to be

$$\omega_2 = \frac{(7.8621)^2}{L^2}\sqrt{\frac{EI}{\rho A}} = 17.84\frac{h}{L^2}\sqrt{\frac{E}{\rho}} \tag{2.3.29}$$

We will see later in §2.4 that this is also a very good approximation for the first higher harmonic. (The exact constant is 17.8034 instead of 17.84).

§2.4. Vibration Modes and the Buckling of a Beam

In the last section, the Rayleigh-Ritz method was developed for the vibration frequencies of beam structures. The Rayleigh-Ritz method can only be used to find an approximate eigenvalue of the vibration system when an approximate shape function can be assumed. However, as the static displacement function is usually a good approximation for the shape function of the fundamental vibration mode, the Rayleigh-Ritz method is very effective in finding the frequency for a fundamental vibration mode. For higher vibration modes, the Rayleigh-Ritz method is also applicable in principle, but it relies somewhat on the speculation on the features of the shape functions.

A more fundamental method for finding the vibration frequencies and the shape functions for all the vibration modes is to establish and solve the differential equation of the vibration system [6]. Generally, it is a quite complicated process for practical problems so that numerical methods are often used for this purpose. However, even with commercially available computer aided design tools (such as ANSYS, etc), much skill and dedication are needed to make use of these tools. Here in this section, we will introduce the differential equation method and the solutions to the vibration problems of beam structures. The beam structure to be used as an example is a double-clamped beam with a rectangular cross section.

§2.4.1. Differential Equation for Free Vibration of a Beam

The small amplitude vibration of a beam can be described by a differential equation. The equation can be derived based on the following discussion.

Consider an elemental section, dx, of the beam as shown in Fig. 2.4.1. The internal shear force applied on the left-hand side of the section is an upward force f and the internal shear force applied on the right-hand side of the section is a downward force $f + df$. The bending moment applied on the left-hand side and right-hand side are M and $M + dM$, respectively. $q(x)$ is the loading force per unit length of the beam. As f and $f + df$ are in opposite directions, the net force df will cause an acceleration of the elemental section. Thus, the relation for force balance is

$$df + q(x)dx = \rho A\frac{\partial^2 w(x,t)}{\partial t^2}dx$$

where $q(x)dx$ is the loading force on dx ($q=\rho gbh$ if the loading is the weight of the beam), $A=bh$ and $w(x,t)$ is the displacement function of the beam. From this relation, we have

$$\frac{df}{dx} + q(x) = \rho A \frac{d^2 w(x,t)}{dt^2} \qquad (2.4.1)$$

From the balance of the bending moments acting on the section, we have

$$M + fdx - \frac{q(dx)^2}{2} = M + dM$$

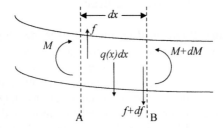

Fig. 2.4.1. Forces and force moments acting on a section of vibrating beam

Therefore, for an infinitesimal dx, we have

$$f = \frac{dM}{dx} \qquad (2.4.2)$$

By substituting Eq. (2.4.2) into Eq. (2.4.1), we find

$$\frac{d^2 M}{dx^2} + q(x) = \rho A \frac{d^2 w}{dt^2}$$

According to Eq. (2.2.10), i.e., $M = -EIw''(x)$, we have

$$EI \frac{\partial^4 w}{\partial x^4} + \rho A \frac{\partial^2 w}{\partial t^2} = q(x) \qquad (2.4.3)$$

This is the linear partial differential equation governing the vibration of the beam. If there is no loading on the beam, the equation is

$$EI \frac{\partial^4 w}{\partial x^4} + \rho A \frac{\partial^2 w}{\partial t^2} = 0 \qquad (2.4.4)$$

It may be mentioned here that, by substituting Eq. (2.2.10) into Eq. (2.4.2), the shearing force can be expressed as

$$f = -EIw'''(x)$$

As the shear force at the free end of a beam is zero, the boundary condition for the free end is $w'''(x) = 0$ in addition to $w''(x) = 0$.

§2.4.2. Vibration Frequencies of a Double-clamped Beam

By using the boundary conditions corresponding to the practical problem, Eq. (2.4.4) can be solved to find the vibration frequencies and corresponding shape functions for different vibration modes [7]. For a double-clamped beam (a bridge) with length L, the boundary conditions are

$$w(0) = 0, w(L) = 0, w'(0) = 0, w'(L) = 0 \tag{2.4.5}$$

The general solution to Eq. (2.4.4) can be considered as the superposition of different vibration modes, i.e.,

$$w(x,t) = \sum_n W_n(\frac{x}{L})(A_n \cos \omega_n t + B_n \sin \omega_n t) \tag{2.4.6}$$

where ω_n is the radial frequency and W_n the shape function of the nth vibration mode. By substituting Eq. (2.4.6) into Eq. (2.4.4) and using the dimensionless variable, $\eta = x/L$, we have the equations for $W_n(\eta)$

$$W_n''''(\eta) - \frac{\rho A}{EI} L^4 \omega_n^2 W_n(\eta) = 0$$

Let $(k_n)^4 = \rho A L^4 \omega_n^4 / EI$ or $\omega_n^2 = EI k_n^4 / (\rho A L^4)$, we have

$$W_n''''(\eta) - k_n^4 W_n(\eta) = 0$$

By using $W_n = e^{\lambda \eta}$ as a trial solution, we have

$$\lambda^4 - k_n^4 = 0$$

The solution to the equation are

$$\lambda = k_n, -k_n, ik_n, -ik_n$$

Therefore, the general form of shape function is

$$W_n(\eta) = Ae^{k_n \eta} + Be^{-k_n \eta} + Ce^{ik_n \eta} + De^{-ik_n \eta}$$

or

$$W_n(\eta) = A\sin(k_n\eta) + B\cos(k_n\eta) + C\sinh(k_n\eta) + D\cosh(k_n\eta) \qquad (2.4.7)$$

As the eiginfunctions of the differential equation (the shape-functions), $W_n(x)s$, are orthogonal, i.e., we have the relations (§51 of [6])

$$\int_0^L W_m W_n dx = 0 \ \text{(for } m \neq n)$$

By substituting Eq. (2.4.7) into Eq. (2.4.5), the equations for A, B, C and D are

$$\begin{pmatrix} 1 & 1 & 1 & 1 \\ k_n & -k_n & ik_n & -ik_n \\ e^{k_n} & e^{-k_n} & e^{ik_n} & e^{-ik_n} \\ k_n e^{k_n} & -k_n e^{-k_n} & ik_n e^{ik_n} & -ik_n e^{-ik_n} \end{pmatrix} \begin{pmatrix} A \\ B \\ C \\ D \end{pmatrix} = 0 \qquad (2.4.8)$$

For nontrivial solutions, the determinant of the matrix in Eq. (2.4.8) must be zero

$$\begin{vmatrix} 1 & 1 & 1 & 1 \\ k_n & -k_n & ik_n & -ik_n \\ e^{k_n} & e^{-k_n} & e^{ik_n} & e^{-ik_n} \\ k_n e^{k_n} & -k_n e^{-k_n} & ik_n e^{ik_n} & -ik_n e^{-ik_n} \end{vmatrix} = 0 \qquad (2.4.9)$$

Eq. (2.4.9) can be reduced to a simple equation

$$\cosh(k_n)\cos k_n - 1 = 0 \qquad (2.4.10)$$

This result indicates that the eigenvalues, $k_n s$, for the shape functions must satisfy the transcendental equation, Eq. (2.4.10). Once a k_n is found from Eq. (2.4.10), the corresponding vibration frequency is determined by

$$\omega_n = \sqrt{\frac{k_n^4 EI}{L^4 \rho A}} \qquad (2.4.11)$$

or

$$f_n = \frac{k_n^2}{2\pi L^2}\sqrt{\frac{EI}{\rho A}}$$

Some of the $k_n s$ found using Eq. (2.4.10) are listed below

$k_o = 0$ does not represent a vibration mode

$k_1 = 4.730$, $k_2 = 7.8532$, $k_3 = 10.996$,

$k_n \approx (n\pi + \pi/2)$ for $n > 3$ (from the approximation of $\cos k_n = 0$)

The radial frequency of the fundamental vibration mode is

$$\omega_1 = \frac{4.730^2}{L^2}\sqrt{\frac{EI}{\rho A}} = 6.4585 \frac{h}{L^2}\sqrt{\frac{E}{\rho}} \tag{2.4.12}$$

and the radial frequency of the first higher harmonic is

$$\omega_2 = \frac{7.8532^2}{L^2}\sqrt{\frac{EI}{\rho A}} = 17.803 \frac{h}{L^2}\sqrt{\frac{E}{\rho}} \tag{2.4.13}$$

By comparing Eqs. (2.4.12) and (2.4.13) with the results of the Raylaigh-Ritz method, i.e., Eqs. (2.3.23) and (2.3.28) in §2.3, we find that the discrepancies between the vibration frequencies found by the Rayleigh-Ritz method and the exact results shown in above are only a fraction of one percent and that those found by the Rayleigh-Ritz method are always a little bit larger.

The main purpose of the above discussion is to make it clear that the natural vibration frequencies of a double-clamped beam form an infinite spectrum of vibration frequencies corresponding to a spectrum of infinite vibration modes. This is generally true for any vibration system and the situation is generally much more complicated if the mechanical system has more degrees of freedom.

The vibration frequency of a cantilever beam can also be discussed by repeating the above procedure. For a cantilever beam, the boundary conditions are [6]

$$w(0) = 0, \; w'(0) = 0, \; w''(L) = 0, \; w'''(L) = 0$$

and the k_ns found are

$k_1 = 1.875$, $k_2 = 4.694$, $k_3 = 7.855$, $k_4 = 10.996$, $k_5 = 14.137$, $k_6 = 17.279....$

Thus, the radial frequency of the fundamental vibration mode is

$$\omega_1 = \frac{1.875^2}{L^2}\sqrt{\frac{EI}{\rho A}} = 1.015 \frac{h}{L^2}\sqrt{\frac{E}{\rho}}$$

When this result is compared with that of Rayleigh-Ritz method as shown in Eq. (2.3.16), the difference is again smaller than 1%.

§2.4.3. Vibration With an Axial Force

In §2.4.1 and §2.4.2, no axial force is considered in the discussion on the free vibration frequencies of a beam. However, the axial force exists in many cases and can have significant influence on vibration. One of its main effects is to cause the vibration frequency of a vibration mode to shift. This effect is utilized for sensing purposes in resonant sensors. Here in this section we will derive the differential equation, discuss the frequency effect of the axial force and give out some useful relations.

(1) Differential Equation of a Beam With an Axial Force

For a beam with an axial force, the free vibration of small amplitude can be described by a linear partial differential equation. The equation can be derived as outlined in the following discussion.

Consider an elemental section, dx, of the beam. The internal forces applied on both ends of the section are shown in Fig. 2.4.2. Fig. 2.4.2 differs from Fig. 2.4.1 by two additional axial forces (denoted by N and $N+dN$) on both ends of the section. The effect of the axial forces causes an additional bending moment $Ndw = Nw'(x) \cdot dx$. The balance of bending moments requires

$$M + fdx - N\frac{dw}{dx}dx = M + dM$$

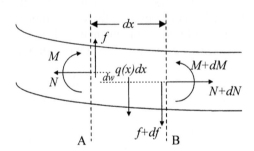

Fig. 2.4.2. Forces and force moments in a section of a beam with an axial force

Therefore, we find

$$f = \frac{dM}{dx} + N\frac{dw}{dx}$$

By substituting the relation into Eq. (2.4.1), we have

$$EI\frac{\partial^4 w(x,t)}{\partial x^4} - N\frac{\partial^2 w(x,t)}{\partial x^2} + \rho A\frac{\partial^2 w(x,t)}{\partial t^2} = q(x) \qquad (2.4.14)$$

If the loading $q(x)$ is negligible, we have

$$EI \frac{\partial^4 w(x,t)}{\partial x^4} - N \frac{\partial^2 w(x,t)}{\partial x^2} + \rho A \frac{\partial^2 w(x,t)}{\partial t^2} = 0 \tag{2.4.15}$$

(2) The Solutions to the Equation

To solve the equation, the method used in the §2.4.2 is applied. By assuming that

$$w(x,t) = \sum_n W_n(\eta)(A_n \cos \omega_n t + B_n \sin \omega_n t)$$

and substituting the $w(x,t)$ into Eq. (2.4.15), for the n'th vibration mode, we have

$$W_n''''(\eta) - \frac{NL^2}{EI} W_n''(\eta) - \frac{\rho A L^4}{EI} \omega_n^2 W(\eta) = 0 \tag{2.4.16}$$

where $\eta = x/L$. Assuming that $W_n(\eta) = A e^{\lambda \eta}$, we have the equation for λ

$$\lambda^4 - \frac{NL^2}{EI} \lambda^2 - \frac{\rho A L^4}{EI} \omega_n^2 = 0 \tag{2.4.17}$$

By letting

$$\beta = \frac{NL^2}{2EI} \tag{2.4.18}$$

and

$$k_n^4 = \frac{\rho A L^4}{EI} \omega_n^2 \tag{2.4.19}$$

The equation for λ is

$$\lambda^4 - 2\beta \lambda^2 - k_n^4 = 0 \tag{2.4.20}$$

Therefore, we have the solutions

$$\lambda = \pm i \sqrt{\sqrt{\beta^2 + k_n^4} - \beta}, \pm \sqrt{\sqrt{\beta^2 + k_n^4} + \beta}$$

By letting

$$k_{n1} = \sqrt{\sqrt{\beta^2 + k_n^4} - \beta} \;; \; k_{n2} = \sqrt{\sqrt{\beta^2 + k_n^4} + \beta} \tag{2.4.21}$$

we have the solutions

$$\lambda_1 = ik_{n1}, \lambda_2 = -ik_{n1}, \lambda_3 = k_{n2}, \lambda_4 = -k_{n2},$$
(2.4.22)

Therefore, the general solution to Eq. (2.4.16) is

$$W_n(\eta) = Ae^{ik_{n1}\eta} + Be^{-ik_{n1}\eta} + Ce^{k_{n2}\eta} + De^{-k_{n2}\eta}$$
(2.4.23)

To satisfy the boundary conditions of the beam and follow the same procedure given in §2.4.2, for nontrivial solutions, k_n must satisfy the following equation [7]

$$\cos k_{n1} \cosh k_{n2} - \frac{\beta}{k_n^2} \sin k_{n1} \sinh k_{n2} - 1 = 0$$
(2.4.24)

The corresponding shape functions are

$$W_n(\eta) = \frac{\cosh(k_{n2}\eta) - \cos(k_{n1}\eta)}{\cosh(k_{n2}) - \cos(k_{n1})} - \frac{\sinh(k_{n2}\eta) - \sin(k_{n1}\eta)}{\sinh(k_{n2}) - \sin(k_{n1})}$$
(2.4.25)

If there is no axial force, i.e. $N=0$, Eq. (2.4.24) returns to Eq. (2.4.10). If $N \neq 0$, β can be determined by Eq. (2.4.18). Then, Eq. (2.4.21) and Eq. (2.4.24) should be used to find k_n. Therefore, k_n is a function of β (i.e., a function of the axial force N). The corresponding vibration frequency ω_n can be found by Eq. (2.4.19)

$$\omega_n(\beta) = k_n^2(\beta)\sqrt{\frac{EI}{\rho A L^4}}$$
(2.4.26)

The procedure for finding k_n for a specific N to satisfy Eqs. (2.4.21) and (2.4.24) is a tedious numerical work. S. Bouwstra found that the solutions can be approximated by an analytical relation [7]

$$\omega_n(N) = \omega_n(0)\sqrt{1 + \gamma_n \frac{NL^2}{12EI}}$$
(2.4.27)

where the γ_ns are coefficients representing the contribution of the axial force. The values of the k_ns (at zero axial force) and the γ_ns are listed in Table 2.4.1. The error caused by the approximation of Eq. (2.4.27) is smaller than 0.5% under the condition that $NL^2/12EI < 1$.

Table 2.4.1 Values for k_n, and γ_n

n	k_n	γ_n
1	4.7300	0.2949
2	7.8532	0.1453
≥3	$(n+\frac{1}{2})\pi$	$\dfrac{12(k_n - 2)}{k_n^3}$

(3) Vibration Frequencies With Axial Forces

For small frequency variations, further approximation can be made for Eq. (2.4.27) by

$$\omega_n(N) = \omega_n(0)(1 + \frac{\gamma_n}{2} \frac{NL^2}{12EI})$$

(2.4.28)

or

$$f_n(N) = f_n(0)(1 + \frac{\gamma_n}{2} \frac{NL^2}{12EI})$$

(2.4.29)

To conclude this section, we consider the following example. For a double-clamped beam of silicon with dimensions of $b=100\mu m$, $h=20\mu m$, $L=2000\mu m$, the fundamental vibration frequency with no axial force is $f_1(0)= 43.90$ kHz. From Eq. (2.4.27), the vibration with an axial force of $N=0.02$N has a frequency

$$f_1(N) = f_1(0)\sqrt{1 + 0.2949 \times 0.1882} = 1.0833 f_1(0) = 47.557 kHz$$

The corresponding frequency shift is 3.657 kHz. If Eq. (2.4.29) is used, we have

$$f_1(N) = f_1(0)(1 + \frac{1}{2} \times 0.2949 \times 0.1882) = 1.0867 f_1(0) = 47.707 kHz$$

The corresponding frequency shift is 3.807 kHz. The difference is about 4%.

§2.4.4. Buckling of a Double-clamped Beam

According to the discussion in §2.4.3, the vibration frequency of the beam decreases with the compressive axial force on the beam. It is well known that the vibration frequency may diminish to zero at a certain value of compressive axial force N_b and the beam buckles as shown in Fig. 2.4.3. N_b is referred to as the buckling force of the beam. For example, a silicon dioxide bridge in a MEMS device may buckle due to the compressive strain or stress in the bridge caused by the thermal effect. The value of the strain or stress that makes the beam start to buckle is referred to as a buckling strain or stress. Now let us analyze the buckling effect quantitatively.

For a compressive axial force N, Eq. (2.4.15) can be rewritten as

$$EI\frac{\partial^4 w(x,t)}{\partial x^4} + N\frac{\partial^2 w(x,t)}{\partial x^2} + \rho A\frac{\partial^2 w(x,t)}{\partial t^2} = 0$$

(2.4.30)

By letting $w(x,t) = X(x)T(t)$, we have

$$EIX''''(x)T(t) + NX''(x)T(t) + \rho AX(x)\ddot{T}(t) = 0$$

If the vibration is sinusoidial, i.e., $T(t) = C\sin(\omega t + \alpha)$, we have the equation for $X(x)$

$$\frac{EI}{\rho A}X''''(x) + \frac{N}{\rho A}X''(x) - \omega^2 X(x) = 0 \qquad (2.4.31)$$

Fig. 2.4.3. The buckling of a double-clamped beam

By letting $X(x) = e^{\lambda\frac{x}{L}}$, we have

$$\frac{EI}{\rho A L^4}\lambda^4 + \frac{N}{\rho A L^2}\lambda^2 - \omega^2 = 0 \qquad (2.4.32)$$

where L is the length of the beam. The solution to Eq. (2.4.32) is

$$\lambda_{1,2} = \pm\sqrt{-\frac{NL^2}{2EI} + \sqrt{\left(\frac{NL^2}{2EI}\right)^2 + \frac{\rho A L^4}{EI}\omega^2}} \ ; \ \lambda_{3,4} = \pm i\sqrt{+\frac{NL^2}{2EI} + \sqrt{\left(\frac{NL^2}{2EI}\right)^2 + \frac{\rho A L^4}{EI}\omega^2}}$$

By letting

$$\alpha = \sqrt{-\frac{NL^2}{2EI} + \sqrt{\left(\frac{NL^2}{2EI}\right)^2 + \frac{\rho A L^4}{EI}\omega^2}} \ ; \beta = \sqrt{\frac{NL^2}{2EI} + \sqrt{\left(\frac{NL^2}{2EI}\right)^2 + \frac{\rho A L^4}{EI}\omega^2}} \qquad (2.4.33)$$

The solutions to Eq. (2.4.32) can be written as

$$\lambda_1 = \alpha, \lambda_2 = -\alpha, \lambda_3 = i\beta, \lambda_4 = -i\beta$$

Therefore, the general solution to Eq.(2.4.31) takes the form of

$$X(x) = A\sinh\left(\alpha\frac{x}{L}\right) + B\cosh\left(\alpha\frac{x}{L}\right) + C\sin\left(\beta\frac{x}{L}\right) + D\cos\left(\beta\frac{x}{L}\right) \qquad (2.4.34)$$

The boundary condition for a double-clamped beam (i.e., a bridge) is

$$X(0) = 0, X(L) = 0, X'(0) = 0, X'(L) = 0 \qquad (2.4.35)$$

By substituting Eq. (2.4.34) into Eq. (2.4.35), we obtain the following equations

$$\begin{cases} 0 \cdot A + B + 0 \cdot C + D = 0 \\ A \sinh \alpha + B \cosh \alpha + C \sin \beta + D \cos \beta = 0 \\ A \dfrac{\alpha}{L} + B \cdot 0 + C \dfrac{\beta}{L} + D \cdot 0 = 0 \\ A \dfrac{\alpha}{L} \cosh \alpha + B \dfrac{\alpha}{L} \sinh \alpha + C \dfrac{\beta}{L} \cos \beta + D \dfrac{\beta}{L} \cos \beta = 0 \end{cases} \tag{2.4.36}$$

For nontrivial solutions, the determinant of the matrix in Eq. (2.4.36) must be zero

$$\begin{vmatrix} 0 & 1 & 0 & 1 \\ \sinh \alpha & \cosh \alpha & \sin \beta & \cos \beta \\ \dfrac{\alpha}{L} & 0 & \dfrac{\beta}{L} & 0 \\ \dfrac{\alpha}{L} \cosh \alpha & \dfrac{\alpha}{L} \sinh \alpha & \dfrac{\beta}{L} \cos \beta & -\dfrac{\beta}{L} \sin \beta \end{vmatrix} = 0 \tag{2.4.37}$$

Eq.(2.4.37) can be developed as

$$\sinh \alpha \sin \beta \left(\frac{\alpha^2 - \beta^2}{L^2} \right) + \frac{\alpha \beta}{L^2} \left(\sin^2 \beta + \cos^2 \beta - \cos \beta \cosh \alpha - \sinh^2 \alpha + \cosh \alpha \cos \beta + \cosh^2 \alpha \right) = 0$$

By the relations of $\beta^2 = \alpha^2 + NL^2 / EI$, $\cosh^2 \alpha - \sinh^2 \alpha = 0$ and $\cos^2 \beta + \sin^2 \beta = 0$, we have

$$\frac{\alpha \beta}{L^2} (1 - \cosh \alpha \cos \beta) = \frac{N}{EI} \sinh \alpha \sin \beta$$

As $\alpha \beta / L^2 = \omega \sqrt{\rho A / EI}$, we have

$$\omega = \sqrt{\frac{N^2}{EI\rho A} \cdot \frac{\sinh(\alpha)\sin(\beta)}{2[1 - \cosh(\alpha)\cos(\beta)]}}$$

As $\alpha \cong \omega L \sqrt{\rho A / N}$ and $\beta \cong L\sqrt{N / EI}$ for $\omega \cong 0$, we have

$$\beta \sin \beta = 2(1 - \cos \beta), \text{ or, } \sin \frac{\beta}{2} \left(\beta \cos \frac{\beta}{2} - 2 \sin \frac{\beta}{2} \right) = 0$$

There are two solutions to the equation: 1) $\beta=2n\pi$ from $\sin(\beta/2)=0$ and 2) $\beta=0$ (i.e., $N=0$) from $\tan(\beta/2)= (\beta/2)$. Obviously, the lowest non-zero solution $\beta=2\pi$ is the buckling state of the beam and we have the buckling force

$$N_b = \frac{4\pi^2 EI}{L^2} = \frac{\pi^2 Ebh^3}{3L^2} = 3.29 \frac{Ebh^3}{L^2} \tag{2.4.38}$$

Thus, the buckling stress is

$$T_b = \frac{\pi^2 Eh^2}{3L^2} \tag{2.4.39}$$

It may be mentioned here that if the buckling force is obtained from Eq. (2.4.24) by the "zero frequency condition", the constant in Eq. (2.4.38) is 3.39 instead of 3.29.

§2.5. Damped and Forced Vibration

§2.5.1. Damping Force

For the free vibration as described in last section, there is no need for the system to do work against resistive forces so that its total energy remains constant at any time. In practice, the vibration of a real system is always resisted by dissipative forces, such as air viscosity, friction, acoustic transmission, internal dissipation, etc. The system thus does positive work. The energy for the work is subtracted from the vibration energy and is usually converted into thermal energy. Therefore. damping is the process whereby energy is taken from the vibration system.

Generally, damping is inevitable in any system. For example, the internal friction in a spring (or flexure) is always dissipative and transforms part of the vibration energy into thermal energy in each cycle. Another common cause of damping is the viscosity of the surrounding fluid such as air. The fluid exerts viscous forces on the moving object and opposes its movement through the fluid.

As a first order approximation, a damping force, F_d, is proportional to the speed of the movement

$$F_d = -c\dot{x} \tag{2.5.1}$$

where c is called the coefficient of damping force and the negative sign indicates that the force, is opposite to the moving direction. A vibration system with damping is usually modeled in Fig. 2.5.1. The damping effect is represented by a damper.

Damping is a destructive factor in maintaining a vibration. In many cases, measures are taken to reduce damping so that the vibration can be maintained with a minimum energy supplement per cycle. However, in many other cases, damping is deliberately introduced into a system to reduce oscillation. For example, for a micro-accelerometer, air damping is necessary and proper air damping should be considered from the very beginning of the design stage.

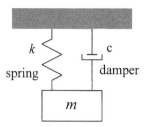

Fig. 2.5.1. A simplified model for a vibration system with damping

§2.5.2. Vibration With Damping

(1) Attenuation of Oscillation

For a damped vibration system as shown in Fig. 2.5.1, the differential equation for the movement is the second-order differential equation

$$m\ddot{x} = -kx - c\dot{x}$$

where m is the mass and c is the coefficient of damping force. Or, we have

$$\ddot{x} + \frac{c}{m}\dot{x} + \frac{k}{m}x = 0 \tag{2.5.2}$$

Using $\omega_o^2 = \frac{k}{m}$ and $n = \frac{c}{2m}$, we have

$$\ddot{x} + 2n\dot{x} + \omega_o^2 x = 0 \tag{2.5.3}$$

where ω_o is the radial frequency of the vibration system if there is no damping (i.e., a free vibration) and n is called the coefficient of damping. By letting $x = Ae^{\lambda t}$, we have

$$\lambda^2 + 2n\lambda + \omega_o^2 = 0 \tag{2.5.4}$$

The solutions to the equation are

$$\lambda_{1,2} = -n \pm \sqrt{n^2 - \omega_o^2} \tag{2.5.5}$$

The performances of the system can be discussed according to the ratio between n and ω_o. The ratio $\zeta = n/\omega_o$ is known as the damping ratio of the system.

(a) Slight Damping (Under-damping)
If $n < \omega_o$, i.e., $\zeta < 1$, we have

$$\lambda_{1,2} = -n \pm i\sqrt{(\omega_o^2 - n^2)}$$

The solution to Eq. (2.5.3) is

$$x = Ae^{-nt}\sin(\sqrt{\omega_o^2 - n^2}\,t + \alpha) \qquad\qquad (2.5.6)$$

Eq. (2.5.6) indicates that the system will have an oscillation, but it differs from the free vibration in that: (i) The vibration frequency is $\omega_d = \sqrt{\omega_o^2 - n^2} = \omega_o\sqrt{1 - \zeta^2}$, which is smaller than the free vibration frequency ω_o, and (ii) The amplitude of the vibration decays exponentially with time.

(b) Heavy Damping (Over-damping)
If $n > \omega_o$, i.e. $\zeta > 1$, we have

$$\lambda_{1,2} = -n \pm \sqrt{(n^2 - \omega_o^2)}$$

and

$$x = e^{-nt}(c_1 e^{\sqrt{n^2 - \omega_o^2}\,t} + c_2 e^{-\sqrt{n^2 - \omega_o^2}\,t}) \qquad\qquad (2.5.7)$$

There is no oscillation of the displacement; the mass returns to its balanced position slowly.

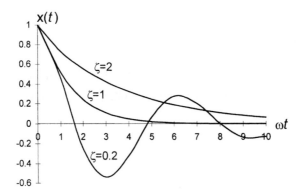

Fig. 2.5.2. Amplitude attenuation of systems with different damping ratio

(c) Critical Damping
If $n = \omega_o$, i.e. $\zeta = 1$, we have $\lambda_1 = \lambda_2 = -n = -\omega_o$ and the special solution to Eq. (2.5.3) is $x_1(t) = c_1 e^{-nt}$. Assuming a general solution of $x = A(t)x_1(t)$ and substituting it into Eq. (2.5.3), we have $d^2A/dt^2 = 0$. Therefore, the general solution to Eq. (2.5.3) is

$$x(t) = c_1 e^{-nt}(c_2 t + c_3) \tag{2.5.8}$$

The result indicates that there is no oscillation in case of critical damping but the time taken for the displacement to become virtually zero is a minimum.

Fig. 2.5.2 shows the $x(t)$ relations for the three damping conditions.

(2) Response to a Step Force

For a spring-mass system, the balanced position of the mass is taken as $x=0$. With a constant force F_o on the mass, the balanced position is $x_o = F_o/k$.

Now let us consider the time response of the system to a step force F_o. When the force is applied on the mass at $t = 0$, the mass moves from it original position, $x=0$, towards its new balanced position, $x_o=F_o/k$. If the damping effect is negligible, the system will possess a total energy of $E=F_o x_o$ when the mass arrives at x_o due to the work done by the force. As the potential energy in this condition is $k x_o^2/2=F_o x_o/2$, the mass must have a kinetic energy $F_o x_o/2$ at x_o. This means that the mass will move on after passing through x_o until it reaches x', where the work done by the applied force equals the elastic potential energy

$$F_o x' = \frac{1}{2} k x'^2$$

As $x'=2x_o$ is not a balanced position, the mass will move back toward $x = 0$. In this way, the mass oscillates between 0 and $2x_o$. When the effect of damping is considered the oscillation will die down and the mass will finally settle at $x=x_o$ when the excessive energy is completely consumed by the damping. Therefore, the response of the system to a step force is strongly dependent on the damping.

The process can be described by the equation

$$m\ddot{x} + c\dot{x} + kx = F_o$$

As $F_o=kx_o$, we have

$$m\ddot{x} + c\dot{x} + k(x - x_o) = 0$$

By letting $x_1 = x-x_o$, we have the differential equation for x_1

$$m\ddot{x}_1 + c\dot{x}_1 + kx_1 = 0 \tag{2.5.9}$$

with the initial conditions

$$x_1|_{t=0} = -x_o, \quad \dot{x}_1|_{t=0} = 0 \tag{2.5.10}$$

Therefore, the solutions to Eq. (2.5.3) are applicable to Eq. (2.5.9) and the initial conditions shown in Eq. (2.5.10) can be used to determine the constants in the solutions.

The analysis gives the following results:

(a) The Under-damping Conditions (i.e., $\zeta < 1$)
In this condition, the system oscillates before it settles down. The expression for $x(t)$ is

$$x = x_o\left(1 - \frac{e^{-\zeta\omega_o t}}{\sqrt{1-\zeta^2}}\sin(\sqrt{1-\zeta^2}\,\omega_o t + \alpha)\right)$$

where $\omega_o = \sqrt{\dfrac{k}{m}}$ and $\alpha = \sin^{-1}\sqrt{1-\zeta^2}$.

(b) The Critical Damping Condition (i.e., $\zeta=1.0$)
In this condition, the system moves to its new balanced position without oscillation. The dependence of position x on time is

$$x = x_o\left(1 - (1+\omega_o t)e^{-\omega_o t}\right)$$

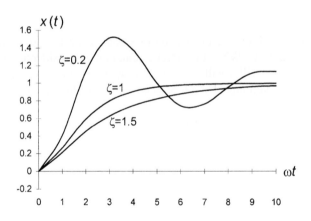

Fig. 2.5.3. Response to a step force for different damping ratios

(c) The Over-damping Conditions (i.e., $\zeta > 1$)
In this condition, the system reaches it new balance position slowly. The expression of $x(t)$ is

$$x = x_o\left(1 - \frac{\zeta+\sqrt{\zeta^2-1}}{2\sqrt{\zeta^2-1}}e^{(-\zeta+\sqrt{\zeta^2-1})\omega_o t} + \frac{\zeta-\sqrt{\zeta^2-1}}{2\sqrt{\zeta^2-1}}e^{(-\zeta-\sqrt{\zeta^2-1})\omega_o t}\right)$$

For example, curves for the three damping conditions are shown in Fig. 2.5.3

§2.5.3. Vibration Driven by Force

Suppose that a spring-mass system is set to continuous oscillation by a periodic driving force acting on the mass. If the force is a sinusoidal one with a frequency of ω and an amplitude of F_o

$$F = F_o \sin(\omega t)$$

the differential equation for the system is

$$m\ddot{x} = -kx - c\dot{x} + F_o \sin(\omega t) \tag{2.5.11}$$

The solution to Eq. (2.5.11) takes the form

$$x = x_1 + x_2$$

where x_2 is a specific solution to Eq. (2.5.11) and x_1 is a general solution to the homogenous differential equation of the damped system, i.e., the equation for x_1 is

$$m\ddot{x}_1 + c\dot{x}_1 + kx_1 = 0 \tag{2.5.12}$$

The solution to Eq. (2.5.12) is known to be a damped oscillation (for $n < \omega_o$)

$$x_1 = A \exp(-nt)\sin(\sqrt{(\omega_o{}^2 - n^2)} \cdot t + \alpha)$$

We assume that x_2 is a stable oscillation

$$x_2 = B\sin(\omega t - \varphi) \tag{2.5.13}$$

where B is the amplitude of the resulting vibration and φ the phase lag of the vibration against the sinusoidal force. By substituting Eq. (2.5.13) into Eq. (2.5.11), we have

$$[B(\omega_o{}^2 - \omega^2) - f\cos\varphi]\sin(\omega t - \varphi) + (2nB\omega - f\sin\varphi)\cos(\omega t - \varphi) = 0$$

where $f = F_o/m$. To satisfy this equation, we have

$$B(\omega_o{}^2 - \omega^2) - f\cos\varphi = 0$$

and

$$2nB\omega - f\sin\varphi = 0$$

Therefore the amplitude B and the phase lag φ are found to be

$$B = \frac{f}{\sqrt{(\omega_o^2 - \omega^2)^2 + 4n^2\omega^2}} \qquad (2.5.14)$$

and

$$\varphi = tg^{-1}\frac{2n\omega}{\omega_o^2 - \omega^2} \qquad (2.5.15)$$

Thus, the solution to Eq. (2.5.11) is

$$x = Ae^{-nt}\sin(\sqrt{\omega_o^2 - n^2}t + \alpha) + B\sin(\omega t - \varphi) \qquad (2.5.16)$$

The solution given in Eq. (2.5.16) indicates that, in the early stages, beats occur due to forced vibration and damped vibration, giving rise to transient oscillations, which are usually short-lived and can be ignored in most cases. We will only consider what happens when conditions are steady.

(1) Vibration Frequency

The frequency of the steady vibration is the same as the frequency of the driving force. However, there is a phase lag that is a function of the free vibration frequency ω_o, the driving frequency ω and the damping coefficient n, as is shown by Eq. (2.5.15).

(2) Amplitude

The amplitude of the steady vibration is given by Eq. (2.5.14). It can be written as

$$B = \frac{B_o}{\sqrt{(1 - \frac{\omega^2}{\omega_o^2})^2 + 4(\frac{n}{\omega_o})^2(\frac{\omega}{\omega_o})^2}}$$

where $B_o = F_o/k$ is the static displacement of the mass caused by a constant force F_o. If ω/ω_o is designated as λ and a relative amplitude $\beta = B/B_o$ is introduced, we have

$$\beta = \frac{1}{\sqrt{(1 - \lambda^2)^2 + 4\zeta^2\lambda^2}} \qquad (2.5.17)$$

For a constant force amplitude F_o, the dependence of β on the driving frequency is:

(a) Low Driving Frequencies ($\omega \ll \omega_o$ or $\lambda \ll 1$)

In this case, $\beta \approx 1$. This means that the amplitude is the same as the displacement caused by the static force F_o.

(b) Medium Frequencies ($\omega \approx \omega_o$ or $\lambda \approx 1$)

The amplitude has a maximum at $\omega = \omega_o \sqrt{1 - 2\zeta^2}$. If the damping is light ($\zeta < 0.7$), ω is very closer to ω_0 and β is approximately equal to $1/2\zeta$, which can be very large for small ζ. The resonance peak disappears at $\zeta \geq 0.7$. In this case, the curve for $\beta \sim \omega$ relation has the largest flat region (i.e., the largest bandwidth). Therefore, $\zeta = 0.7$ is often referred to as an optimum damping condition.

(c) High Frequencies ($\omega \gg \omega_o$ or $\lambda \gg 1$)

In this case, we have $\beta \approx 1/\lambda^2$. This means that the amplitude decreases very fast with frequency.

Example curves showing the frequency dependence of the relative amplitude are given in Fig. 2.5.4, which shows that the amplitude of a forced vibration is a function of driving frequency and the damping ratio.

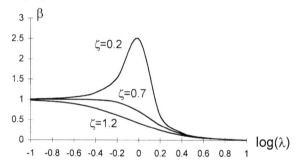

Fig. 2.5.4. Amplitude-frequency relations for different damping ratios

(3) Phase Lag.

The forced vibration takes on the frequency of the driving force, but it has a phase lag as shown in Eq. (2.5.15). For very small ω, φ is close to zero. φ increases with ω, passes $\pi/2$ at the natural frequencies of the system and approaches π at very high frequency, as shown in Fig. 2.5.5.

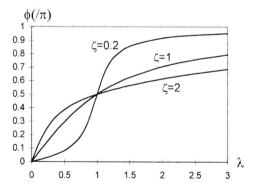

Fig. 2.5.5. Phase lag of forced vibrations for different damping ratios

§2.5.4. Resonance and Quality Factor

(1) Resonant Frequency

According to Eq. (2.5.17) or Fig. 2.5.4, for slight damping, the amplitude has a maximum at a frequency near the free vibration frequency. This phenomenon is known as resonance and the corresponding frequency is called resonant frequency. Using the condition of $d\beta/d\lambda = 0$, we find that the resonant frequency of a forced vibration is at $\lambda_r = \sqrt{1-2\zeta^2}$ or $\omega_r = \omega_o\sqrt{1-2\zeta^2}$. As the natural vibration frequency (without damping) of the system is ω_o and the damped vibration frequency of the system is $\omega_d = \omega_o\sqrt{1-\zeta^2}$, we find that, for slight damping, ω_o, ω_d and ω_r are all close to each other.

At resonant frequency the relative amplitude reaches a maximum value

$$\beta_r = \frac{1}{2\zeta\sqrt{1-\zeta^2}} \tag{2.5.18}$$

For slight damping, we have the amplitude at resonance: $\beta_r \doteq 1/2\zeta$.

(2) Phase Lag at Resonance

By substituting ω_r into Eq. (2.5.15), we have the phase lag at resonance

$$\varphi_r = \tan^{-1}\frac{\sqrt{1-2\zeta^2}}{\zeta} \tag{2.5.19}$$

For slight damping, $\varphi_r \doteq \tan^{-1}(1/\zeta) \doteq \pi/2$ or $\zeta = 1/\tan\varphi_r$.

(3) Quality Factor

For many sensor applications, the sharpness of the resonant peak is important for high resolution or high accuracy of measurement. The parameter to indicate the sharpness of a resonance curve is the Quality factor or Q-factor of the system. There are several definitions for the Q factor. They are equivalent for slight damping.

(a) Definition 1

Q is mathematically defined as the peak value of the relative amplitude at resonant frequency, i.e.,

$$Q = \beta_r = \frac{1}{2\zeta\sqrt{1-\zeta^2}} \tag{2.5.20}$$

It is reasonable to find that Q is related to the damping ratio ζ. The smaller the damping ratio, the larger the Q.

(b) Definition 2

In physics, Q is defined as the ratio between the total system energy and the average energy loss in one radian at resonant frequency, i.e.,

$$Q = 2\pi \frac{E}{\Delta E} \tag{2.5.21}$$

where E is the total energy of the vibration system and ΔE is the energy dissipated by damping in one cycle (equals 2π radian) of oscillation. Eq. (2.5.21) can be justified by the following discussion.

For a system vibrating at its resonant frequency ω_r, the displacement is

$$x = A \sin \omega_r t$$

The total energy of the system is

$$E = \frac{1}{2} m \dot{x}^2 = \frac{1}{2} m A^2 \omega_r^2$$

The energy dissipation in one cycle is

$$\Delta E = -\int_0^T F_d \dot{x} dt$$

where F_d is the damping force, i.e., $F_d = -c\dot{x}$. Therefore, we have

$$\Delta E = \int_0^T c \dot{x}^2 dt = \pi c A^2 \omega_r$$

and

$$Q = 2\pi \frac{E}{\Delta E} = \frac{m \omega_r}{c}$$

By using $n = \dfrac{c}{2m}$ and $\zeta = \dfrac{n}{\omega_o} \cong \dfrac{n}{\omega_r}$, we have $Q \cong \dfrac{1}{2\zeta}$.

(c) Definition 3

The third definition for Q is similar to the definition of Q for an electrical resonant circuit. The energy of a vibration system is proportional to the square of the vibration amplitude. At resonance, the system has its maximum energy

$$E_r = C\beta_r^2$$

The relative amplitude for a half maximum energy is $\beta_r / \sqrt{2}$. There are two frequencies for the amplitude: one is smaller than ω_r and another larger than ω_r. If the two frequencies are ω_1 and ω_2, the corresponding relative frequencies, λ_1 and λ_2, can be found by the equation

$$\frac{1}{\sqrt{(1-\lambda^2)^2 + 4\zeta^2\lambda^2}} = \frac{1}{\sqrt{2}}\beta_r$$

λ_1 and λ_2 are found to be

$$\lambda_{1,2} = \sqrt{1-2\zeta^2}\,(1 \pm \frac{\zeta}{1-2\zeta^2})$$

This leads to

$$\Delta\lambda = \lambda_1 - \lambda_2 = \sqrt{1-2\zeta^2}\,\frac{2\zeta}{1-2\zeta^2} = \frac{2\zeta}{\sqrt{1-2\zeta^2}}$$

Therefore, for slight damping, we have

$$\frac{\lambda_r}{\Delta\lambda} = \frac{1-2\zeta^2}{2\zeta} \doteq \frac{1}{2\zeta}$$

This means that the Q factor can be equally defined as

$$Q = \frac{\lambda_r}{\Delta\lambda} \tag{2.5.22}$$

It is clear that the three definitions are equivalent for small ζ, i.e., for a large Q value. For silicon micro resonators in air, the Q value is usually of the order of 10^2. It may go up to over one hundred thousand in a vacuum.

(4) The Measurement of Q
Q factor is a very important parameter to characterize a vibration system. As the damping ratio that determines the Q factor is usually very small and difficult to estimate theoretically, the Q factor of a system is usually found through experimental measurements.

(a) Method 1
According to the basic definition, Q is equal to the relative amplitude of the system at resonance, i.e., β_r. Therefore, if we find amplitude-frequency relation for a constant driving force for frequencies from dc to a frequency higher than the resonant frequency, $Q=\beta_r$ can be found at the resonant frequency. As the measurement must cover large frequency range

and the amplitude may change by several orders of magnitude, the measurement is often time-consuming and inaccurate.

(b) Method 2

According to Eq. (2.5.22), Q can be found by the $A-f$ measurement in a small frequency range near the resonant frequency. First, we find the resonant frequency f_r and the amplitude A_r at the frequency. Then the amplitude-frequency relation around the resonant frequency is measured so that the frequencies f_1 and f_2 with amplitude of $A_r / \sqrt{2}$ are found. The Q can be found by the relation $Q = f_r /(f_2 - f_1)$.

(c) Method 3

According to Eq. (2.5.20) and Eq. (2.5.19), Q is related to the damping ratio ζ, which is related to the phase lag at resonance. Therefore, we first find the resonant frequency ω_r and then find the phase lag φ_r at resonant frequency. Q can then be calculated by $Q = \tan \varphi_r / 2$.

§2.5.5. Vibration Driven by Vibration

Forced vibration of a spring-mass system was discussed in §2.5.3, where a sinusoidal force with constant amplitude was used to excite the system. The force may be an electrostatic force, an electromagnetic force, etc. However, in many other cases, a mass-spring system is excited into vibration by a vibration. For example, an accelerometer is often mounted on a vibrator (or, an excitor) in a pig-back form for the calibration of the accelerometer or on a machine to measure its vibration. In these cases, the inertia force applied on the seismic mass of the accelerometer is related to the frequency of the vibration as well as the vibration amplitude. Therefore, the situation will be a little different from that shown in §2.5.3.

Consider that a micro mechanical accelerometer is mounted on a large machine for measurement. Suppose that the mass of the accelerometer is much smaller than that of the machine so that the vibration of the machine is not affected by the measurement. The setup is schematically shown in Fig. 2.5.6, where the accelerometer is shown as a spring-mass system with a damper.

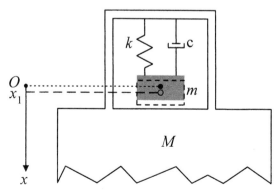

Fig. 2.5.6. Vibration driven by a vibration

With reference to the inertia coordinate system Ox, if the vibration of the machine is

$$x_M = d_o \sin \omega t \tag{2.5.23}$$

where d_0 is the amplitude of the vibration and ω the vibration frequency. If the displacement of the seismic mass is x_m, the elastic force of the spring on the seismic mass is $-k(x_m-x_M)$ as the package of the accelerometer is moving with the machine. Therefore, the differential equation for the seismic mass is

$$m\ddot{x} + c(\dot{x}_m - \dot{x}_M) + k(x_m - x_M) = 0 \tag{2.5.24}$$

By defining the relative displacement of the seismic mass against the machine (or the package) as $\xi = x_m - x_M$, Eq. (2.5.24) can be written as

$$m\ddot{\xi} + c\dot{\xi} + k\xi = -m\ddot{x}_M = ma_o \sin \omega t \tag{2.5.25}$$

where $a_o = d_o\omega^2$ is the acceleration amplitude of the vibration.

Equation (2.5.25) is a differential equation for a forced vibration by a periodic force $ma_o \sin \omega t = md_o\omega^2 \sin \omega t$. For a slight damping condition, the solution to the equation is

$$\xi = B\sin(\omega t - \varphi); B = \frac{d_o(\omega^2 / \omega_o^2)}{\sqrt{\left(1 - \omega^2 / \omega_o^2\right)^2 + (\omega^2 / \omega_o^2)/Q^2}}; \varphi = \tan^{-1}\frac{\omega\omega_o}{Q(\omega_o^2 - \omega^2)} \tag{2.5.26}$$

where ω_0 is the natural vibration frequency of the accelerometer and φ the phase lag. Now we discuss some typical situations based on Equation (2.5.26) as follows.

(a) Low Vibration Frequency

For low vibration frequency, i.e., $\omega \ll \omega_o$ and $(1-\omega^2/\omega_o^2) \gg \omega/Q\omega_o$, we have $\varphi \approx 0$ and

$$B = \frac{d_o\omega^2 / \omega_o^2}{1 - \omega^2 / \omega_o^2} \doteq d_o\left(\frac{\omega^2}{\omega_o^2} + \frac{\omega 4}{\omega_o^4}\right) \tag{2.5.27}$$

The displacement of the seismic mass with reference to the machine is

$$\xi = \frac{md_o\omega^2}{k}\left(1 + \omega^2 / \omega_o^2\right)\sin \omega t \tag{2.5.28}$$

For the calibration of an accelerometer, the amplitude of acceleration $d_o\omega^2$ is often kept constant but the vibration frequency is changed. As the output of an accelerometer is usually proportional to ξ, Eq. (2.5.28) means that the sensitivity of the accelerometer is a function of vibration frequency

$$S(\omega) = \frac{md_o\omega^2}{k}\left(1 + \omega^2/\omega_o^2\right) \equiv S_o\left(1 + \omega^2/\omega_o^2\right)$$

where $S_o = md_o\omega^2/k$ is the sensitivity at very low frequency and $S(\omega)$ is the sensitivity at frequency ω. It can be shown that sensitivity $S(\omega)$ is higher than S_o by 10% at a frequency of $\omega = 0.316\omega_o$ for an accelerometer with slight damping.

(b) Resonant Frequency ω_{res}
If the vibration frequency ω equals the resonant frequency ω_{res} of the accelerometer, we have $B = Qd_o$. This means that the amplitude of the seismic mass can be much larger than that of the driving vibration if the accelerometer is not properly damped.

(c) High Frequency $\omega \gg \omega_o$
In this case, we have $B = d_o$ and $\varphi = 180°$. Therefore, we have $\xi = x_m - x_M = -d_o\sin\omega t$, or $x_m \approx 0$. This means that, at high frequency, the seismic mass do not follow the vibration at all. As the seismic mass keeps static when it is observed in an inertia coordination system, its relative displacement against its package can be used as a "mirror" to observe the vibration of the machine. In this case, the amplitude, instead of the acceleration, of the vibration can be measured.

§2.6. Basic Mechanics of Diaphragms

§2.6.1. Long Rectangular Diaphragm

(1) One Dimensional Model
Consider a long rectangular diaphragm with clamped edges. Suppose that the length of the diaphragm, L, is much larger than its width, $2a$, and both L and $2a$ are much larger than its thickness, h, as shown in Fig. 2.6.1(a). If diaphragm is loaded with a pressure, p, on top, the diaphragm bends and strain and stress are induced in the diaphragm.

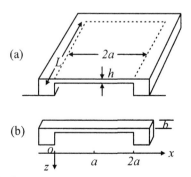

Fig. 2.6.1. Long rectangular diaphragm (a) perspective view (b) a section of the structure

As the length of the diaphragm is much larger than its width, one-dimensional approximation can be used except for the areas near the two far ends of the diaphragm. For a simplified analysis we cut a narrow strip of width b across the diaphragm as shown in Fig. 2.6.1(b). Now the problem is similar to the one of double-clamed beam with uniform loading.

According to Hooke's law

$$\varepsilon_{xx} = \frac{T_{xx}}{E} - v\frac{T_{yy}}{E} \tag{2.6.1}$$

$$\varepsilon_{yy} = \frac{T_{yy}}{E} - v\frac{T_{xx}}{E} \tag{2.6.2}$$

As $L \gg 2a$, the condition of $\varepsilon_{yy} = 0$ has to be satisfied. From Eq. (2.6.2), we have $T_{YY} = vT_{XX}$. From Eq. (2.6.1), we have

$$T_{XX} = \frac{E}{1-v^2}\varepsilon_{xx} = -\frac{E}{1-v^2}zw''(x) \tag{2.6.3}$$

where $w(x)$ is the displacement function of the beam. Except for the replacement of E by $E_{eff} = E/(1-v^2)$, all the relations for a double-clamped beam are true for the strip.

In this case, the bending moment of the strip of width b is

$$M_b = -E_{eff}Iw''(x) = -\frac{bh^3 E}{12(1-v^2)}w''(x) \tag{2.6.4}$$

By defining the flexure rigidity of diaphragm

$$D = \frac{Eh^3}{12(1-v^2)}$$

the bending moment for the strip is

$$M_b = -bDw''(x) \tag{2.6.5}$$

(2) Displacement and Stress

With a pressure p, the total loading force on the top of the strip is $2pab$. Therefore, the supporting forces on both clamped edges are $F_o = pab$ as shown in Fig. 2.6.2. If the weight of the strip is negligible, the bending moment at point x on the strip is

$$M_b(x) = bapx - bm_o - \frac{1}{2}bpx^2$$

where m_o is the restrictive bending moment per unit width of the diaphragm edge. Therefore, the equation for $w(x)$ is

$$-bDw''(x) = bapx - bm_o - \frac{1}{2}bpx^2$$

By this equation and the boundary conditions $w(0) =0$, $w'(0)=0$, $w(2a)=0$, $w'(2a)=0$ we find

$$w(x) = \frac{p}{24D}x^4 - \frac{ap}{6D}x^3 + \frac{a^2 p}{6D}x^2 = \frac{p}{24D}(l-x)^2 x^2 \qquad (2.6.6)$$

where $l=2a$. Therefore, the stress on top of the beam is

$$T_{XX}(x) = \frac{3p}{h^2}(x^2 - 2ax + \frac{2}{3}a^2) \qquad (2.6.7)$$

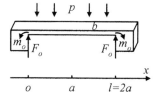

Fig. 2.6.2. A strip of the diaphragm with pressure loading

The stress distribution along the strip is shown in Fig. 2.6.3. Note that the maximum stress is at the edge of the beam with a value of $2pa^2/h^2$ and a negative maximum of $-pa^2/h^2$ is at the center of the diaphragm. The stress in the y-direction is always $T_{YY}= \nu T_{XX}$ in the diaphragm.

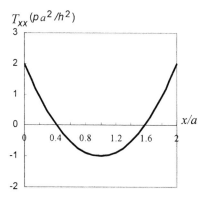

Fig. 2.6.3. T_{XX} on top of the diaphragm

§2.6.2. Differential Equations of Diaphragm

(1) Strain and Stress in diaphragm

For a diaphragm in the x-y plane of a Cartesian coordinate system, with a bending deformation, two normal strain components in the diaphragm are

$$\varepsilon_{xx}(z) = \frac{z}{r_x} = -z\frac{\partial^2 w(x.y)}{\partial x^2}, \varepsilon_{yy}(z) = \frac{z}{r_y} = -z\frac{\partial^2 w(x,y)}{\partial y^2} \tag{2.6.8}$$

where r_x and r_y are the radii of curvature of the diaphragm along the x- and y-directions, respectively, and $w(x,y)$ is the displacement function of the diaphragm in the z-direction. From Eq. (2.1.8) and the relations of $u= -z(\partial w/\partial x)$ and $v= -z(\partial w/\partial y)$ we have the shearing strain component

$$\varepsilon_{xy} = 2e_{xy} = \frac{\partial u}{\partial y} + \frac{\partial v}{\partial x} = -2z\frac{\partial^2 w(x,y)}{\partial x \partial y} \tag{2.6.9}$$

From Hooke's Law and the principle of superposition, we have

$$\varepsilon_{xx} = \frac{T_{XX}}{E} - v\frac{T_{YY}}{E}$$

$$\varepsilon_{yy} = \frac{T_{YY}}{E} - v\frac{T_{XX}}{E}$$

Therefore, we have

$$T_{XX} = \frac{E}{1-v^2}(\varepsilon_{xx} + v\varepsilon_{yy}); T_{YY} = \frac{E}{1-v^2}(\varepsilon_{yy} + v\varepsilon_{xx})$$

By substituting Eq. (2.6.8) into these equations, the two normal stress components are

$$T_{XX} = -\frac{Ez}{1-v^2}(\frac{\partial^2 w}{\partial x^2} + v\frac{\partial^2 w}{\partial y^2}); T_{YY} = -\frac{Ez}{1-v^2}(\frac{\partial^2 w}{\partial y^2} + v\frac{\partial^2 w}{\partial x^2}) \tag{2.6.10}$$

Also, according to Eqs. (2.1.14) and (2.6.9), the shearing stress component can be found as

$$T_{XY} = G\varepsilon_{xy} = -\frac{Ez}{1+v} \cdot \frac{\partial^2 w(x,y)}{\partial x \partial y} \tag{2.6.11}$$

(2) Differential Equation of a Diaphragm

For a two dimensional diaphragm with a pressure load, p, the differential equation for the displacement of the diaphragm can be derived by analyzing the balance conditions for forces and bending moments in an elemental area of diaphragm, $dxdy$. The procedure is similar to that shown in §2.4 for a beam, but of course with some more complexity. The general equation for displacement $w(x,y,t)$ is found to be [8]

$$D\left[\frac{\partial^4 w}{\partial x^4} + 2\frac{\partial^4 w}{\partial x^2 \partial y^2} + \frac{\partial^4 w}{\partial y^4}\right] + h\rho\frac{\partial^2 w}{\partial t^2} = p(x,y) \qquad (2.6.12)$$

As the time dependence of the displacement is considered, this equation is also useful for vibration analysis. Clearly, Eq. (2.6.12) is the two-dimensional version of Eq. (2.4.3).

If only a steady displacement $w(x,y)$ caused by a uniform pressure, p, is considered, Eq. (2.6.12) is reduced to

$$D\left[\frac{\partial^4 w}{\partial x^4} + 2\frac{\partial^4 w}{\partial x^2 \partial y^2} + \frac{\partial^4 w}{\partial y^4}\right] = p \qquad (2.6.13)$$

This equation will be used for the stress distribution of a diaphragm in §2.6.3 and §2.6.4.

If a free vibration is considered, Eq. (2.6.12) is reduced to

$$D\left[\frac{\partial^4 w}{\partial x^4} + 2\frac{\partial^4 w}{\partial x^2 \partial y^2} + \frac{\partial^4 w}{\partial y^4}\right] + h\rho\frac{\partial^2 w}{\partial t^2} = 0 \qquad (2.6.14)$$

This equation will be used for the vibration problems of a diaphragm in §2.6.5.

§2.6.3. Circular Diaphragm

For a circular diaphragm with a clamped edge at $r=a$, the displacement of the diaphragm caused by a pressure, p, can be found by solving the differential equation for the diaphragm. As the displacement is circular symmetric, it will be convenient to use a polar coordinate system with its origin at the center of the diaphragm as shown in Fig. 2.6.4. The displacement is a function of r, $w=w(r)$.

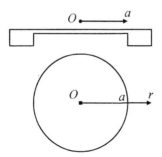

Fig. 2.6.4. Geometries of a circular diaphragm

Eq. (2.6.13) can be written as

$$D\left(\frac{\partial^2}{\partial x^2} + \frac{\partial^2}{\partial y^2}\right)\left(\frac{\partial^2}{\partial x^2} + \frac{\partial^2}{\partial y^2}\right)w(x,y) = p$$

In a polar coordinate system, the equation can be written as

$$\frac{1}{r}\frac{d}{dr}r\frac{d}{dr}\left[\frac{1}{r}\frac{d}{dr}r\frac{d}{dr}w(r)\right]=\frac{p}{D}$$ (2.6.15)

The solution to Eq. (2.6.15) is quite straightforward. To satisfy the boundary conditions of $w(a)=0$, $w'(a)=0$, $w'(0)=0$ and the finite value of w'' at $r=0$, the solution is found to be

$$w(r)=\frac{pa^4}{64D}\left(1-\frac{r^2}{a^2}\right)^2\equiv w(0)\left(1-\frac{r^2}{a^2}\right)^2$$ (2.6.16)

where $w(0)=pa^4/64D$ is the maximum displacement at the center of the diaphragm. The equation for stress in the diaphragm can be derived from Eq. (2.6.10) by replacing x and y by r and t, respectively, i.e.,

$$T_r\equiv T_{rr}=-\frac{Ez}{1-v^2}\left(\frac{\partial^2 w}{\partial r^2}+v\frac{\partial^2 w}{\partial t^2}\right);T_t\equiv T_{tt}=-\frac{Ez}{1-v^2}\left(\frac{\partial^2 w}{\partial t^2}+v\frac{\partial^2 w}{\partial r^2}\right)$$ (2.6.17)

where r and t indicate radial and tangential directions, respectively. As w is not a function of angular coordinate θ, the shearing stress T_{rt} is always zero in the polar coordinate system. For the top of the diaphragm, i.e., $z=-h/2$, $T_{rt}=0$ and the normal stress components are

$$T_r=\frac{3a^2}{8h^2}p\left[(3+v)\frac{r^2}{a^2}-(1+v)\right];T_t=\frac{3a^2}{8h^2}p\left[(1+3v)\frac{r^2}{a^2}-(1+v)\right]$$ (2.6.18)

The stresses T_r and T_t on the top of the diaphragm are shown by the curves in Fig. 2.6.5.

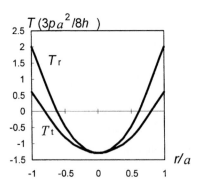

Fig. 2.6.5. Normal stress T_r and T_t on the top of a circular diaphragm

The stress components T_r and T_t at the edges are

$$T_r(a) = \frac{3}{4}\frac{a^2}{h^2}p, \ T_t(a) = \frac{3}{4}v\frac{a^2}{h^2}p \tag{2.6.19}$$

The stresses at the center are

$$T_r(0) = T_t(0) = -\frac{3}{8}(1+v)\frac{a^2}{h^2}p \tag{2.6.20}$$

§2.6.4. Square and Rectangular Diaphragms

The calculation of displacement for square or rectangular diaphragms with clamped edges is a rather complicated problem. The accurate solutions can only be achieved by numerical analysis such as finite-element method or finite-differential method. There are many commercially available software tools for this purpose and we will not go into further detail in this regard.

There are also many approximate analytical expressions for the relations between pressure and displacement for square and rectangular diaphragms. They are usually obtained by an assumed function with some constants to be decided. Obviously, the function should satisfy the boundary conditions of the diaphragm. The constants in the assumed function are then found by minimizing the total energy in the diaphragm for a pressure loading, p. For a rectangular diaphragm with a length of $2b$ and a width of $2a$, the total energy of the diaphragm under pressure p is [8]

$$E = \int_{-b}^{b}\int_{-a}^{a}\frac{D}{2}\left\{(\frac{\partial^2 w}{\partial x^2}+\frac{\partial^2 w}{\partial y^2})^2 - 2(1-v)\left[\frac{\partial^2 w}{\partial x^2}\frac{\partial^2 w}{\partial y^2}-(\frac{\partial^2 w}{\partial x \partial y})^2\right]\right\}dxdy - \int_{-b}^{b}\int_{-a}^{a}pwdxdy \tag{2.6.21}$$

The simplest displacement function for a rectangular diaphragm has only one constant, k, to be determined

$$w(x,y) = k(a^2-x^2)^2(b^2-y^2)^2 \tag{2.6.22}$$

By substituting Eq. (2.6.22) into Eq. (2.6.21) and by letting $\partial E / \partial k = 0$, we find

$$k = \frac{7p}{128(a^4+b^4+\frac{4}{7}a^2b^2)D} \tag{2.6.23}$$

(1) Square Diaphragm

For a square diaphragm with a side length $2a$ as shown in Fig.2.6.6, the simplest expression of displacement for a pressure p is

$$w(x,y) = 0.0213 p \frac{a^4}{D} \left(1 - \frac{x^2}{a^2}\right)^2 \left(1 - \frac{y^2}{a^2}\right)^2 \cong \frac{1}{47} p \frac{a^4}{D} \left(1 - \frac{x^2}{a^2}\right)^2 \left(1 - \frac{y^2}{a^2}\right)^2 \qquad (2.6.24)$$

The maximum displacement of the diaphragm is at the center of the square with a value of $w(0,0)=pa^4/47D$. Obviously, this value is larger than that of the circular diaphragm with a diameter of $2a$ (cf. The maximum displacement is $w(0,0)=pa^4/64D$).

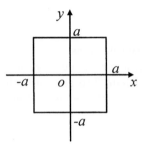

Fig. 2.6.6. Structure geometries of a square diaphragm

Using Eqs. (2.6.10), (2.6.11) and the notification of $\tilde{x} = x/a, \tilde{y} = y/a$, the stresses on the surface of the diaphragm are found to be

$$T_{XX} = -0.51 p \frac{a^2}{h^2} \left[(1-\tilde{y}^2)^2(1-3\tilde{x}^2) + v(1-\tilde{x}^2)^2(1-3\tilde{y}^2)\right.$$

$$T_{YY} = -0.51 p \frac{a^2}{h^2} \left[(1-\tilde{x}^2)^2(1-3\tilde{y}^2) + v(1-\tilde{y}^2)^2(1-3\tilde{x}^2)\right] \qquad (2.6.25)$$

$$T_{XY} = 2.045(1-v) p \frac{a^2}{h^2} (1-\tilde{x}^2)(1-\tilde{y}^2)\tilde{x}\tilde{y}$$

The maximum stresses are at the centers of the edges, e.g., at points $(\pm a, 0)$. The stress components found there are

$$T_{XX} = 1.02 p \frac{a^2}{h^2}; T_{YY} = v T_{XX}, \; T_{XY} = 0 \qquad (2.6.26)$$

At the center of the diaphragm (0,0), we have

$$T_{XX} = T_{YY} = -0.51(1+v) p \frac{a^2}{h^2} \text{ and } T_{XY} = 0$$

A more complicated approximation for the displacement of square diaphragm is

$$w(x,y) = p \frac{a^4}{D} \left(1 - \frac{x^2}{a^2}\right)^2 \left(1 - \frac{y^2}{a^2}\right)^2 \left[0.02023 + 0.00535 \frac{x^2+y^2}{a^2} + 0.00625 \frac{x^2 y^2}{a^4}\right] \qquad (2.6.27)$$

By using this expression, the stresses at edge centers ($\pm a$, 0) are

$$T_{XX} = 1.23p\frac{a^2}{h^2}, T_{YY} = vT_{XX}, T_{XY} = 0 \tag{2.6.28}$$

(2) Rectangular Diaphragm

For a rectangular diaphragm with a width $2a$ and a length $2b$ as shown in Fig. 2.6.7, the approximate displacement of the diaphragm under a pressure of p is easily found from Eqs. (2.6.22) and (2.6.23).

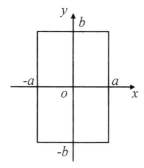

Fig. 2.6.7. Structure geometries of a rectangular diaphragm

Another approximate expression frequently used is

$$w(x, y) = \frac{p(1-v^2)}{2Eh^3}\frac{(a^2-x^2)^2(b^2-y^2)^2}{a^4+b^4} \tag{2.6.29}$$

This is an empirical expression known as Grashof equation [9]. From Eq. (2.6.29), we have

$$T_{XX} = \frac{pa^2}{h^2}\frac{b^4}{a^4+b^4}\left\{(1-3\tilde{x}^2)(1-\tilde{y}^2)^2 + v\frac{a^2}{b^2}(1-3\tilde{y}^2)(1-\tilde{x}^2)^2\right\}$$

$$T_{YY} = \frac{pa^2}{h^2}\frac{b^4}{a^4+b^4}\left\{v(1-3\tilde{x}^2)(1-\tilde{y}^2)^2 + \frac{a^2}{b^2}(1-3\tilde{y}^2)(1-\tilde{x}^2)^2\right\} \tag{2.6.30}$$

$$T_{XY} = -4p(1-v)\frac{a^2}{h^2}\frac{a}{b}(1-\tilde{x}^2)(1-\tilde{y}^2)\tilde{x}\tilde{y}$$

At the center near an edge of the diaphragm (e.g., $x=a$, $y=0$, or, $\tilde{x}=1$, $\tilde{y}=0$), we have

$$T_{XX} = 2\frac{pa^2}{h^2}\frac{b^4}{a^4+b^4}, T_{YY} = vT_{XX}, T_{XY} = 0 \tag{2.6.31}$$

The stress components at the center of the diaphragm are

$$T_{XX} = \frac{pa^2}{h^2}\frac{1+\nu a^2/b^2}{1+a^4/b^4}, T_{YY} = \frac{pa^2}{h^2}\frac{a^2/b^2+\nu}{1+a^4/b^4}, T_{XY} = 0 \tag{2.6.32}$$

§2.6.5. Vibration Frequencies of Diaphragms

The partial differential equation for the free vibration of diaphragm is Eq. (2.6.14). It can be written as

$$\frac{Eh^2}{12(1-\nu^2)}\left\{\frac{\partial^4 w}{\partial x^4}+2\frac{\partial^4 w}{\partial x^2\partial y^2}+\frac{\partial^4 w}{\partial y^4}\right\}+\rho\frac{\partial^2 w(x,t)}{\partial t^2}=0 \tag{2.6.33}$$

The free vibration frequencies of a diaphragm are calculated using the equation and the boundary conditions of the diaphragm. For a rectangular diaphragm with clamped edges at $(\pm a, y)$ and $(x, \pm b)$ the boundary conditions are

$$w(x=\pm a, y)=0, \frac{\partial w}{\partial x}(x=\pm a, y)=0, w(x,y=\pm b)=0, \frac{\partial w}{\partial y}(x, y=\pm b)=0 \tag{2.6.34}$$

As a diaphragm is a two dimensional system, the vibration frequencies of the diaphragm are usually denoted by two indices, i.e., in the form of ω_{mn} and the number of the vibration modes is infinite. However, only the lowest frequencies are usually important in practical problems.

To determine the values of ω_{mn}, it is necessary to find the analytical expression of $w(x,y,t)$ by solving Eq. (2.6.33) with the boundary conditions shown in Eq. (2.6.34). In fact, only approximate values can be obtained from numerical or semi-numerical methods.

The approximate vibration frequencies for square and rectangular diaphragms have been given by Pons [10] for normal vibration for (100) silicon diaphragms with edges in the <110> directions

$$\omega_{mn} = C_{mn}\sqrt{\frac{E}{12(1-\nu^2)\rho}}\frac{h}{A} \tag{2.6.35}$$

where $A=4ab$ is the area of the diaphragm. If $E/12(1-\nu^2)=1.42\times10^{10}$Pa and $\rho=2330$Kg/m^3, the constants C_{mn}s are given in Table 2.6.1 with respect to the ratio $r=b/a$ for some lowest vibration modes. The results have been verified by experiments.

Table 2.6.1. The C_{mn}'s for some lowest vibration modes of rectangular diaphragm ($r=b/a$)

C_{mn}	$r=1$	$r=2$	$r=3$
00	35.16	48.57	69.19
01	71.91	61.90	76.16
10	71.91	127.17	187.34
11	104.35	139.29	194.51

For a circular diaphragm with a clamped edge, the vibration problem was investigated by Timoshenko [6]. Using the Rayleigh-Ritz method, he found that the radial frequencies of vibration took the form of

$$\omega_{mn} = C_{mn}\sqrt{\frac{E}{12(1-\nu^2)\rho}}\,\frac{h}{d^2} \tag{2.6.36}$$

where $d=2a$ is the diameter of the diaphragm, ρ the density of the diaphragm material and C_{mn}'s are the constants corresponding to a given number m of nodal circles and a given number n of nodal diameters. Some values of C_{mn} for the lowest vibration frequencies are listed in Table 2.6.2.

As an example, for a circular silicon diaphragm with $d=2$mm, $h=10\mu$m, we find $f_{00}=40.09$kHz.

When the C_{00} in Table 2.6.1 for $r=1$ and the C_{00} for $n=0$ in Table 2.6.2 are compared, the constant $C_{00} = 40.8$ for a circular diaphragm is larger than the constant $C_{00}=35.16$ for a square diaphragm. This is reasonable as a circular diaphragm is stiffer than a square diaphragm if they have the same thickness, and the side length of the square diaphragm is equal to the diameter of the circular diaphragm.

Table 2.6.2. The C_{mn}'s for some lowest vibration modes of circular diaphragm

m	$n=0$	$n=1$	$n=2$
0	40.8	84.9	139.4
1	159.1	--	--
2	355.6	--	--

Timoshenko [6] also considered the vibration frequencies of a circular diaphragm with mass density of ρ in a medium with a mass density of ρ'. Due to the effect of medium, the lowest vibration frequency changes to

$$\omega = \frac{40.8h}{d^2\sqrt{1+\beta}}\sqrt{\frac{E}{12(1-\nu^2)\rho}} \tag{2.6.37}$$

where $d=2a$ is the diameter of the diaphragm, $\beta=0.6689(\rho'/\rho)\cdot(a/h)$. For a silicon diaphragm in the air of an atmospheric pressure, we have $\beta=3.72\times10^{-4}\cdot(a/h)$. If $a=1$mm and $h=50\mu$m, we have $\beta=0.00743$ and $\omega'=0.996\omega_0$. The frequency variation is less than one percent. But for a silicon diaphragm in water, we find $\beta=0.287\cdot(a/h)$. For the same geometric structure, we have $\beta=5.47$ and $\omega'=0.385\omega_0$. Clearly, the vibration frequency is significantly reduced by the surrounding medium.

§2.7. Problems

Problems on General Relations of Stress and Strain

1. Consider a silicon beam as shown in the figure below. The width, the thickness and the length of the beam are w=100μm, h=20μm and L=500μm, respectively. The top surface of the beam is a (001)-plane and the length of the bean is in the [100] direction of the silicon crystal. If the beam is stretched by a pair of forces, F=2N (N: Newton), on both ends, find the changes in length, width and thickness of the beam.

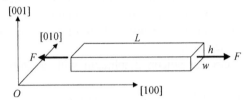

2. Consider a silicon beam as shown in the figure below. The width, the thickness and the length of the beam are w=100μm, h=20μm and L=500μm, respectively. The top surface of the beam is a (001)-plane and the length of the beam is in the [110] direction of the silicon crystal. If the beam is stretched by a pair of forces, F=2N, on both ends, find the change in length, width and thickness of the beam.

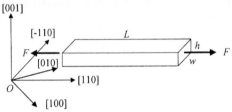

Problems on Stress and Strain of Beam Structures

3. The cross section of a silicon beam is shown in the figure below. The width of the beam is b and the thickness is h. The x-y plane is the neutral plane of the beam and Oy' is in parallel with the Oy and the distance between Oy and Oy' is r. If the moment of inertia of the beam against Oy is I and the moment of inertia against Oy' is I', find the relation between I and I'.

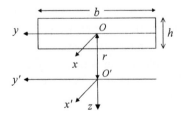

4. The cross section of a composite beam consisting of two layers is shown in the figure below. The top layer of the composite beam is SiO_2 and the substrate is silicon. Find

the position of the neutral plane for free bending. ($E_{ox}=7\times10^{10}$Pa, $E_{Si}=1.7\times10^{11}$Pa, $t_{ox}=0.5\mu$m and $t_{Si}=2\mu$m)

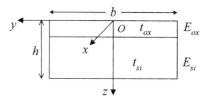

5. For a beam with a circular cross section as shown in the figure below, find the moment of inertia for a pure bending. (In terms of the radius a and the Young's modulus E)

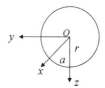

6. Consider a beam with a T-shaped cross-section as shown in the figure below. The cross section can be divided into two rectangles. The top one has a width of b_1 and a thickness of h_1, while the lower one has a width of b_2 and a thickness of h_2. (The centers of the two rectangles are O_1 and O_2 in the figure.). Find the position of the neutral plane z_0 for pure bending and the moment of inertia against the neutral plane.

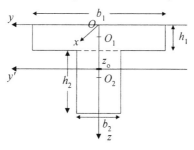

7. A section of optical fiber of length $L=10$mm is free-standing horizontally as shown in the figure below. Find the displacement of the free end due to its own weight, assuming that the diameter of the optical fiber is $d=100\mu m$, the density of the glass $\rho=2200$Kg/m^2 and the Young's modulus of glass $E=7\times10^{10}$Pa.

8. A polysilicon cantilever beam overhangs over a silicon substrate as shown in the figure below. The thickness of the beam is 2μm and the original distance between the beam and the substrate is 2μm. Find the maximum length the cantilever beam may have without

contacting the substrate due to the displacement caused by its own weight. (Note: the Young's modulus of polysilicon is $E=1.7\times10^{11}$Pa.)

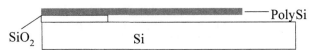

9. A silicon beam has a width b, a thickness h and a length L. The beam is simple-supported on its both ends as shown in the figure below. Find the displacement function $w(x)$ and the maximum displacement of the beam due to its own weight.

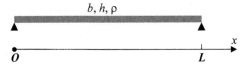

10. A double-supported beam-mass silicon structure is shown in the figure below, where $a_1=500\mu$m, $b=50\mu$m, $h=10\mu$m, $A=B=4$mm, $H=300\mu$m. Find (1) the displacement of the mass under its own weight, (2) the maximum stress on the beam surface.

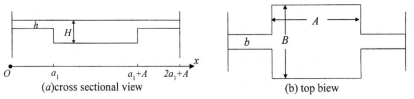

(*a*)cross sectional view   (b) top biew

11. For the same double-supported beam-mass silicon structure as shown in problem 10 above, find the maximum acceleration in the normal direction the structure can tolerate without rupture (assuming the rupture stress for silicon is $T_R=3\times10^9$Pa).

12. For the same double-supported beam-mass silicon structure as shown in problem 10 above, find the maximum displacement of the mass (assuming that the mass of the beam is negligible but the bending of the mass is considered).

13. A cantilever silicon beam with a mass M at its free end as shown in the figure below, where b, h, L are the width, the thickness and the length of the beam, respectively. If the mass of the beam is considered, find the displacement of the beam under gravity.

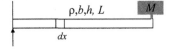

Problems on Vibration Frequency –I

14. A polysilicon cantilever beam has a thickness of $h=2\mu$m and a length of $L=500\mu$m. Find its basic natural vibration frequency ($E=1.7\times10^{11}$Pa, $\rho=2330$Kg/m^3).

15. A double-clamped polysilicon beam (bridge) has a thickness of $h=2\mu$m and a length of $L=500\mu$m. Find its basic natural vibration frequency.

16. A silicon structure is shown in the figure below, where a_1=500μm, b =50μm, h=20μm, A=B=4mm and H=300μm. If the mass of the beam and the bending of the mass are negligible, find the vibration frequencies of the following vibration mode: (1) the vibration normal to the mass plane; (2) the lateral vibration in the mass plane: (3) the angular vibration of the mass around the beam axis.

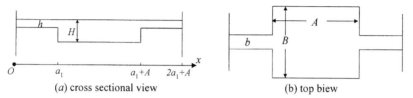

(a) cross sectional view   (b) top biew

17. For two silicon cantilever beams with cross sections as shown in the figure below, where b=5μm, h=10μm, and L=1000μm. (Note that the two beams have the same cross sectional area). Compare two beams on
(1) the maximum displacements at the free ends of the beams under their own weight;
(2) the maximum stresses on top of the beams;
(3) the basic natural vibration frequencies of the beams (in the z-direction).

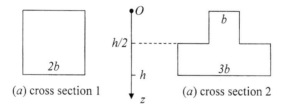

(a) cross section 1   (a) cross section 2

18. A silicon beam-mass structure is shown in the figure below, where a_1=100μm, b_1=20μm, a_2=150μm, b_2=30μm, A=B=4mm, h=10μm and H=300μm, find the basic vibration frequency of the mass in its normal direction.

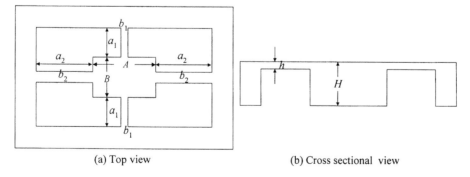

(a) Top view   (b) Cross sectional view

19. A cantilever silicon beam with a concentrated mass M at its free end as shown in the figure below, where b, h, L are the width, the thickness and the length of the beam, respectively. If the mass of the beam and the mass M are comparable, show that the basic vibration frequency found from Rayleigh method is Eq. (2.3.22) in the text.

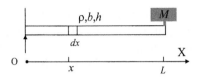

Problems on Vibration Frequency –II

20. For a beam in a steady state, Eq. (2.4.3) reduces to $EIw''''=q(x)$. from this differential equation, derive Eq. (2.2.28) for a cantilever beam, the static displacement function under its own weight. (Hint: the boundary conditions are $w(0)=w'(0)=w''(L)=w'''(L)=0$)

21. Starting from the differential equation shown in problem 20, derive Eq. (2.2.36) for a bridge, the static displacement function under its own weight.

22. Starting from the differential equation given in problem 20, derive Eq. (2.2.44), the static displacement function under its own weight, for a double-clamped beam-mass structure as shown in the figure in problem 16.

23. For a silicon bridge with dimensions of $b=100\mu m$, $h=5\mu m$ and $L=1000\mu m$, how large is the applied tensile stress to raise its basic vibration by 1%.

24. For a double-clamped silicon beam with dimensions of $b=100\mu m$, $h=5\mu m$ and $L=1000\mu m$, how large is the applied compressive stress to make the beam to buckle.

25. The structure of a SiO_2 bridge is shown in the figure below. The dimensions of the bridge are $b=100\mu m$, $h=0.5\mu m$ and $L=100\mu m$. The SiO_2 was grown at 900°C and cooled down to room temperature (20°C) rapidly. If the SiO_2 has no intrinsic strain at its growth temperature, (1) determine the thermal strain in the SiO_2 bridge at room temperature and (2) decide if the bridge buckles or not at a 20°C.

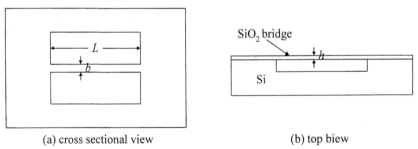

(a) cross sectional view   (b) top biew

Problems on Damped and Forced Vibration

26. A silicon beam-mass structure is shown in the figure in problem 16. The dimensions of the structure are $a_1=500\mu m$, $b=50\mu m$, $h=20\mu m$, $A=B=4mm$ and $H=300\mu m$. If the mass of the beam is negligible, find (1) the vibration frequency of damped vibration for a coefficient of damping force of $c=0.044kg/sec$, (2) the vibration amplitude at low frequency and at resonance for a driving force of $F=(1\times10^{-4}N)\sin\omega t$.

27. Plot the amplitude – frequency relation and the phase lag – frequency relation for a vibration structure with a quality factor of $Q=20$ (Note: using of Excel or other appropriate software tools is required.)

28. If a vibration system has a phase lag of 89.98° at resonance, find the quality factor of the system.

29. An amplitude-frequency relation for a vibration system is shown in the figure below. Find the damping ratio of the system.

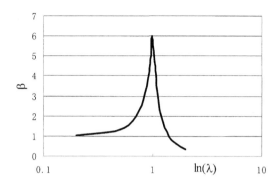

Problems on Diaphragm Structures

30. A long rectangular silicon diaphragm with clamped edges has a width of $2a$=1mm and a thickness of h=10μm. The length of the diaphragm is much larger than its width. Find (1) the maximum stress on the diaphragm surface for an applied pressure of p=1atm on the surface (if E=1.7×10^{11} Pa, v=0.3), (2) the maximum pressure, p_{max}, the diaphragm can support (if the rupture stress of silicon is T_R=3×10^{9}Pa).

31. From differential equation Eq. (2.6.15) and appropriate boundary conditions for a circular plate with clamped edge, derive the displacement function Eq. (2.6.16) for the circular diaphragm.

32. The edge length of a square diaphragm with clamped edges is 2mm and the thickness of the diaphragm is 20μm. If a pressure of 1 atm is applied on the top surface of the diaphragm, find the stress at the edge center of the diaphragm and the stress at the point 0.1mm away from the edge center.

33. The edge length of a square diaphragm with clamped edges is 2mm and the thickness of the diaphragm is 20μm. Find the basic natural vibration frequency of the diaphragm.

34. The diameter of a circular diaphragm with clamped edges is d=2a=2mm and the thickness of the diaphragm is 20μm. Find the basic natural vibration frequencies of the diaphragm in air and in water.

References

[1] S.P. Timoshenko and J.N. Goodier, "Theory of Elasticity", Third Edition, McGraw-Hill Book Company, 1970

[2] E.P. Popov, "Introduction to Mechanics of Solid", 1968 Practice-Hall, Inc., Englewoods, NJ

[3] S. Clark, K. Wise, Pressure sensitivity in anisotropically etched thin-diaphragm pressure sensors, IEEE Trans. on Electron Devices, Vol. ED-26 (1979) 1887-1896

[4] P. Townsend, D. Barnett, T. Brunner, Elastic relationships in layered composite media with approximation for the case of thin films on a thick substrate, J. Appl. Phys. Vol. 62 (1987) 4438-4444

[5] G. Lutz, "Semiconductor Radiation Detectors", Springer-Verlag, Berlin Heidelberg, 1999

[6] S.P. Timoshenko, D. H. Young, W. Weaver, "Vibration Problems in Engineering", 4th edition, John Wiley and Sons, 1974

[7] S. Bouwstra, B. Geijselaers, On the resonance frequencies of microbridge, Digest of Technical Papers, The 6th Proc. Int. Conf. Solid-State Sensors and Actuators, San Francisco, CA, USA, June24-27, 1991 (Transducers'91) 538-542

[8] S.P. Timoshenko and S. Woinowsky-Krieger, "Theory of Plates and Shells", 2nd Edition, McGraw-Hill Book Company, 1959

[9] A.E.H. Love, "Mathematical Theory of Elasticity", Dover Publications, New York, 4th edition, 1944

[10] P. Pons, G. Blasquez, Natural vibration frequencies of silicon diaphragms, Digest of Technical Papers, The 6th Int. Conf. on Solid-State Sensors and Actuators, San Francisco, CA, USA, June 24-27, 1991 (Transducers' 91) 543-546

Chapter 3

Air Damping

For a conventional mechanical machine, the damping caused by the surrounding air can generally be ignored. This is because the energy dissipated by air damping is much smaller than the energy dissipated by other mechanisms.

As air damping is related to the surface area of the moving parts, air damping may become very important for micro mechanical devices and systems in determining their dynamic performance due to the large surface area to volume ratio of the moving parts. For some micro mechanical devices, the energy consumed by air damping must be minimized so that the motion of mechanical parts can be maximized with a limited energy supply. For some other situations, air damping has to be controlled to a proper level so that the system has an optimum dynamic performance. Therefore, estimating the damping effect of the system is one of the most important steps in the analysis and design process of micro mechanical devices.

In this chapter, the basic concept of air damping is introduced and the air damping by different mechanisms are analyzed.

§3.1. Drag Effect of a Fluid

§3.1.1. Viscosity of a Fluid

(1) The Coefficient of Viscosity of a Fluid

Although a fluid at rest cannot permanently resist the attempt of a shear stress to change its shape, viscous force appears to oppose the relative motion between different layers of the fluid. Viscosity is thus an internal friction between adjacent layers moving with different velocities.

The internal shear force in a steady flow of a viscous fluid is proportional to the velocity gradient. If the flow is in the x-direction and the speed of the flow is distributed in the y-direction, i.e., the flow velocity in the x-direction, u, is a function of y, the shear force τ_{yx} is

$$\tau_{yx} = \mu \frac{du(y)}{dy} \qquad (3.1.1)$$

where μ is the coefficient of viscosity of the fluid. If the coefficient of viscosity of a fluid is a constant for a steady flow, the fluid is called a Newtonian fluid.

According to Eq. (3.1.1), the coefficient of viscosity has a unit of Pa·sec or Pa·s. At room temperature (20°C), air has a coefficient of viscosity of 1.81×10^{-5} Pa·s and the coefficient of viscosity of water is 1.0×10^{-3} Pa·s.

(2) The Mechanisms of Viscosity
    Though both liquid and gas show viscosity, they have different properties due to different mechanisms in physics.
    For a steady liquid, the relative positions of adjacent molecules in the same layer are basically stable, but the relative positions of molecules in adjacent layers of a laminar flow change due to the flow. Fig. 3.1.1 shows the change of the relative position between molecules A and B in adjacent layers with different flow velocities, where the molecule A has a higher velocity than molecule B. A is catching up B and takes over it. The approach of A to B is accompanied by a decrease of intermolecular potential energy and an increase in molecular kinetic energy. While the molecular kinetic energy becomes disordered, a temporary bond is formed. The external force must do work if the molecules are later to be separated. The work done by the external force becomes random energy.

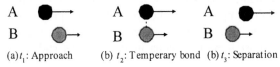

(a)t_1: Approach   (b) t_2: Temperary bond (b)t_3: Separation

Fig. 3.1.1. Mechanism of viscosity in liquid

    According to the mechanism described for a liquid, a temperature increase means that the molecules have a greater thermal speed. Thus, less energy is needed to de-bond the molecular pair). Therefore, the viscosity of most liquids decreases with temperature.

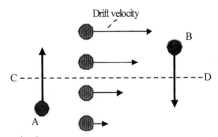

Fig. 3.1.2. Mechanism of viscosity in gas

    For gases, the thermal motion of a molecule is much larger than its drift motion related to the flow of the gas. In Fig. 3.1.2, the molecule A with a smaller drift velocity moving up across the boundary CD (due to the thermal motion) acquires a larger drift velocity, i.e., gains drift momentum, and experiences a force to the right. This means that the molecule has exerted a force to the left on the upper layer, which tends to retard the faster layer.
    Similarly, the molecule B in the faster layer moving down across the boundary CD (due to the thermal motion) exerts a force to the right on the slower layer into which it moves.

Thus, the molecules in the upper layer exert a force to the right on the lower layer, which tends to speed up the lower layer.

(3) The Temperature Dependence of Gas Viscosity

Due to the mechanism described above, a temperature increase means that molecules have a greater thermal speed, which increases the rate at which they cross the layers. Therefore, the viscosity of a gas increases with temperature. A quantitative analysis by a simple model based on the kinetic theory of gas [1] predicts that

$$\mu = \frac{1}{3}\rho\bar{v}\lambda \tag{3.1.2}$$

where ρ is the gas density, $\bar{v}$ is the average velocity of the molecules and λ is the mean free path of the molecules. According to the Kinetic Theory of gas, $\bar{v}$, λ and ρ are

$$\bar{v} = \sqrt{\frac{8RT}{\pi M_m}}, \quad \lambda = \frac{1}{\pi\sqrt{2}nd^2} \quad \text{and} \quad \rho = n\frac{M_m}{N_{av}}$$

respectively, where R is the Universal Molar Gas constant ($R=8.31$ kg·m^2/sec^2/°K), M_m the molar mass, d the effective molecular diameter of the gas, T the absolute temperature and N_{av} the Avogadro constant ($N_{av}=6.022\times10^{23}$/mol). Therefore, we have

$$\mu = \frac{2\sqrt{M_m R}}{3\sqrt{\pi^3 d^2 N_{AV}}}\sqrt{T} \tag{3.1.3}$$

Eq. (3.1.3) suggests that μ is independent of pressure P. Maxwell confirmed experimentally that this result is true over a wide range of pressure around an atmospheric pressure. Eq. (3.1.3) also indicates that μ increases in direct proportion to $\sqrt{M_m}$ and $\sqrt{T}$. Experiments have confirmed that μ increases with temperature but the power slightly exceeds 1/2.

The temperature and molecular dependence of μ can be expressed by an empirical relation known as Sutherland Equation [2]

$$\mu = \mu_o \frac{1+T_S/T_o}{1+T_S/T}\sqrt{\frac{T}{T_o}} \tag{3.1.4}$$

where $T_o=273.16$K, μ_o is the coefficient of viscosity at T_o and T_S is a constant. μ_o and T_S are dependent on the specific gas considered. μ_o and T_S for some gases are listed in Table 3.1.1.

Table 3. 1.1. μ_o and T_S for some gases

gas	air	N_2	H_2	CO_2
$\mu_o/10^{-6}$ (Pa·sec)	17.2	16.6	8.40	13.8
$T_S/°K$	124	104	71	254

Usually, the coefficient of viscosity of liquid is much more sensitive to temperature than that of gas. The data for the coefficient of viscosity of water at one atmospheric pressure are listed in Table 3.1.2. For comparison, the data for air are also listed.

Table 3.1.2. Temperature dependence of coefficient of viscosity for water and air (in 10^{-3} Pa·sec for water and in 10^{-6} Pa·sec for air)

t /°C	0	10	20	30	40	50	60	70	80	90	100
H_2O	1.79	1.30	1.02	0.80	0.65	0.55	0.47	0.41	0.36	0.32	0.28
air	17.2	17.8	18.1	18.7	19.2	19.6	20.1	20.4	21.0	21.6	21.8

§3.1.2. Viscous Flow of a Fluid

(1) Equations for Viscous Flow

Consider an elemental cubic in a fluid as shown in Fig. 3.1.3. There are six shear force components on its surface caused by the velocity gradient of the flow: $\tau_{xy}(x)dydz$, $\tau_{xy}(x+dx)dydz$, $\tau_{yz}(y)dxdz$, $\tau_{yz}(y+dy)dxdz$, $\tau_{zx}(z)dxdy$, $\tau_{zx}(z+dz)dxdy$. There are also six normal force components on its surface caused by pressure: $P(x)dydz$, $P(x+dx)dydz$, $P(y)dxdz$, $P(y+dy)dxdz$, $P(z)dxdy$ and $P(z+dz)dxdy$.

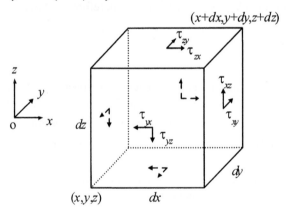

Fig. 3.1.3. Shearing stresses on the surfaces of an element cube in liquid

For a steady flow, assuming the weight of the fluid is negligible, the force balance for the cube in the z-direction is

$$[P(z) - P(z + dz)]dxdy + [\tau_{xz}(x + dx) - \tau_{xz}(x)]dydz + [\tau_{yz}(y + dy) - \tau_{yz}(y)]dxdz = 0$$

Therefore, we have

$$\frac{\partial P}{\partial z} = \frac{\partial \tau_{xz}}{\partial x} + \frac{\partial \tau_{yz}}{\partial y}$$

As $\tau_{xz} = \mu\partial w/\partial x$ and $\tau_{xz} = \mu\partial w/\partial y$, we have

$$\frac{\partial P}{\partial z} = \mu(\frac{\partial^2 w}{\partial x^2} + \frac{\partial^2 w}{\partial y^2})$$
(3.1.5)

where w is the velocity component in z-direction. For the same reason, we have

$$\frac{\partial P}{\partial y} = \mu(\frac{\partial^2 v}{\partial x^2} + \frac{\partial^2 v}{\partial z^2})$$
(3.1.6)

and

$$\frac{\partial P}{\partial x} = \mu(\frac{\partial^2 u}{\partial y^2} + \frac{\partial^2 u}{\partial z^2})$$
(3.1.7)

where u and v are velocity components in the x- and y-directions, respectively. Eqs. (3.1.5), (3.1.6) and (3.1.7) are equations for viscous flow of a fluid caused by a pressure, P.

(2) Viscous Flow in a Pipe

Let the length of the pipe be L and the radius of the circular cross section a, and $L \gg a$, as shown in Fig. 3.1.4. If z-axis is taken along the centroid of the pipe, Eq. (3.1.5) is the only equation to be used to determine the flow.

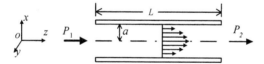

Fig. 3.1.4. Fluid flow in a long pipe

As the length of pipe, L, is much larger than its radius, a, the flow in the pipe is in the z-direction and the velocity distribution is symmetric radially. By using a polar coordinate in the x-y plane and putting the origin at the center of the cross section of the pipe, Eq. (3.1.5) can be written as

$$\frac{\partial P}{\partial z} = \mu \frac{1}{r} \frac{\partial}{\partial r}\left(r \frac{\partial}{\partial r} w(r) \right)$$

After integration, we have

$$r \frac{\partial}{\partial r} w(r) = \frac{1}{2\mu} \frac{\partial P}{\partial z} r^2 + C_1$$

As $\partial w(r)/\partial r = 0$ at $r=0$, we have $C_1=0$. Thus, a second integration yields

$$w(r) = \frac{1}{4\mu}\frac{\partial P}{\partial z}r^2 + C_2$$

According to the "non-slippage" boundary condition of

$$w(a) = 0 \qquad\qquad\qquad\qquad (3.1.8)$$

we find

$$w(r) = -\frac{1}{4\mu}\frac{\partial P}{\partial z}(a^2 - r^2)$$

The negative sign indicates that the velocity is opposite to the pressure gradient. If the pressure difference between the two ends of the pipe is P, i.e. $\partial P/\partial z = P/L$, we have

$$w(r) = -\frac{1}{4\mu}\frac{P}{L}(a^2 - r^2)$$

The flow rate, i.e., the volume of fluid passing through the pipe per unit time, is

$$Q = \int_0^a |w(r)|2\pi r dr$$

By simple calculation, we have

$$Q = \frac{\pi a^4}{8\mu}\frac{P}{L} \qquad\qquad\qquad\qquad (3.1.9)$$

This is referred to as Poiseuille equation. By Eq. (3.1.9), the average velocity of the flow is

$$\bar{w} = \frac{Q}{\pi a^2} = \frac{a^2}{8\mu}\frac{P}{L} \qquad\qquad\qquad\qquad (3.1.10)$$

It may be mentioned here that, for gases, the above results are accurate only when the diameter of the pipe is much larger then the mean free path of the gas molecules. Otherwise, the non-slippage boundary condition might have to be modified.

(3) Reynolds' Number

The flow pattern described in the above is an orderly flow that is called a streamline flow or laminar flow. A streamline flow occurs only when the speed of the flow is small.

The flow will become turbulent if the speed of the flow exceeds a certain limit. The criterion for turbulence is usually given by the value of the Reynolds' number, *Re*. *Re* is a dimensionless number, which, for a tube, takes the form of

$$Re = \frac{\bar{v}\rho d}{\mu}$$

where ρ is the mass density of the fluid, $\bar{v}$ the average velocity of the fluid and d the diameter of the tube.

Re is a convenient parameter for measuring the stability of flow. However, the critical value of *Re* that causes instability of fluid flow depends strongly on the shape of the tube and can only be determined by experiments. For tubes with circular cross-section, we have:

(i) *Re*< 2200, the flow is laminar

(ii) *Re*~ 2200, the flow is unstable

(iii) *Re*≥ 2200, the flow is turbulent

The Reynolds' number is also useful in measuring the stability of fluid flowing through a solid object inside the fluid (or, the stability of an object moving through a fluid at rest). In this case, the general form of the Reynolds' number is

$$Re = \frac{v\rho l}{\mu}$$

where l is a characteristic dimension of the object. For example, l is the diameter of a sphere and, for a column with a circular cross section moving through the fluid laterally, l is the diameter of the cross section, etc. The critical value of the Reynolds' number that causes instability depends on the shape of the object and can only be determined by experiments. As Re is usually small for a micro machine, only laminar flow condition will be considered in our study.

§3.1.3 Drag Force Damping

A drag force will be applied on a body if the body is held steadily in a fluid flow (or the body is dragged through a steady fluid) because there exists a velocity gradient between the boundary layer and the more distant points in the viscous fluid. As the analysis for the drag force is quite complicated, the drag forces for some simple body structures moving through an infinitive viscous fluid are given here [3].

(a) Sphere With a Radius *r*:

$$F = 6\pi\mu r v \qquad (3.1.11)$$

(b) Circular Dish With a Radius of *r* Moving in its Normal Direction:

$$F = 16\mu r v \qquad (3.1.12)$$

(c) Circular Dish With a Radius of *r* Moving in its Plane Direction:

$$F = \frac{32}{3} \mu r v \qquad (3.1.13)$$

where v is the speed of the sphere or the circular dish relative to the distant fluid.

When Eqs. (3.1.11), (3.1.12) and (3.1.13) are compared, we can find that the dependence of drag forces on different cross sections or on the moving direction are not significant. All three drag forces for low speed motion can be written in the same form as

$$F = 6\pi\alpha\mu r v \qquad (3.1.14)$$

where the value of α for a sphere, a dish moving in its normal direction and a dish moving in its plane direction are 1.0, 0.85 and 0.567, respectively. Note that drag forces are independent of the mass density of the fluid, ρ (known as Stokes' law).

However, this conclusion is not true for higher moving speeds [3]. The force on a sphere with a radius, r, oscillating in a fluid is given by

$$F = -\beta_1 v - \beta_2 \frac{dv}{dt} \qquad (3.1.15)$$

with

$$\beta_1 = 6\pi\mu r + 3\pi r^2 \sqrt{2\rho\mu\omega}$$

and

$$\beta_2 = \frac{2}{3}\pi\rho r^3 + 3\pi r^2 \sqrt{\frac{2\rho\mu}{\omega}}$$

where ω the radial frequency of the oscillation. Note that both β_1 and β_2 are dependent on the mass density of the fluid.

§3.1.4. Damping by Acoustic Transmission

For a plate vibrating in a fluid, it radiates acoustic energy into the fluid through sound waves and the fluid reacts on the plate with a resistant damping force. For a circular plate with a radius of a vibrating in the direction normal to the plate (i.e., for a piston-like movement), the damping force on the plate is found to be [4]

$$F_{acou} = \rho c A \left\{ 1 - 2 J_1\left(\frac{4\pi a}{\lambda}\right) \middle/ \left(\frac{4\pi a}{\lambda}\right) \right\} \dot{x} \qquad (3.1.16)$$

where ρ is the density of the fluid, c the speed of sound wave in the fluid, A the area of the plate ($A = 2\pi a^2$) and λ the wave length of the sound. $J_1(x)$ is the first order Bessel function.

Equation (3.1.16) shows that the acoustic damping force is relatively larger only when $2\pi a \gg \lambda$. In this case, we have

$$F_{acou} = \rho c A \dot{x} \tag{3.1.17}$$

However, acoustic damping is usually insignificant for most micro mechanical devices in air. It might be significant for micro mechanical devices in water or other liquids.

§3.1.5. The Effects of Air Damping on Micro-Dynamics

As seen in §3.1.3, the drag force applied to a sphere moving in a viscous fluid at a speed of v is $F = 6\pi\mu r v$, where μ is the coefficient of viscosity of the fluid and r the radius of the sphere. The ratio between the drag force F and the mass of the body, M, is

$$\frac{F}{M} = \frac{6\pi\mu r v}{4\pi r^3 \rho / 3} = \frac{4.5\mu v}{\rho r^2} \tag{3.1.18}$$

where ρ is the specific density of the body. It is obvious that the smaller the dimension of the body, the larger the effect of the drag force on the body. For example, for a silicon ball of radius $r=1$cm moving in air with a velocity of 1 cm/sec, F/M is 3.5×10^{-6} m/sec^2, while, for a silicon ball of radius 10 microns, F/M is 3.5 m/sec^2, one million times larger. Therefore, the drag force caused by the viscosity of the surrounding air is usually negligible for conventional mechanical structure but it may play an important role for the motion of micro machines.

Now let us look at a practical example. The differential equation for a beam-mass (spring-mass) accelerometer is

$$m\ddot{x} = -kx - c\dot{x}$$

where k is the spring constant of the beam and c is the coefficient of damping force caused by the surrounding air. A very important dynamic parameter of the accelerometer is the damping ratio of the system, ζ. The definition of ζ is

$$\zeta = \frac{c}{2m\omega_o} = \frac{c}{2\sqrt{mk}}$$

where ω_0 is the free vibration frequency of the system. The damping ratio, ζ, for an accelerometer is usually required to be around 0.7 so that the system shows the best frequency response to an input signal (not shown in the equation). Quite often, the quality factor, Q, is used to characterize the mechanical system. For small damping, the relation between the quality factor and the damping ratio is: $Q = 1/2\zeta$.

According to Eq. (3.1.14), the coefficient of damping force, c, is proportional to the dimensions of the mechanical structure and the coefficient of viscosity of the surrounding fluid. As m is quite large for an accelerometer made of conventional mechanical structures,

ζ is usually very small in air. It is quite difficult to raise the damping ratio, ζ, to around 0.7 even if the structure is filled with oil of high viscosity. But for an accelerometer made of micro mechanical structures, the damping ratio, ζ, can be easily raised to around 0.7 in air by using some mechanical structure to increase the damping force in a controlled way. An additional advantage of air damping over oil damping include a much lower temperature coefficient and the ease of packaging the device. The basic mechanisms of air damping for micromechanical structures are squeeze-film air damping and slide-film air damping in addition to the drag force damping discussed above. The basic principles and relations for these mechanisms will be analyzed in §3.2, §3.3 and §3.4. The damping of microstructures in rarefied air will be discussed in §3.5.

§3.2. Squeeze-film Air Damping

§3.2.1. Reynolds' Equations for Squeeze-film Air Damping

(1) Squeeze-film Air Damping

When a plate is placed in parallel to a wall and moving towards the wall, the air film between the plate and the wall is squeezed so that some of the air flows out of the gap as shown in Fig. 3.2.1. Therefore, an additional pressure Δp develops in the gap due to the viscous flow of the air. On the contrary, when the plate is moving away from the wall, the pressure in the gap is reduced to keep the air flowing into the gap.

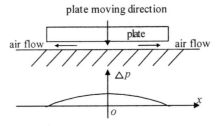

Fig. 3.2.1. Pressure built-up by squeeze-film motion

In both moving directions of the plate, the forces on the plate caused by the built-up pressure are always against the movement of the plate. The work done by the plate is consumed by the viscous flow of the air and transformed into heat. In other words, the air film acts as a damper and the damping is called squeeze-film air damping.

Obviously, the damping force of squeeze-film air damping is dependent on the gap distance; the smaller the gap, the larger the damping force. When the plate is very far away from the wall, the pressure build-up is negligible and the damping force is mainly due to the drag force discussed in §3.1.

The squeeze-film air damping has significant effects on the dynamic behavior of microstructures. In many cases, it should be reduced and, in some other cases, it should be controlled to an expected level.

(2) Reynolds' Equation

Suppose that we have a pair of plates in parallel with the *x-y* plane of the Cartesian coordinates as shown in Fig. 3.2.2 and the dimensions of the plates are much larger than the distance between them so that the gas flow between the plates caused by the relative motion of the plates is lateral (in the *x-* and *y-*direction but not in the *z-*direction).

Let us consider an elemental column, *hdxdy* (where $h = h_2 - h_1$), as shown in Fig. 3.2.2, where q_x is the flow rate in the *x-*direction per unit width in the *y-*direction and q_y is the flow rate in the *y-*direction per unit width in the *x-*direction.

The balance of mass flow for the elemental column requires

$$\left(\rho q_x\right)_x dy - \left(\rho q_x\right)_{x+dx} dy + \left(\rho q_y\right)_y dx - \left(\rho q_y\right)_{y+dy} dx = \left(\frac{\partial \rho h_2}{\partial t} - \frac{\partial \rho h_1}{\partial t}\right) dxdy$$

By making use of the relations $(\rho q_x)_{x+dx} = (\rho q_x)_x + [\partial(\rho q_x)/\partial x]dx$, $(\rho q_y)_{y+dy} = (\rho q_y)_y + [\partial(\rho q_y)/\partial y dy$ and $h=h_2-h_1$, we have

$$\frac{\partial\left(\rho q_x\right)}{\partial x} + \frac{\partial\left(\rho q_y\right)}{\partial y} + \frac{\partial\left(\rho h\right)}{\partial t} = 0 \tag{3.2.1}$$

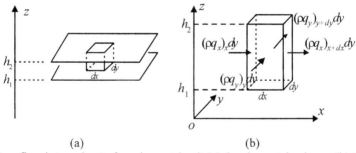

(a)   (b)

Fig. 3.2.2. Mass flow into and out of an elemental unit (a) the elemental column; (b) the flow rates

To find q_x and q_y, we have to find the speed distribution in the *z-*direction first. As the dimensions of the plate is much larger than the gap and the flow is lateral, the velocity components *u* and *v* are functions of *z* only. From Eq. (3.1.7), we have

$$\frac{\partial P}{\partial x} = \mu \frac{\partial^2 u}{\partial z^2}$$

For a small gap, *P(x,y)* is not a function of *z*. By integrating the equation twice, we have

$$u(z) = \frac{1}{2\mu}\frac{\partial P}{\partial x}z^2 + C_1 \frac{1}{\mu}z + C_2 \tag{3.2.2}$$

If the plates do not move laterally and we put the origin of the coordinates on the bottom plate, the boundary conditions for Eq. (3.2.2) are

$$u(0) = 0 , \ u(h) = 0$$

Therefore

$$u(z) = \frac{1}{2\mu} \frac{\partial P}{\partial x} z(z - h)$$

(3.2.3)

The flow rate in the x-direction for a unit width in the y-direction is

$$q_x = \int_0^h u dz = -\frac{h^3}{12\mu} \left(\frac{\partial P}{\partial x} \right)$$

(3.2.4)

The negative sign in the equation indicates that the flow is in the direction with decreasing pressure.

Similarly, we have

$$q_y = -\frac{h^3}{12\mu} \left(\frac{\partial P}{\partial y} \right)$$

(3.2.5)

By substituting Eqs. (3.2.4) and (3.2.5) into (3.2.1), we find

$$\frac{\partial}{\partial x} \left(\rho \frac{h^3}{\mu} \frac{\partial P}{\partial x} \right) + \frac{\partial}{\partial y} \left(\rho \frac{h^3}{\mu} \frac{\partial P}{\partial y} \right) = 12 \frac{\partial(h\rho)}{\partial t}$$

(3.2.6)

Eq. (3.2.6) is referred to as Reynolds' equation. Eq. (3.2.6) can also be derived from the much more complicated Navier-Stokes equation under the condition that the Modified Reynolds' Number for a squeeze film, R_S, is much smaller than unity [5, 6], i.e., the condition of

$$R_S = \frac{\omega h^2 \rho}{\mu} \ll 1$$

where ω is the radial frequency of the oscillating plate. This condition is satisfied for typical silicon microstructures. For example, an accelerometer with an air film thickness of 10 microns, oscillating at a frequency of 1 kHz, would have a modified Reynolds' number of $R_S = 0.045$.

As h is assumed to be uniform both in the x- and y-directions, we have

$$\frac{\partial}{\partial x} \left(\rho \frac{\partial P}{\partial x} \right) + \frac{\partial}{\partial y} \left(\rho \frac{\partial P}{\partial y} \right) = \frac{12\mu}{h^3} \frac{\partial(h\rho)}{\partial t}$$

For an isothermal film, the air density, ρ, is proportional to pressure P, i.e., $\rho = \dfrac{P}{P_o} \rho_o$. The above equation can be written as

$$\frac{\partial}{\partial x}\left(P \frac{\partial P}{\partial x} \right) + \frac{\partial}{\partial y}\left(P \frac{\partial P}{\partial y} \right) = \frac{12\mu}{h^3} \frac{\partial(h\rho)}{\partial t} \tag{3.2.7}$$

Eq. (3.2.7) can be developed into

$$\left(\frac{\partial P}{\partial x} \right)^2 + \left(\frac{\partial P}{\partial y} \right)^2 + P\left(\frac{\partial^2}{\partial x^2}P + \frac{\partial^2}{\partial y^2}P \right) = \frac{12\mu}{h^3}\left(P\frac{dh}{dt} + h\frac{\partial P}{\partial t} \right)$$

Assuming that $h=h_o+\Delta h$ and $P=P_o+\Delta P$, for small motion distance of the plate, we have $\Delta h \ll h_0$ and $\Delta P \ll P_0$. Under these conditions, the equation can be approximated as

$$P_o\left(\frac{\partial^2 \Delta P}{\partial x^2} + \frac{\partial^2 \Delta P}{\partial y^2} \right) = \frac{12\mu}{h^3} P_o h_o \left(\frac{1}{h_o}\frac{d\Delta h}{dt} + \frac{1}{P_o}\frac{\partial \Delta P}{\partial t} \right) \tag{3.2.8}$$

If $\Delta P / P_o \ll \Delta h / h_o$, we have

$$\frac{\partial^2 \Delta P}{\partial x^2} + \frac{\partial^2 \Delta P}{\partial y^2} = \frac{12\mu}{h^3}\frac{d\Delta h}{dt} \tag{3.2.9}$$

or,

$$\frac{\partial^2 P}{\partial x^2} + \frac{\partial^2 P}{\partial y^2} = \frac{12\mu}{h^3}\frac{dh}{dt} \tag{3.2.10}$$

In Eq. (3.2.10), P is equivalent to ΔP. Thus, letter P in Eq. (3.2.10) is sometimes read as ΔP. However, attention must be given to the difference in the boundary conditions for P and ΔP : $P = P_o$ but $\Delta P = 0$ at the periphery of the plate.

Before ending this section, let us discuss once more the condition for Eq. (3.2.10). Suppose that the typical dimension of the plate is l (e.g. the radius of a disk or the half width of a rectangle) and the motion of the plate is a sinusoidal vibration with amplitude δ, i.e., $h=h_o+\delta\sin\omega t$. From Eq. (3.2.10), we can make a rough and ready estimation of ΔP

$$\frac{\Delta P}{l^2} = \frac{12\mu}{h^3}\delta\omega\cos\omega t$$

or

$$\frac{\Delta P}{P_o} = \frac{12\mu l^2\omega}{P_o h^2}\frac{\delta\cos\omega t}{h} = \sigma\frac{\delta\cos\omega t}{h}$$

where $\sigma \equiv 12\mu l^2\omega/P_o h_o^2$ is referred to as a "squeeze number". As $\delta\cos(\omega t)$ has the same order of magnitude as $\Delta h = \delta\sin(\omega t)$, we obtain

$$\frac{\Delta P}{P_o} \cong \sigma\frac{\Delta h}{h_o}$$

Therefore, the condition for the validity of Eq. (3.2.10), i.e. $\Delta P/P_o \ll \Delta h/h_o$, is equivalent to a small squeeze number $\sigma \ll 1$, or

$$\frac{l}{h_o} \ll \sqrt{\frac{P_o}{12\mu\omega}}$$

As we know that the air pressure P in a closed cylinder as shown in Fig. 3.2.3 is inversely proportional to the distance between the piston and the bottom of the cylinder, i.e., $\Delta P/P = \Delta h/h$ (for $\Delta h \ll h$). However, the air pressure between two parallel plates will be much more difficult to build-up due to the open border, i.e., $\Delta P/P \ll \Delta h/h$ in most practical conditions.

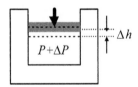

Fig. 3.2.3. Gas compression by a piston

The squeeze number implies the efficiency of squeezing in establishing the pressure. For a piston sealed in its peripheral as shown in Fig. 3.2.3, we have

$$\frac{\Delta p}{p_o} = 1\cdot\frac{\Delta h}{h_o}$$

which means that the relative pressure change equals to the squeeze rate. Thus we have a squeeze number of unity and the squeeze number for a plate with open peripheral is always smaller than unity.

As an example, let us assume that $\omega = 2\pi\times10^3/\text{sec}$, $P_o = 10^5$ Pa (i.e., 1 atm.) and $\mu = 1.8\times10^{-5}$ Pa·sec (for air at 20°C). The requirement for l for the validity of Eq. (3.2.10) is $l \ll 271\,h_o$ (e.g., $l \ll 5.42\text{mm}$ for $h_o = 20\mu\text{m}$). For the same conditions but the oscillating frequency is $\omega = 2\pi\times10^4/\text{sec}$, the condition becomes $l \ll 85.6 h_o$ (e.g. $l \ll 1.71\text{mm}$ for $h_o = 20\mu\text{m}$).

§3.2.2. Long Rectangular Plate

(1) Damping Pressure and Damping Force

Consider a pair of rectangular plates with length, L, much larger than width, B. The origin of the Cartesian coordinates is at the center of the lower plate and the x-axis is along the width direction, as schematically shown in Fig. 3.2.4. As the problem is virtually one-dimensional, Eq. (3.2.10) is written as

$$\frac{d^2 P}{dx^2} = \frac{12\mu}{h^3}\frac{dh}{dt} \tag{3.2.11}$$

The boundary conditions are

$$P\left(\pm\frac{1}{2}B\right) = 0 \tag{3.2.12}$$

By integrating Eq. (3.2.11) twice, we obtain the damping pressure

$$P(x,t) = \frac{6\mu}{h^3}\frac{dh}{dt}x^2 + C_1 x + C_2$$

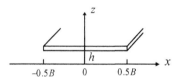

Fig. 3.2.4. Squeeze-film air damping of a long rectangular plate

By using the boundary conditions, we obtain

$$P(x,t) = -\frac{6\mu}{h^3}\left(\frac{B^2}{4} - x^2\right)\frac{dh}{dt} \tag{3.2.13}$$

$P(x,t)$ is positive when the air film is squeezed ($dh/dt<0$), and vice versa. The maximum damping pressure is at the center of the plate ($x=0$) where $P(0,t)=-(3\mu B^2/2h^3)dh/dt$. The schematic distribution of the damping pressure is shown in Fig. 3.2.5.

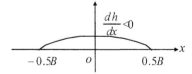

Fig. 3.2.5. Pressure distribution under a long rectangular plate

The damping force F on the plate is

$$F_{lr} = \int_{-B/2}^{B/2} P(x)L\,dx = -\frac{\mu B^3 L}{h^3}\frac{dh}{dt} \equiv -\frac{\mu B^3 L}{h^3}\dot{h}$$

According to the definition of $F = -c\dot{x}$, the coefficient of damping force for a long rectangular plate is

$$c_{lr} = \frac{\mu B^3 L}{h^3} \tag{3.2.14}$$

Note that Eq. (3.2.14) is only valid for rectangular plates whose length, L, is much larger than their width, B. For a rectangular plate with a comparable L and B, the squeeze-film air damping will be discussed in §3.2.4.

(2) Example

Suppose that the width of the plates is B=2mm, the length of the plates 10 mm, the gap distance h_o=20μm and the motion of the upper plate $h=h_o+\delta\sin\omega t$, where δ=1μm and ω=2π×10³/s. The environment is air of one atmospheric pressure. Thus, we have

$$\frac{dh}{dt} = \delta \cdot \omega\cos\omega t$$

and

$$P(x,t) = -\frac{6\mu}{h^3}\left(\frac{B^2}{4} - x^2\right)\delta \cdot \omega\cos\omega t = -\frac{3\mu B^2}{2h^3}\left[1 - \left(\frac{2x}{B}\right)^2\right]\delta\omega\cos\omega t$$

As the coefficient of viscosity of air is μ=1.8×10⁻⁵ Pa·s (at 20°C), the damping pressure at x=0 is

$$P(0,t) = -85\cos\omega t \text{ (Pa)}$$

Therefore, the maximum damping pressure is 8.5×10^{-4} atm. As $\delta/h = 0.05$, the result satisfies the condition $\Delta P/P_o \ll \delta/h_o$. The pressure is not easily to build up due to the open boundary condition and the low viscosity of gas. This phenomenon is often described as "incompressible gas" approximation.

According to Eq. (3.2.14) the coefficient of damping force is c_{lr}=0.182kg/s.

§3.2.3. Circular and Annular Plates

(1) Circular Plate

For a circular plate moving against a substrate as schematically shown in Fig. 3.2.6, the equation for air damping can be written in a polar coordinate system as

$$\frac{1}{r}\frac{\partial}{\partial r}\left(r\frac{\partial}{\partial r}P(r)\right)=\frac{12\mu}{h^3}\frac{dh}{dt} \tag{3.2.15}$$

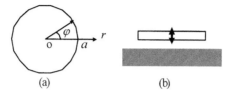

(a)   (b)

Fig. 3.2.6. A circular plate (a) top view and the polar coordinate system; (b) cross sectional view

The boundary conditions are

$$P(a)=0\,,\ \ \frac{dP}{dr}(0)=0 \tag{3.2.16}$$

where a is the radius of the plate. By integrating Eq. (3.2.15) and using the boundary conditions in Eq. (3.2.16), we find the damping pressure

$$P(r)=-\frac{3\mu}{h^3}\left(a^2-r^2\right)\frac{dh}{dt} \tag{3.2.17}$$

The damping force on the circular plate is

$$F_{cir}=\int_{0}^{a}P(r)2\pi rdr=-\frac{3\pi}{2h^3}\mu a^4\frac{dh}{dt}$$

or

$$F_{cir}=-\frac{3}{2\pi}\frac{\mu A^2}{h^3}\frac{dh}{dt}=-0.4775\frac{\mu A^2}{h^3}\frac{dh}{dt} \tag{3.2.18}$$

where $A=\pi a^2$ is the area of the plate. The coefficient of damping force is

$$c_{cir}=\frac{3\pi}{2h^3}\mu a^4 \tag{3.2.19}$$

(2) Annular Plate

For an annular plate moving against a substrate, the equation for air damping is the same as Eq. (3.2.15), but the boundary conditions are

$P(a)=0,\ P(b)=0$

where a and b are the outer and inner radii of the annular plates, respectively, as shown in Fig. 3.2.7. By solving Eq. (3.2.15) with the boundary conditions, the damping pressure is

$$P(r) = \left(-\frac{3\mu}{h^3}a^2\left(1-\frac{r^2}{a^2}\right) + \frac{3\mu}{h^3}a^2\left(1-\frac{b^2}{a^2}\right)\frac{\ln(r/a)}{\ln(r/b)} \right)\frac{dh}{dt} \tag{3.2.20}$$

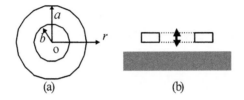

(a)   (b)

Fig. 3.2.7. An annular plate (a) top view and the polar coordinate system; (b) cross sectional view

If the ratio of b/a is designated as β, we have the damping force for the annular plate

$$F_{ann} = \int_b^a P(r)2\pi r dr = -\frac{3\pi\mu a^4}{2h^3}\left(1-\beta^4+\frac{(1-\beta^2)^2}{\ln\beta}\right)\dot{h}$$

The damping force can be written as

$$F_{ann} = -\frac{3\pi\mu a^4}{2h^3}G(\beta)\dot{h} = -\frac{3\mu A^2}{2\pi h^3}G(\beta)\dot{h}$$

where $A=\pi a^2$ and $G(\beta)$ is

$$G(\beta) = 1-\beta^4+\frac{(1-\beta^2)^2}{\ln\beta}$$

The coefficient of damping force for an annular plate is

$$c_{ann} = \frac{3\mu a^2 A}{2h^3}G(\beta) \tag{3.2.21}$$

§3.2.4. Rectangular Plate

In this section, we will discuss the squeeze-film air damping for a rectangular plate in a general form. If the side lengths in the x- and y-directions of the plate are $B=2a$ and $L=2b$, respectively, as shown in Fig. 3.28, and a and b are comparable, the differential equation for pressure in the air film is Eq. (3.2.10) and the boundary conditions are

$$P(\pm a, y) = 0, \ P(x, \pm b) = 0 \tag{3.2.22}$$

The solution to Eq. (3.2.10) can be divided into two parts: $P = p_1 + p_2$, where p_1 is a specific solution to Eq. (3.2.10), i.e., p_1 is a solution to equation

$$\frac{\partial^2 p_1}{\partial x^2} + \frac{\partial^2 p_1}{\partial y^2} = \frac{12\mu}{h^3}\dot{h}$$ (3.2.23)

and p_2 is a general solution to the Laplace equation

$$\frac{\partial^2 p_2}{\partial x^2} + \frac{\partial^2 p_2}{\partial y^2} = 0$$ (3.2.24)

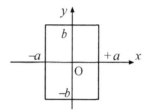

Fig. 3.2.8. Rectangular plate with comparable edge lengths

(1) Solution of p_1

By assuming that $p_1 = A + Bx + Cx^2$ and p_1 meets the boundary conditions of $p_1(\pm a)=0$, we find

$$p_1 = -\frac{6\mu}{h^3}\frac{dh}{dt}(a^2 - x^2)$$ (3.2.25)

(2) Boundary Condition of p_2

From the definition of $P = p_1 + p_2$ and the boundary conditions of $P(\pm a,y)=0$, we have

$$p_1(\pm a, y) + p_2(\pm a, y) = 0$$

According to Eq. (3.2.25), the boundary conditions for p_2 at $x = \pm a$ can be shown to be

$$p_2(\pm a, y) = 0$$ (3.2.26)

According to Eq. (3.2.22), the boundary conditions for P at $y = \pm b$ should be $P(x, \pm b) = 0$, i.e.,

$$p_1(x) + p_2(x, \pm b) = 0$$

Therefore, the boundary conditions for p_2 at $y = \pm b$ are

$$p_2(x,\pm b) = -p_1(x) = \frac{6\mu}{h^3}\frac{dh}{dt}(a^2 - x^2)$$ (3.2.27)

The complete boundary conditions for p_2 are Eqs. (3.2.26) and (3.2.27).

(3) Solution of p_2

To find the solution of p_2, we separate the variables by assuming that

$$p_2 = X(x)Y(y)$$ (3.2.28)

By using Eq. (3.2.28), Eq. (3.2.24) becomes

$$X''(x)Y(y) + Y''(y)X(x) = 0$$

or

$$\frac{X''(x)}{X(x)} = -\frac{Y''(y)}{Y(y)} = \lambda$$

Therefore, we have two independent equations

$$X''(x) - \lambda X(x) = 0$$

and

$$Y''(y) + \lambda Y(y) = 0$$

We assume that $X(x) = A_1\cos\alpha x + A_2\sin\alpha x$. As $X(\pm a)=0$, we have $A_2=0$ and

$$X(x) = A_1\cos(\frac{2n\pi x}{a}) \qquad (n=1,3,5,...)$$ (3.2.29)

Similarly, we assume that $Y(y) = C_1\cosh(\gamma y) + C_2\sinh(\gamma x)$. By using the boundary condition for $Y(y)$, $Y(b)=Y(-b)$, we have $C_2=0$ and

$$Y(y) = C_1\cosh(\frac{2n\pi y}{a}) \quad (n=1,3,5,...)$$ (3.2.30)

By Eqs. (3.2.28), (3.2.29) and (3.2.30), $P_2(x, y)$ can be written as

$$p_2(x,y) = \sum_{n=1,3,5,}^{\infty} a_n\cosh\frac{n\pi y}{2a}\cos\frac{n\pi x}{2a}$$ (3.2.31)

To satisfy the boundary conditions shown in Eq. (3.2.27), we have

$$\sum_{n=1,3,5,}^{\infty} a_n \cosh \frac{n\pi b}{2a} \cos \frac{n\pi}{2a} x = \frac{6\mu}{h^3} \dot{h}(a^2 - x^2) \qquad (3.2.32)$$

The constant a_n's are thus found to be

$$a_n = \frac{\int_{-a}^{a} \frac{6\mu}{h^3} \dot{h}(a^2 - x^2) \cos \frac{n\pi x}{2a} dx}{\cosh \frac{\pi n b}{2a} \int_{-a}^{a} \cos^2 \frac{n\pi x}{2a} dx} = \frac{192\mu \dot{h} a^2}{n^3 \pi^3 h^3} \frac{\sin \frac{n\pi}{2}}{\cosh \frac{n\pi b}{2a}} \qquad (n=1,3,5, \text{etc.})$$

Therefore, we have

$$p_2(x,y) = \frac{192\mu a^2}{h^3 \pi^3} \dot{h} \sum_{n=1,3,5,}^{\infty} \frac{\sin \frac{n\pi}{2}}{n^3 \cosh \frac{n\pi b}{2a}} \cosh \frac{n\pi y}{2a} \cos \frac{n\pi x}{2a}$$

The final solution to the problem is

$$P = p_1 + p_2 = -\frac{6\mu \dot{h}}{h^3}(a^2 - x^2) + \frac{192\mu \dot{h} a^2}{h^3 \pi^3} \sum_{n=1,3,5,}^{\infty} \frac{\sin \frac{n\pi}{2}}{n^3 \cosh \frac{n\pi b}{2a}} \cosh \frac{n\pi y}{2a} \cos \frac{n\pi x}{2a} \qquad (3.2.33)$$

(4) Damping Force
   The damping force on the rectangular plate is

$$F_{rec} = \int_{-a}^{a} dx \int_{-b}^{b} P(x,y) dy \equiv -\frac{\mu L B^3}{h^3} \dot{h} \beta \left(\frac{B}{L} \right) \qquad (3.2.34)$$

where the factor $\beta(B/L)$ is

$$\beta \left(\frac{B}{L} \right) = \left\{ 1 - \frac{192}{\pi^5} \left(\frac{B}{L} \right) \sum_{n=1,3,5,}^{\infty} \frac{1}{n^5} \tanh \left(\frac{n\pi L}{2B} \right) \right\} \qquad (3.2.35)$$

The dependence of β on B/L is shown by the curve in Fig. 3.2.9. For a very long plate, $\beta=1$, and for a square plate (i.e., $a=b$), $\beta=0.42$. The coefficient of the damping force is

$$c_{rec} = \frac{\mu L B^3}{h^3} \beta \left(\frac{B}{L} \right) \qquad (3.2.36)$$

This result shows that the coefficient of damping force for a rectangular plate is similar to that of a long strip plate except for a correction factor β that is related to the aspect ratio of the plate. For a long strip plate the correction factor equals unity.

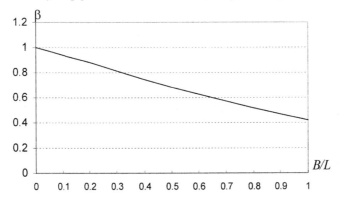

Fig. 3.2.9. The dependence of factor β on the aspect ratio B/L

§3.2.5. Perforated Infinite Thin Plate

Moving plates in microstructures are sometimes perforated (1) to facilitate the etch of sacrificial layer to release the moving parts, and (2) to reduce the damping effect to a certain level for some applications such as accelerometers, microphones,etc. Therefore, the estimation of air damping force for a perforated plate is important in designing the devices.

Suppose that a plate is perforated with circular holes of radius r_o and the holes are uniformly distributed in a simple square array or a hexagon close-packed pattern as shown in Fig. 3.2.10. If the density of hole is n, the area (the cell) allocated to a hole is $A_1 = 1/n$. If the cell is approximated as a circular plate with a hole in the center (i.e., the cell is approximated as an annulus), the outer radius of the annulus is

$$r_c = \frac{1}{\sqrt{\pi n}}$$

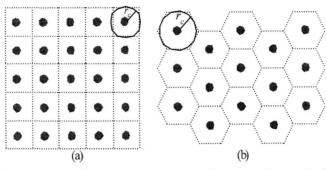

Fig. 3.2.10. Schematic drawing of perforations (a) square array; (b) hexagon close-packed array

The damping force on the whole plate is the summation of the damping force of each cell. So let us first consider the damping force on a cell. The equation for the damping pressure caused by the parallel motion of an annular cell is Eq. (3.2.15) in §3.2.3.

$$\frac{1}{r}\frac{\partial}{\partial r}r\frac{\partial}{\partial r}P(r) = \frac{12\mu}{h^3}\dot{h}$$

Under the approximation conditions that (1) the plate is much larger than a cell, so the air-flow between the cells is negligible (the "infinite plate" approximation), (2) the diameter of the hole is much larger than the thickness of the plate so there is no pressure build-up in the hole (the "thin plate" approximation), the boundary conditions for the cell are

$$P(r_o) = 0, \frac{\partial P}{\partial r}(r_c) = 0 \qquad (3.2.37)$$

By solving Eq. (3.2.15) with boundary conditions, the damping pressure is

$$P(r) = \frac{3\mu r_c^2}{h^3}\dot{h}\left(2\ln\frac{r}{r_o} - \frac{r^2 - r_o^2}{r_c^2}\right)$$

The damping force on a cell is

$$F_1 = \int_{r_o}^{r_c}\frac{3\mu r_c^2}{h^3}\dot{h}\left(2\ln\frac{r}{r_o} - \frac{r^2 - r_o^2}{r_c^2}\right)2\pi r dr = -\frac{\mu(\pi r_c^2)^2}{h^3}\dot{h}\frac{3}{\pi}\left[2\ln\frac{r_c}{r_o} - \left(1 - \frac{r_o^2}{r_c^2}\right)^2 - \frac{1}{2}\left(1 - \frac{r_o^4}{r_c^4}\right)\right]$$

If r_o/r_c is designated as β, we have

$$F_1 = -\frac{3\mu A_1^2}{2\pi h^3}\dot{h}\left(4\beta^2 - \beta^4 - 4\ln\beta - 3\right) = -\frac{3\mu}{2\pi h^3 n^2}\dot{h}\left(4\beta^2 - \beta^4 - 4\ln\beta - 3\right) \qquad (3.2.38)$$

Thus, the total damping force on the perforated plate is approximately

$$F_p = \frac{A}{A_1}F_1 = -\frac{3\mu A}{2\pi n h^3}\dot{h}k(\beta) = -\frac{3\mu A^2}{2\pi h^3 N}\dot{h}k(\beta) \qquad (3.2.39)$$

where N is the total hole number of the plate and $k(\beta)$ is

$$k(\beta) \equiv 4\beta^2 - \beta^4 - 4\ln\beta - 3 \qquad (3.2.40)$$

This is the same result as that given in reference [7]. The dependence of k on β is shown by the curve in Fig. 3.2.11.

For a finite plate area A, the damping force is over estimated by Eq. (3.2.39) (especially for small holes), as the boundary effect of the plate has not been considered. An empirical approximation [8] assumes that the damping force given by Eq. (3.2.39) and the damping force of a non-perforated plate of the same shape and size are considered to act in parallel.

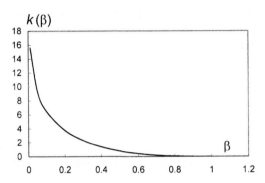

Fig. 3.2.11. The dependence of factor k on β

For example, for a rectangular hole-plate, the squeeze-film damping force, F_{rec}, of a non-perforated rectangular plate can be found using Eq. (3.2.35). Then the resultant damping force of the perforated plate, F_R, is given by the following relation

$$F_R = \frac{F_p F_{rec}}{F_p + F_{rec}} \tag{3.2.41}$$

It may be mentioned here that the relations obtained in this subsection is based on Reynolds' equation for a solid plate (Eq.3.2.10). Therefore, the discussion leaves much to be improved:

(a) The Boundary Effect
Eq. (3.2.41) for boundary effect is pure intuitive. Theoretically, the boundary effect should be treated using a differential equation for the hole-plate with appropriate boundary conditions.

(b) The Consideration of Thick Hole-plate
With the progress of technologies, deep reactive ion etching (deep RIE) technology is getting popular. More and more devices are fabricated out of thick hole-plates. In these cases, the damping effect of air flow in the hole has to be considered.

(c) The Consideration of the End Effect of Hole
For a long and thin pipe, the end effect is insignificant. However, The end effect might be significant for short pipes.
To improve the understanding of squeeze-film air damping of hole-plate, a modified Reynolds' equation specifically for hole-plate has been developed. The establishment of the equation and the applications of the equation on hole-plate will be given in §3.3.

§3.2.6. Damping of Oscillating Beams

Micro beams are widely used in micro mechanical sensors and actuators, especially in resonant type devices, for which air damping is a major design factor. Therefore, knowledge of air damping effects on oscillating micro beams is important. As the motion of the oscillating beam structure is not uniform, there is no closed form solution for beam vibration problems involving damping force. Though numerical analysis is possible, it is not convenient as a design tool. Therefore, two simplified models are introduced in this section to obtain approximate solutions in closed form.

(1) Squeeze-film Damping

Suppose that a beam is in parallel with a substrate and oscillates in its normal direction in air. Assume the length of the beam, L, is much larger than its width, B, while its width is much larger than its thickness, h. If the gap between the beam and the substrate, d_o, is small when compared to the beam width, the main damping mechanism is the squeeze-film air damping. As the moving speed of the oscillating beam is not uniform and its distribution is dependent on the vibration mode, the air damping force can hardly be simplified by a lumped model. As air flow caused by the vibration is mainly lateral, according to Eq. (3.2.14), the squeeze-film air damping force per unit length of beam is

$$\widetilde{F}_d = \frac{\mu B^3}{d_o^{\,3}} \dot{w}(x,t) \tag{3.2.42}$$

where $\dot{w}(x,t)$ is the moving speed of the beam sector considered. With reference to Eq. (2.4.3) and with the damping effect considered, the differential equation for a forced vibration of the beam is

$$\rho B h \ddot{w}(x,t) + \widetilde{c}\,\dot{w}(x,t) + EI \frac{\partial^4 w(x,t)}{\partial x^4} = \widetilde{F}(x)\sin \omega t \tag{3.2.43}$$

where ρ is the mass density, E the Young modulus of the beam material, I the moment of inertia of the beam, $\widetilde{F}(x)$ the external driven force on unit length of beam and $\widetilde{c}$ the coefficient of squeeze-film air damping force per unit length of the beam. According to Eq. (3.2.42), we have

$$\widetilde{c} = \frac{\mu B^3}{d_o^{\,3}} \tag{3.2.44}$$

Displacement function $w(x,t)$ in Eq. (3.2.43) can be developed as

$$w(x,t) = \sum_{i=1}^{\infty} \varphi_i(t) W_i(x) \tag{3.2.45}$$

where $W_i(x)$'s are the shape-functions of the free vibration of the beam structure, i.e., they are eigenfunctions of the following equations

$$EI \frac{d^4 W_i(x)}{dx^4} = \rho Bh\omega_i^2 W_i(x)$$

By substituting Eq. (3.2.45) into Eq. (3.2.43), we have

$$\rho Bh \sum_{i=1}^{\infty} \ddot{\varphi}_i(t)W_i(x) + \tilde{c} \sum_{i=1}^{\infty} \dot{\varphi}_i(t)W_i(x) + EI \sum_{i=1}^{\infty} \omega_i^2 \varphi_i(t)W_i(x) = \tilde{F}(x)\sin\omega t \qquad (3.2.46)$$

By multiplying $W_n(x)$, taking integration and making use of the orthogonal characteristics of $W_n(x)$'s (i.e., $\int_0^L W_i(x)W_n(x)dx = 0$ for $i \neq n$ and $\int_0^L W_n^2(x)dx = 1$), we have

$$\rho Bh\ddot{\varphi}_n(t) + \tilde{c}\dot{\varphi}_n(t) + \rho Bh\omega_n^2 \varphi_n(t) = \tilde{F}_n \sin\omega t \qquad (3.2.47)$$

where

$$\tilde{F}_n = \int_0^L \tilde{F}(x)W_n(x)dx \qquad (3.2.48)$$

The effectiveness of an external driving force for a specific mode is determined by Eq. (3.2.48). This is very useful in designing the driving electrode for an electrostatic actuator.

According to the result given in §2.5.3, for a system with equation

$$m\ddot{x} + c\dot{x} + m\omega_o^2 x = F_o \sin\omega t$$

the damping ratio of the forced vibration is $\zeta = c/2m\omega_o$ and the quality factor is $Q = m\omega_o/c$.

Therefore, according to Eq. (3.2.47), the damping ratio of the n'th vibration mode is

$$\zeta_n = \frac{\tilde{c}}{2\rho Bh\omega_n} = \frac{\mu B^2}{2\rho h d_o^3 \omega_n} \qquad (3.2.49)$$

Thus, the quality factor of the n'th vibration mode (for small damping) is

$$Q_n = \frac{\rho h d_o^3 \omega_n}{\mu B^2} \qquad (3.2.50)$$

This indicates that the Q factor of a beam resonator is related to the frequency of the specific vibration mode, as well as the geometries of the beam. Note that the higher the vibration mode (i.e., the higher the vibration frequency), the higher the Q factor.

(2) Drag Force Damping

If a beam is far away from any surrounding object (as in the case where a beam is driven into vibration by a piezoelectric method), the main damping mechanism is the drag force of air flow. As there is no closed form solution to the damping force for the beams, a simplified dish-string model is proposed in this section. Similar to the bead model [9], the dish-model replaces the beam with a string of dishes as shown in Fig. 3.2.12. The diameter of the dishes is equal to the beam width, B, and the interference between neighboring dishes is neglected. According to Eq. (3.1.12), the air damping force on the i'th dish is

$$\widetilde{F}_{ai} = 8\mu B\dot{w}_i \tag{3.2.51}$$

where w_i is the displacement of the i'th dish. Since the number of dishes per unit length of beam is $1/B$, the damping force per unit length of beam is

$$\widetilde{F}_{al} = 8\mu\dot{w} \tag{3.2.52}$$

The equation for a forced vibration of the beam with drag force damping is

$$\rho bh\ddot{w}(x,t) + \widetilde{c}_{al}\dot{w}(x,t) + EI\frac{d^4 w(x,t)}{dx^4} = \ddot{F}(x,t) \tag{3.2.53}$$

where $\widetilde{c}_{al}$ is the coefficient of air-flow damping force per unit length of beam. According to Eq. (3.2.52), we have

$$\widetilde{c}_{al} = 8\mu \tag{3.2.54}$$

Following similar argumentation as made for squeeze-film damping, the damping ratio for the n'th vibration mode (with a radial frequency of ω_n) is

$$\zeta_{an} = \frac{\widetilde{c}_{al}}{2\rho Bh\omega_n} = \frac{4\mu}{\rho Bh\omega_n} \tag{3.2.55}$$

and the Q factor for the vibration mode is

$$Q_{sn} = \frac{\rho Bh\omega_n}{8\mu} \tag{3.2.56}$$

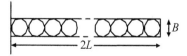

Fig. 3.2.12 Dish-string model for air damping of beam (top view)

In the above discussion on damping force, the dishes are considered as being isolated from one another and the dishes do not fill the beam area completely. As a matter of fact, no air flow is allowed between two neighboring dishes. Therefore, the damping force given by Eq. (3.2.52) is under-estimated and the Q factor given in (3.2.56) is overestimated.

§3.2.7. Effects of Finite Squeeze Number

In §3.2.1, Reynolds' equation for squeeze-film air damping was derived. For convenience, Eq. (3.2.7) is given again below

$$\frac{\partial^2 P^2}{\partial x^2} + \frac{\partial^2 P^2}{\partial y^2} = \frac{24\mu}{h^3}\frac{\partial(hP)}{\partial t}$$

Under the condition that the squeeze number σ is small, (the definition of squeeze number σ is given in §3.2.1.), Eq. (3.2.7) can be simplified to Eq. (3.2.10)

$$\frac{\partial^2 P}{\partial x^2} + \frac{\partial^2 P}{\partial y^2} = \frac{12\mu}{h^3}\frac{dh}{dt}$$

Based on Eq. (3.2.10), discussions on squeeze-film air damping for some typical structures were carried out in §3.2.2 to §3.2.5.

However, if σ is small but not negligible, Eq. (3.2.7) instead of Eq. (3.2.10) must be used for squeeze-film air damping problems. The effect of finite squeeze-number has been discussed by Sadd and Stiffer [10] for plates of typical shapes. Among them the simplest one is the long rectangular plate with a length L much larger than its width B. They found that the damping force on the plate for a varying gap distance $h=h_0(1+\delta\sin\omega t)$ is

$$F = P_o LB \int_0^1 (\tilde{P}-1)d\tilde{x} = P_o LB\left[-\frac{1}{3}\frac{h_o^2\dot{h}}{h^3\omega}\sigma + \frac{2}{15}\frac{h_o^4\ddot{h}}{h^5\omega^2}\sigma^2 - \frac{1}{3}\frac{h_o^4\dot{h}^2}{h^6\omega^2}\sigma^2\right] \qquad (3.2.57)$$

where P_0 is the air pressure. Based on Eq. (3.2.57), two conditions are considered

(1) Small Amplitude

By using the small amplitude condition of $\varepsilon<<1$ so that $h \approx h_o$ and using the relations $\dot{h} = h_o\varepsilon\omega\cos\omega t$, $\ddot{h} = -h_o\varepsilon\omega^2\sin\omega t$, Eq. (3.2.57) can be approximated as

$$F = P_o LB\left[-\frac{1}{3}\sigma\varepsilon\cos\omega t - \frac{2}{15}\sigma^2\varepsilon\sin\omega t - \frac{1}{6}\sigma^2\varepsilon^2 - \frac{1}{6}\sigma^2\varepsilon^2\cos 2\omega t\right] \qquad (3.2.58)$$

$$\equiv F_D + F_k + F_R + F_{2\omega}$$

The meaning of the four terms are explained as follows:

(a) Damping effect

The damping effect is shown by the first term in Eq. (3.2.58)

$$F_D = -\frac{1}{3}P_oLB\sigma\epsilon\cos\omega t = -\frac{\mu LB^3}{h_o^3}h_o\epsilon\omega\cos\omega t = -\frac{\mu LB^3}{h_o^3}\dot{h}$$

This is exactly the same result as given in Eq. (3.2.14), i.e., it is the result of the first order approximation.

The energy loss of the system due to this damping force in one cycle is

$$\Delta E = \int_0^T -\frac{\mu LB^3}{h_o^3}h_o\epsilon\omega\cos\omega t \dot{h}dt = \int_0^{2\pi} -\frac{\mu LB^3\omega}{h_o^3}h_o^2\epsilon^2\cos^2\omega t d\omega t = -\frac{\pi\mu LB^3\omega}{h_o}\epsilon^2$$

Therefore, we have

$$Q = \frac{2\pi E_T}{\Delta E} = \left(2\pi\frac{1}{2}k(h_o\epsilon)^2\right)\bigg/\left(2\pi\frac{\mu LB^3\omega}{2h_o}\epsilon^2\right) = \frac{kh_o^3}{\mu LB^3\omega}$$

where E_T is the total energy of the system and k is the elastic constant of the mechanical spring supporting the plate.

(b) Elastic effect
The elastic effect is given by the second term in Eq. (3.2.58)

$$F_k = -\frac{2}{15}P_oLB\sigma^2\epsilon\sin\omega t = -\frac{2}{15}LB\frac{9\mu^2\omega^2B^4}{P_oh_o^5}h_o\epsilon\sin\omega t = -\frac{6}{5}LB\frac{\mu^2\omega^2B^4}{P_oh_o^5}\Delta h$$

F_k is significant at high squeeze numbers, when the air flow into and out of the gap fails to keep up with the motion of plate. Since the trapped air acts as a spring, which does not cause energy losses. Therefore, the effect is referred to as the "elastic effect".

(c) Rectification effect
The third term F_R in Eq. (3.2.58) is a constant force stemmed from the quadratic term of p. Therefore, it is referred to as the rectification force due to the nonlinear relationship between the pressure and the displacement.

(d) The force of higher harmonics
The last term in Eq. (3.2.58), $F_{2\omega}$, represents a force component whose frequency is twice as large as that of the oscillating plate. This component is also caused by the quadratic term of p_2.

(2) Large Amplitude
If the amplitude of the oscillating plate is not negligible, then the value of $h=h_o(1+\epsilon\sin\omega t)$ (instead of $h\approx h_o$) has to be used in Eq. (3.2.57). To examine the effect of large amplitude, Eq. (3.2.57) is developed

$$F = A_o + A_1\cos\omega t + B_1\sin\omega t + A_2\cos 2\omega t + B_2\sin 2\omega t + \cdots \tag{3.2.59}$$

where A_o is the rectification force, and A_1 and B_1 are the amplitudes of the damping and elastic forces, respectively. They are found to be

$$A_o = \frac{1}{2\pi}\int_0^{2\pi} F \cdot d(\omega t) = P_o LB \frac{\varepsilon^2(4+3\varepsilon^2)\sigma^2}{24(1-\varepsilon^2)^{9/2}} \qquad (3.2.60)$$

$$A_1 = \frac{1}{\pi}\int_0^{2\pi} F \cdot \cos\omega t\, d(\omega t) = -P_o LB \frac{\varepsilon\sigma}{3(1-\varepsilon^2)^{3/2}} \qquad (3.2.61)$$

and

$$B_1 = \frac{1}{\pi}\int_0^{2\pi} F \cdot \sin\omega t\, d(\omega t) = -P_o LB \frac{\left(\dfrac{3}{4}\varepsilon^4 + 6\varepsilon^2 + 2\right)\sigma^2\varepsilon}{15(1-\varepsilon^2)^{9/2}} \qquad (3.2.62)$$

Eqs. (3.2.60) to (3.2.62) show that the rectification force, the damping force and the elastic forces increase with an increase in the amplitude indicated by ε.

From Eq. (3.2.61) the damping force is

$$F_D = A_1 \cos\omega t = -\frac{\mu LB^3}{h_o^3}\frac{1}{(1-\varepsilon^2)^{3/2}}\dot{h} \qquad (3.2.63)$$

The coefficient of damping force is

$$c = \frac{\mu LB^3}{h_o^3}\frac{1}{(1-\varepsilon^2)^{3/2}} \qquad (3.2.64)$$

Obviously, for small amplitudes, the coefficient is the same as that given by Eq. (3.2.14).

§3.3. Damping of Perforated Thick Plates

In §3.2.5, the squeeze-film air damping of a perforated thin plate was analysed using Reynolds' equation. The analysis is effective to a hole-plate with lateral dimensions much larger than its gap distance and thickness. In this case, the resistant force caused by the air-flow in holes is negligible and the boundary effect of the plate can be neglected or estimated by Eq. (3.2.41), an equation by intuitive imagination. These conditions can be satisfied for some MEMS structures fabricated using surface micromachining technology.

With the development of MEMS technology, deep reactive ion etching (deep RIE, or ICP) technologies have been widely used for micromachining. Nowadays, thick plate can be perforated with thin holes easily. For thick hole-plate, the analytical results given in §3.2.5 are no more accurate. Naturally, people tend to rely on numerical methods.

Many methods have been proposed to calculate the squeeze-film air damping of perforated thick plates. The finite element method (FEM) is the most popular one for this purpose. However, as the FEM analysis is usually time-consuming and non-transparent, it is not convenient for the design optimisation. Therefore, an analytical model is still strongly desired for design consideration. Fortunately, progress has been made recently by Bao et al by modifying Reynolds' equation for hole-plate with finite thickness [11]. The method is discussed as follows.

§3.3.1. Modified Reynolds' Equation for Hole-plate

Consider a uniformly perforated plate with high hole density. The plate can be divided into cells with each cell having a hole at centre. As the cell is much smaller than the plate, the pressure can be considered as a smooth function of position under the whole plate. For the same reason, the air flowing out through a hole (the flow is in the z-direction) can be considered as penetrating through the whole cell area uniformly. Thus, for the plate moving relative to the substrate, the continuity equation can be modified as

$$\frac{\partial(\rho q_x)}{\partial x} + \frac{\partial(\rho q_y)}{\partial y} + \rho Q_z + \frac{\partial(\rho h)}{\partial t} = 0 \qquad (3.3.1)$$

where $q_x = -\dfrac{h^3}{12\mu}\dfrac{\partial P}{\partial x}$, $q_y = -\dfrac{h^3}{12\mu}\dfrac{\partial P}{\partial y}$ and Q_z is the penetrating rate, which is a function of ΔP, the additional pressure in the gap caused by air damping effect.

The dependence of Q_z on ΔP for a hole-plate can be found based on a simplified model as shown in Fig. 3.3.1. The plate is uniformly perforated with circular holes in an array (The one in the figure is a hexagonal array though a cubic array is also widely used). The radius of the holes is r_0 and the area allocated to a cell is $A_1 = \pi r_c^2$, where r_c is referred to as the cell radius. The thickness of the plate is H and the distance from the plate to the substrate (or, electrodes) is h.

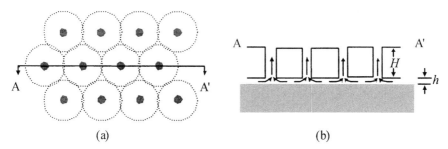

(a)   (b)

Fig. 3.3.1. Schematic structure of a hole-plate (a) top view (b) cross-sectional view

According to Poiseuille Equation (Eq. 3.1.9) for a pipe with radius r_0 and length H, the volume of air passing through the hole in a unit time (the flow rate Q) is

$$Q = \frac{\pi r_o^4}{8\mu}\frac{P_H}{H} \qquad (3.3.2)$$

where P_H is the pressure difference between the two ends of the hole. In the other words, the pressure across the hole established by the flow rate Q is

$$P_H = \frac{8\mu H}{\pi r_o^4} Q \qquad (3.3.3)$$

According to Eq. (3.2.38) and $Q = A_1 dh/dt$, the damping force in a cell area of an infinite thin hole-plate by the lateral air-flow from the border of a cell to the hole of the cell is

$$F_{l1} = \frac{3\mu A_1}{2\pi h^3} k(\beta) Q \qquad (3.3.4)$$

where $k(\beta) = 4\beta^2 - \beta^4 - 4\ln\beta - 3$ and $\beta = r_o / r_c$. From Eqs. (3.3.2) and (3.2.4), the average damping pressure in the cell caused by the flow is

$$\Delta P = P_H + \frac{F_{l1}}{A_1} = P_H \left(1 + \frac{3r_o^4 k(\beta)}{16 H h^3}\right) \equiv \eta(\beta) P_H \qquad (3.3.5)$$

or $P_H = \Delta P / \eta(\beta)$, where $\eta(\beta) \equiv [1 + 3r_o^4 k(\beta)/16 H h^3]$. The average penetrating rate of air through the plate (volume for a unit area in a unit time) is

$$Q_z = \frac{Q}{A_1} = \frac{\beta^2 r_o^2}{8\mu H} \frac{\Delta P}{\eta(\beta)} \qquad (3.3.6)$$

After the substitution of Eq. (3.3.6) into Eq. (3.3.1) and using the derivation steps similar to those for Reynolds' equation in §3.2.1, we have

$$\frac{\partial^2 \Delta P}{\partial x^2} + \frac{\partial^2 \Delta P}{\partial y^2} - \frac{3\beta^2 r_o^2}{2h^3 H} \frac{1}{\eta(\beta)} \Delta P = \frac{12\mu}{h^3} \frac{\partial h}{\partial t} \qquad (3.3.7)$$

This is the modified Reynolds' equation for the squeeze-film air damping of hole-plate. Note that the damping pressure ΔP is sometimes designated as P for simplicity. During the derivation, we assumed that the gas flow in the holes is a fully developed Poiseuille flow. The assumption is, however, not true for many micromachined structures if the radius of the holes, r_o, is comparable to the thickness of the plate, H. In these cases, Sharipov et al [12] proposed that the Poiseuille equation be modified as

$$Q = \frac{\pi r_o^4}{8\mu} \frac{P_H}{H + 3\pi r_o / 8} \qquad (3.3.8)$$

By comparing Eq. (3.3.2) and (3.3.8), we can define the effective hole length

$$H_{eff} = H + \frac{3\pi r_o}{8} \tag{3.3.9}$$

Also, the end effects are considered by replacing H with H_{eff} in other related equations.

§3.3.2. Long Rectangular Hole-plate

To demonstrate the nature of the squeeze-film damping of thick hole-plate, we consider a rectangular hole-plate with a length much larger than its width (i.e., b is much larger than a as shown in Fig. 3.3.2). Thus, Eq. (3.3.7) becomes one-dimensional.

$$\frac{\partial^2 P}{\partial x^2} - \frac{3\beta^2 r_o^2}{2h^3 H_{eff}} \frac{1}{\eta(\beta)} P = \frac{12\mu}{h^3} \frac{\partial h}{\partial t} \tag{3.3.10}$$

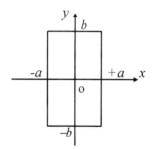

Fig. 3.3.2. Geometries of a long rectangular hole-plate (holes are not shown)

By defining $R = -\frac{12\mu}{h^3} \frac{\partial h}{\partial t}$ and $l = \sqrt{\frac{2h^3 H_{eff}\eta(\beta)}{3\beta^2 r_o^2}}$ (l is a characteristic length or attenuation length of the hole-plate), Eq. (3.3.10) can be simplified as

$$\frac{\partial^2 P}{\partial x^2} - \frac{P}{l^2} + R = 0 \tag{3.3.11}$$

With boundary conditions of $P(\pm a) = 0$, the distribution of damping pressure is

$$P(x) = Rl^2\left(1 - \cosh\frac{x}{l} \Big/ \cosh\frac{a}{l}\right) \tag{3.3.12}$$

The dependence of pressure distribution on the ratio of a/l is shown in Fig. 3.3.3, where the damping pressure has been normalized to Rl^2.

We can learn from Eq. (3.3.12) or Fig. 3.3.3 that the larger the ratio a/l, the larger the effect of hole on damping and the smaller the border effect of the plate, and vice verse. Note that the maximum pressure at centre, $P(0)$, increases with l for a constant a. This might not be easily seen as the pressure in Fig. 3.3.3 have been normalized by Rl^2.

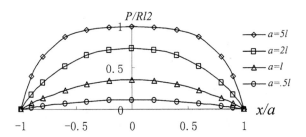

Fig. 3.3.3. Dependence of normalized pressure distribution on a/l

Now let us consider two extreme conditions. If the holes are very thin and the plate is very thick so that the attenuation length l is much larger than the half width a, Eq. (3.3.12) returns to

$$P(0) = -\frac{6\mu}{h^3}(a^2 - x^2)\frac{dh}{dt}$$

This is the same result as given by Eq. (3.2.13) in §3.2.2 (with $B=2a$).

To another extreme, if the plate is very thin so that $a \gg l$, Eq. (3.3.12) shows that the pressure for most area of the plate has a constant value of

$$P = Rl^2 = -\frac{3\mu r_c^2}{2h^3}\dot{h}k(\beta) \tag{3.3.13}$$

This equation agrees with Eq. (3.2.39) for the damping pressure of an infinite thin hole-plates. The above results for the two extremes also justify the modified Reynolds' equation.

Generally, l is indeed much smaller than a, but has a finite value so that its effect on the damping pressure in the areas near the borders cannot be neglected. In these cases, for region $x>0$ (The pressure distribution in the region of $x<0$ is symmetric), Eq. (3.3.12) can be approximated as

$$P(x) = Rl^2(1 - e^{-\frac{a-x}{l}} - e^{-\frac{a+x}{l}}) \tag{3.3.14}$$

This means that the pressure in most of the area under the plate is $P = Rl^2$. However, in the area near the borders, where $a - x$ is comparable with or smaller than l, the pressure drop exponentially

$$P(x) = Rl^2(1 - e^{-\frac{a-x}{l}}) \tag{3.3.15}$$

As an example, the pressure distribution across the plate is plotted in Fig. 3.3.4 for the condition of $a = 10 \cdot l$, where $P_0 = Rl^2$.

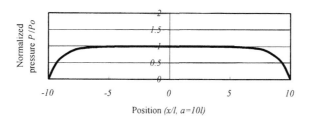

Fig. 3.3.4. Pressure distribution for a long rectangular hole-plate with $a = 10 \cdot l$

From Eq. (3.3.12), the damping force on the long rectangular plate is

$$F_d = 2aLRl^2 \left[1 - \frac{l}{a} \tanh\left(\frac{a}{l}\right) \right] \tag{3.3.16}$$

According to the relation of $F_d = c\dot{h}$, the coefficient of damping force is

$$c = 2aL \frac{8\mu H}{\beta^2 r_o^2} \left(1 + \frac{3r_o^4 K(\beta)}{16 H h^3} \right) \left[1 - \frac{l}{a} \tanh\left(\frac{a}{l}\right) \right] \tag{3.3.17}$$

If the plate is the seismic mass of an accelerometer, the damping ratio of the system is $\zeta = c/(2m\omega_0)$, where ω_0 is the radial frequency of free vibration and m the mass of the plate.

In the condition of $a \gg l$ so that $\tanh(a/l) \cong 1$, we have the following equations

$$F_d \cong 2(a - l)LRl^2 \tag{3.3.18}$$

and

$$c = \frac{8\mu H}{\beta^2 r_o^2} \left(1 + \frac{3r_o^4 K(\beta)}{16 H h^3} \right) 2L \cdot 2(a - l) \tag{3.3.19}$$

§3.3.3. "Effective Damping Area" Approximation

From the above analysis, we can find that, under the condition of $a \gg l$ (in this condition the hole effect is significant) the damping pressure in most internal area is constant Rl^2 but the damping pressure near the two borders drop exponentially. According to Eq. (3.3.18), the damping force is equivalently caused by the pressure distribution shown by the curve in Fig. 3.3.5. Therefore, we may consider that the plate is only effective for damping in an area of width $2(a–l)$, excluding the two border regions, each of width l.

Based on the approximation of "effective damping area", for $a, b \gg l$, the damping force of a rectangular hole-plate can be approximated as

$$F_d \cong 4Rl^2(a-l)(b-l) \tag{3.3.20}$$

Similarly, the damping force on a circular hole-plate can be approximated as

$$F_d \cong Rl^2\pi(a-l)^2 \tag{3.3.21}$$

where a is the radius of the circular hole-plate. This approach may be applicable to some hole-plate with irregular shape by just excluding ineffective border areas of width l.

Let us now consider the conditions for effective damping area approximation. According to the definition of the attenuation length l, if the hole is large enough so that $r_o^4 \gg H_{eff}h^3$, we have $\eta(\beta) \approx \dfrac{3r_o^4 K(\beta)}{16Hh^3}$ and $l \approx r_c\sqrt{\dfrac{k(\beta)}{8}}$. Thus, we have $l < r_c$ if only $\beta > 0.06$. Therefore, the condition $a,b \gg l$ for effective damping area approximation is equivalent to $a,b \gg r_c$ or $a,b > 3r_c$. This means that the effective damping area approximation is a good one if only there are more than five holes across the plate in any directions.

The analytical results by the modified Reynolds' equation have been compared with those by numerical analyses and the experimental data. The agreement is very good. Readers who are interested in the details are referred to reference [11].

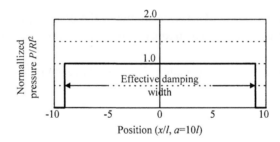

Fig.3.3.5. Approximate pressure distribution for a strip hole-plate with $a = 10 \cdot l$

§3.4. Slide-film Air Damping

§3.4.1. Basic Equations for Slide-film Air Damping

(1) Simplified Model of Slide-film Air Damping

Micromechanical devices fabricated by surface micromachining technology feature thin movable plates (about 2 μm thick) suspended with a small gap over a substrate by flexures. This basic structure facilitates the lateral movement of the plates for such applications as resonators, actuators, accelerometers, etc.

As the dimensions of the moving plates are usually much larger than their thickness and the gap distance, the viscous damping by the ambient air plays a major role in energy dissipation of the dynamic system; the air film behaves as a slide-film damper to the moving structure. To investigate the basic features of slide-film damping, a mechanical model is considered: an infinite plate, immersed in an incompressible viscous fluid, moving in a lateral direction at a constant distance from the substrate as shown in Fig. 3.4.1(a) [13]. The simplified model is shown in Fig. 3.4.1(b).

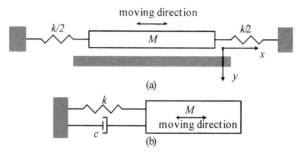

Fig. 3.4.1. Slide-film air damping (a) schematic structure; (b) a simplified model

(2) Basic Equations

The general differential equation for the steady flow of an incompressible fluid is the well-known Navier-Stokes equation [3]

$$\rho\left[\frac{\partial \vec{v}}{\partial t} + (\vec{v} \cdot \nabla)\vec{v}\right] = \vec{F} - \nabla p + \mu \nabla^2 \vec{v} \tag{3.4.1}$$

where ρ is the density of the fluid, $\vec{F}$ the force applied, p the pressure in the fluid, μ the coefficient of viscosity and $\vec{v}$ the velocity of the fluid

$$\vec{v} = u\vec{i} + v\vec{j} + w\vec{k}$$

The notations ∇ and ∇^2 are gradient and Laplace operators, respectively,

$$\nabla = \vec{i}\frac{\partial}{\partial x} + \vec{j}\frac{\partial}{\partial y} + \vec{k}\frac{\partial}{\partial z} \quad \text{and} \quad \nabla^2 = \frac{\partial^2}{\partial x^2} + \frac{\partial^2}{\partial y^2} + \frac{\partial^2}{\partial z^2}$$

For example, the Navier-Stokes equation for the x-direction is

$$\rho\left[\frac{\partial u}{\partial t} + \left(u\frac{\partial}{\partial x} + v\frac{\partial}{\partial y} + w\frac{\partial}{\partial z}\right)u\right] = F_x - \frac{\partial}{\partial x}p + \left(\frac{\partial^2}{\partial x^2} + \frac{\partial^2}{\partial y^2} + \frac{\partial^2}{\partial z^2}\right)u$$

Suppose that the plate is in the x-y plane of the coordinate system and infinite in dimensions. If the movement of the plate is in the x-direction, and there is no external force or pressure gradient in the fluid, we have $u \gg v, w$ and $\frac{\partial^2 u}{\partial z^2} \gg \frac{\partial^2 u}{\partial x^2}, \frac{\partial^2 u}{\partial y^2}$. Thus, Eq. (3.4.1) reduces to

$$\frac{\partial u}{\partial t} + u\frac{\partial u}{\partial x} = \frac{\mu}{\rho}\frac{\partial^2 u}{\partial z^2} \tag{3.4.2}$$

For an infinite plate (i.e., the plate is much larger than the gap distance between the plate and the substrate and its oscillation amplitude), the second term on the left side of Eq. (3.4.2) may be neglected, resulting in

$$\frac{\partial u}{\partial t} = \frac{\mu}{\rho}\frac{\partial^2 u}{\partial z^2} \tag{3.4.3}$$

The non-slippage boundary conditions for the equation are: u equals zero at the substrate surface and u equals the velocity of the moving plate on the surface of the plate.

Now let us discuss more on the approximation conditions for Eq. (3.4.3). Suppose that the motion of plate with reference to its balanced position is a simple harmonic oscillation

$$x(t) = a_o \sin \omega t$$

where a_o is the amplitude of the simple harmonic oscillation, i.e.,

$$u(t) = a_o \omega \cos \omega t \equiv u_o \cos \omega t$$

where $u_o = a_o \omega$. Thus, the first term on the left of Eq. (3.4.2) is

$$\frac{\partial u}{\partial t} = -u_o \omega \sin \omega t \tag{3.4.4}$$

If the typical dimension of the plate is l, the second term on the left of Eq. (2.4.2) is

$$u\frac{\partial u}{\partial x} \approx \frac{u_o^2}{l} = \frac{a_o^2 \omega^2}{l} \tag{3.4.5}$$

and the term on the right is

$$\frac{\mu}{\rho}\frac{\partial^2 u}{\partial z^2} \approx \frac{\mu}{\rho}\frac{a_o \omega}{d^2} \tag{3.4.6}$$

where d is the gap distance between the substrate and the moving plate.

Therefore, the approximation conditions for Eq. (3.4.3) are

a) $|\partial u / \partial t| >> |u \partial u / \partial x|$. This requires small amplitude, i.e., $a_o << l$,

b) $(\mu/\rho)(\partial^2 u / \partial z^2) >> u(\partial u / \partial x)$. From Eqs. (3.4.5) and (3.4.6), this condition becomes $l >> 2(\rho\omega / 2\mu)d^2 a_o$. By defining a characteristic "effective decay distance" $\delta = \sqrt{2\mu / \rho\omega}$, we have

$$l >> 2\frac{d^2}{\delta^2}a_o \tag{3.4.7}$$

We will find later in §3.4.3 that the effective decay distance, δ, is the distance in the z-direction over which the velocity decays from the plate by a factor of e (=2.718...). The curve in Fig. 3.4.2 shows the dependence of δ on the frequency in air at 1 atm. at 20°C.

In the condition of $(\mu/\rho)(\partial^2 u/\partial z^2) >> (\partial u/\partial t)$, i.e., $d << \delta$, Eq. (3.4.3) is further simplified as

$$\frac{\partial^2 u}{\partial z^2} = 0 \tag{3.4.8}$$

In the following sections, two damping models will be considered: (1) a Couette-flow model governed by Eq. (3.4.8) when frequency ω and the distance d are small so that δ is much larger than d, and (2) a Stokes-flow model governed by Eq. (3.4.3) for more general conditions [13].

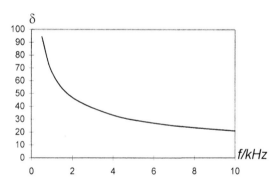

Fig. 3.4.2. The effective decay distance, δ, as a function of frequency

§3.4.2. Couette-flow Model

Suppose that a large plate oscillates laterally over a substrate as shown in Fig. 3.4.3. If the oscillating frequency is low so that $\delta>>d$, the flow pattern of the air around the plate is called a Couette-flow. We will consider the damping force on the plate caused by the viscous fluid by the Couette-flow model with the non-slippage boundary conditions

$$u(0) = u_o \cos \omega t, u(d) = 0 \tag{3.4.9}$$

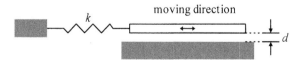

Fig. 3.4.3. Laterally oscillating plate over a substrate

According to Eqs. (3.4.8) and (3.4.9), the velocity distribution of the fluid is

$$u(z) = u(0)(1 - \frac{z}{d})$$ (3.4.10)

where $u(0)$ is the velocity of the moving plate. The resistive shearing force applied to the plate is

$$F = -\mu \frac{u(0)}{d} A$$ (3.4.11)

where A is the area of the plate. According to the Couette-flow model, the velocity gradient on the open side (top) of the plate is zero. Therefore, there is no damping force on the top of the plate and the Q factor of the lateral vibration system is determined only by the damping force described in Eq. (3.4.11). (As a matter of fact, if d on the top is large, the condition for Couette-flow, $\delta \gg d$, is no longer valid, but we will just assume here that the damping force on the top of the plate is negligible.)

The energy dissipated by the damping force in one cycle is

$$\Delta E_{Cd} = \int_0^T A\mu \frac{u(0)}{d} u(0) dt$$

As $u(0) = u_o \cos \omega t$, we have

$$\Delta E_{Cd} = \int_0^{2\pi} A\mu \frac{u_o^2 \cos^2 \omega t}{d} \frac{1}{\omega} d\omega t = \frac{\pi}{\omega} u_o^2 \frac{\mu}{d} A$$

According to the second definition of Q factor in §2.5.4.

$$Q_{Cd} = \frac{\pi m u_o^2}{\Delta E_{Cd}} = \frac{m\omega d}{\mu A}$$ (3.4.12)

If the mass density of the plate is ρ_p and the thickness of the plate is H, Eq. (3.4.12) can be written as

$$Q_{Cd} = \frac{\rho_p H \omega d}{\mu}$$

Note that Q_{Cd} is not dependent on the area of the plate, A.

§3.4.3. Stokes-flow Model

In Couette-flow model, the effective decay distance, δ, is much larger than the gap distance d. Therefore, the velocity profile between the plate and the substrate is linear. The model becomes invalid when the gap distance is large. If the effective decay distance, δ, is not much larger than d, Eq. (3.4.3) has to be used.

By solving Eq. (3.4.3) with the non-slippage boundary conditions given in Eq. (3.4.9), the velocity profile of the fluid is

$$u = u_o \frac{-e^{-\tilde{d}+\tilde{z}}\cos(\omega t+\tilde{z}-\tilde{d}-\theta)+e^{\tilde{d}-\tilde{z}}\cos(\omega t-\tilde{z}+\tilde{d}-\theta)}{\sqrt{e^{2\tilde{d}}+e^{-2\tilde{d}}-2\cos(2\tilde{d})}} \quad (3.4.13)$$

where $\tilde{d} \equiv d/\delta, \tilde{z} \equiv z/\delta$ and θ is a phase lag against the oscillation of plate ($u(0)=u_o\cos\omega t$). The expression for θ is

$$\theta = \arctan\frac{(e^{\tilde{d}}+e^{-\tilde{d}})\sin\tilde{d}}{(e^{\tilde{d}}-e^{-\tilde{d}})\cos\tilde{d}} \quad (3.4.14)$$

Thus, the damping force on the plate (on one side) is

$$F_{Sd} = A\mu\frac{\partial u}{\partial z}\bigg|_{z=0} = \frac{A\mu u_o}{\delta\sqrt{e^{2\tilde{d}}+e^{-2\tilde{d}}-2\cos 2\tilde{d}}}\bigg(-e^{-\tilde{d}}\cos(\omega t-\tilde{d}-\theta)$$
$$+e^{-\tilde{d}}\sin(\omega t-\tilde{d}-\theta)-e^{\tilde{d}}\cos(\omega t+\tilde{d}-\theta)+e^{\tilde{d}}\sin(\omega t+\tilde{d}-\theta)\bigg) \quad (3.4.15)$$

With the damping force, the energy dissipation in one cycle of oscillation is

$$\Delta E_{Sd} = \int_0^T F_{Sd} u_o dt = \frac{\pi A\mu u_o^2}{\omega\delta}\frac{\sinh(2\tilde{d})+\sin(2\tilde{d})}{\cosh(2\tilde{d})-\cos(2\tilde{d})} \quad (3.4.16)$$

and the Q factor is

$$Q_{Sd} = \frac{m\omega\delta}{A\mu}\cdot\frac{\cosh(2\tilde{d})-\cos(2\tilde{d})}{\sinh(2\tilde{d})+\sin(2\tilde{d})} \quad (3.4.17)$$

For the extreme condition of $d<<\delta$, we have $\theta=\pi/4$, $F_{Sd}=F_{Cd}=-A\mu u(0)/d$, $\Delta E_{Sd}=\Delta E_{Cd}$ and $Q_{Sd}=Q_{Cd}$, i.e., the results of Stokes-flow model coincide with those of Couette-flow model.

For another extreme condition of $d>>\delta$, from Eq. (3.4.13), we have

$$u = u_o e^{-\tilde{z}}\cos(\omega t+\tilde{d}-\theta)$$

This shows that the fluid around the plate oscillates with the same frequency as the plate but the oscillation amplitude in the fluid decays exponentially away from the plate. δ is the distance over which the amplitude decreases by a factor of e (=2.718).

Under this condition ($d \gg \delta$), the energy dissipation in one cycle is

$$\Delta E_{S\infty} = \frac{\pi}{\omega} u_o^2 \frac{\mu}{\delta} A$$

and

$$Q_{S\infty} = \frac{m\omega\delta}{\mu A} = \frac{\rho_p H\omega\delta}{\mu}$$

If $Q_{S\alpha}$ is compared with Q_{Cd} in Eq. (3.4.12), we conclude that the damping force now is

$$F_{S\infty} = \frac{\mu A u(0)}{\delta} \tag{3.4.18}$$

As the condition of $d \gg \delta$ means that the effect of substrate is negligible for the oscillating plate, the plate can be considered as an isolated object in the fluid. Now let us compare the result here with the drag force on an isolated object given in §3.1.3.

According to §3.1.3, the drag force on a circular dish moving in its plane direction is

$$F_d = \frac{32}{3} \mu r v = \frac{32}{3\pi} \frac{\mu A v}{r} \tag{3.4.19}$$

where v is the velocity of the plate, equivalent to the $u(0)$ in Eq. (3.4.18), and r is the radius of the dish (i.e., the characteristic dimension). Obviously, the boundary effect has not been considered in $F_{S\alpha}$ and the frequency effect has not been considered in F_d. As the damping effect is underestimated in both models, the larger one is considered to be more accurate. Or, they are just added together for the damping force for an isolated object.

§3.4.4. Air Damping of a Comb Drive Resonator

Many types of silicon micro-resonators have so far been developed. Among them, the lateral driving comb resonator formed by surface micromachining technology is the one of the widest applications.

Fig. 3.4.4 shows schematically the basic structure of the lateral driving comb resonator. The dark areas are fixed fingers (or, fixed electrodes) or anchors of the movable parts. The movable parts include the flexures (narrow beams), moving plate (with etching holes) and fingers (movable electrodes).

The moving parts of the structure can be driven into lateral oscillation by applying an alternating voltage (often with a dc bias, see Chapter 4) between the movable and the fixed electrodes. Quite often, the frequency of the driving force coincides with the resonant frequency of the structure so that the structure is driven into a resonant state. Thus, the structure is referred to as a comb resonator.

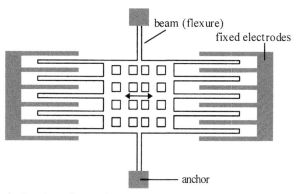

Fig. 3.4.4. A schematic drawing of a comb resonator

One of the most important characteristics of the comb resonator is its mechanical quality factor, Q. For a comb resonator operating in an atmospheric environment, air damping is the dominant factor for the Q factor of the resonator. The air damping force for a comb resonator consists of many components related to the geometries of the structure and different damping mechanisms [14]. The main damping force components are

(a) Slide-film Damping Force on the Bottom
As the gap distance, d_p, between the moving plate and the substrate is much smaller than δ, the damping force is of Couette-flow type and can be expressed as

$$F_1 = \mu \frac{A_p}{d_p} \dot{x} = c_1 \dot{x}$$

where A_P is the effective plate area for the damping calculation, including the areas of the plates, fingers and beams. The etch hole region is also included in A_P as the dimension of the etch hole is usually small so that no lateral flow of air exists in the hole. This is usually the most important force component of all the force components.

(b) Slide-film Damping on the Top
Suppose the structure is far away from any objects above it. The damping force above the moving parts of the structure is of Stokes-flow type. The damping force component is

$$F_2 = \mu \frac{A_p}{\delta} \dot{x} = c_2 \dot{x}$$

where δ is the effective decay distance $\delta = \sqrt{2\mu/\rho\omega}$ and ρ the density of air. For a resonant frequency of 1 kHz, we find $\delta = 67 \mu\text{m}$.

(c) Slide-film Damping of the Sidewalls

The slide damping force of the side walls is

$$F_3 = \mu \frac{A_s}{d_s} \dot{x} = c_3 \dot{x}$$

where A_s is the area of the sidewalls of the movable fingers and d_s is the gap distance between the sidewalls and the fixed fingers. Here we have assumed that: $d_s \ll \delta$.

(d) Air Drag Force

The air drag force on the moving plate is difficult to estimate. Referring to Eq. (3.4.19), the force can be approximated by

$$F_4 \approx \frac{32}{3} \mu l \dot{x} = c_4 \dot{x}$$

where l is the characteristic dimension of the moving structure that can be taken as half the width of the plate.

The total damping force on the moving plate consists of F_1, F_2 and F_3, or F_4. As both F_3, and F_4 are usually quite small, for convenience, the total damping force is expressed as

$$F = F_1 + F_2 + F_3 + F_4 = (c_1 + c_2 + c_3 + c_4)\dot{x} \equiv c\dot{x}$$

and the quality factor is

$$\frac{1}{Q} = 2\zeta = \frac{c}{m\omega}$$

or

$$\frac{1}{Q} = \frac{\mu}{m\omega} \left((\frac{A_p}{d_p} + \frac{A_p}{\delta} + \frac{A_s}{d_s} + 10.7l) \right) \tag{3.4.20}$$

where m is the mass of the plate and ω the radial frequency.

Due to the finite dimension of the structure and the fringe effect at the edges and corners, Eq. (3.4.20) is only a semi-quantitative approximation to the accurate value. However, it does provide useful information in designing a comb resonator.

The resonant frequency ω in Eq. (3.4.20) can be written as

$$\omega = \sqrt{\frac{2Ehb^3}{mL^3}}$$

where L is the effective length of the beams, b the width and h the thickness of the beam flexures. As $m \cong A_p h \rho_p$, we have

$$\frac{1}{Q} = \frac{\mu}{h}\sqrt{\frac{L^3}{2E\rho_p A_p b^3}}\left((\frac{A_p}{d_p} + \frac{A_p}{\delta} + \frac{A_s}{d_s} + 10.7l)\right) \qquad (3.4.21)$$

For most situations, the Couette-flow slide-film damping term, the first term in Eq. (3.4.21), is the dominant factor of air damping. If only the Couette-flow slide-film damping is considered in estimating the quality factor of the structure, we have

$$Q \approx \frac{hd_p}{\mu}\sqrt{\frac{E\rho_p}{A_p}\left(\frac{b}{L}\right)^3} \qquad (3.4.22)$$

This means that, for a high quality factor, the structure should be thick and far away from the substrate. Also, the flexures should have as large a flexure rigidity as possible in their moving direction.

It may be mentioned here that the discussion given in the above is based on 1-D Stokes model that ignores the boundary effects at the edges of the plate. As is well known, the comb-drive structures made of typical surface micro machining technology has thin and fine fingers. Therefore, the ignorance of the boundary effects might cause significant error.

Due to the difficulties in computation, a three-dimensional stokes model has not been considered until recently. W. Ye, et al [15] investigated the air damping of comb drive structures using a 3-D model and compared the theoretical results with experimental data. They found that the 1-D Stokes model underestimated the damping force by a factor of about 2, while the 3-D Stokes model matches the experimental data to within 10% for typical devices.

§3.5. Damping in Rarefied Air

§3.5.1. Effective Viscosity of Rarefied Air

As described in §3.1, gas can be considered as a viscous fluid at an atmospheric pressure due to the frequent collisions among gas molecules. Therefore, the energy dissipation mechanism of air damping is due to the viscous flow of air caused by the movement of the vibrating structure (typically, a micro-bridge or a plate-flexure structure). The coefficient of viscosity of gas, μ, derived based on a simple model [1] is Eq. (3.1.2)

$$\mu = \frac{1}{3}\rho_a \lambda \bar{v}$$

where ρ_a is the mass density of gas, λ the mean free path of the molecules and $\bar{v}$ the average velocity of the molecules. As ρ_a is proportional to the pressure while λ inversely proportional to the pressure, the coefficient of viscosity, μ, is independent of pressure p.

Experimental results show that the effect of air damping is indeed quite constant when the air pressure is near the atmospheric pressure. However, when the air is rarefied to a pressure well below one atmospheric pressure, the air damping is reduced appreciably.

There have been two basic approaches so far in considering the damping in rarefied air: the "effective coefficient of viscosity" and the free molecular model. The first approach suggested that the equations for squeeze-film air damping remain effective in rarefied air, but the "coefficient of viscosity" should be replaced by an "effective" one, μ_{eff}, which is dependent on the pressure via Knudsen number, K_n, of the system. Knudsen number K_n is defined as

$$K_n = \frac{\lambda}{d} \tag{3.5.1}$$

where λ is the mean free path of molecules and d the typical dimension of the damping structure (For example, for the damping of an isolated object d is the typical dimension of the body and, for squeeze-film damping, d is the gap distance between the plate and the wall).

Based on the work of Fukui et al [16], Veijola, et al [17] derived a simple approximate equation for the effective coefficient of viscosity

$$\mu_{eff} = \frac{\mu_o}{1 + 9.658 K_n^{1.159}} \tag{3.5.2}$$

where μ_o is the coefficient of viscosity at atmospheric pressure.

Based on Andrews' experimental data [18], Li [19] derived a similar empirical equation for the effective coefficient of viscosity

$$\mu_{eff} = \frac{\mu_o}{1 + 6.8636 K_n^{0.9906}} \tag{3.5.3}$$

With the effective coefficient of viscosity, the squeeze-film damping effect in rarefied air is related to the dimensions of the plate and the gap distance between the moving plate and the nearby wall as it is in an atmospheric pressure.

The concept of effective viscosity coefficient for rarefied air is reasonable when the air pressures is not very low so that the gas can still be considered as a continuum though the non-slippage boundary conditions might have to be replaced by a slippage one. However, for a pressure much lower than an atmospheric pressure, the collisions among the gas molecules are so reduced that the gas can hardly be considered as a viscous fluid. In this case, the concept of effective coefficient of viscosity would become questionable.

Kinematically, the nature of interaction between a solid-state body and ambient gas is described using Knudsen number K_n. For $K_n \ll 1$, i.e., the typical dimension of micro structures is much larger than the mean free path of the gas molecules, the gas can be considered as a viscous fluid. For $K_n \gg 1$, i.e., the gap distance is much smaller than the mean free path of the gas molecules, the viscous flow model is no more valid and a free molecular model has to be considered.

For example, the mean free path of gas molecules is on the order of 0.1 μm in an atmospheric pressure. However, for a vacuum with a pressure of 1 Pa, the mean free path increases to about 1 cm, which is much larger than the typical dimension of microstructures. In this case, the use of free molecular model would be more reasonable.

§3.5.2. Christian's Model for Rarefied Air Damping

Experiments show that the air damping force on a microstructure reduces significantly when the air pressure is reduced to about 1kPa and below. It is believed that the gas molecules are so far apart in low pressure that the interaction between the gas molecules can be neglected. Therefore, the free molecule model could be used.

Christian proposed a free molecular model for damping in low vacuum [20, 21]. In his model, the resistive damping force on an oscillating plate is found by the momentum transfer rate from the vibrating plate to the surrounding air through the collisions between the plate and the molecules. In his model, no "coefficient of viscosity" in any form has been used and the quality factor of the oscillating plate is found directly.

Consider the air damping force acting on a plate oscillating in its normal direction (x-direction) as shown in Fig. 3.5.1. If the interaction between gas molecules can be neglected, the damping force acting on the oscillating plate is caused by the momentum transformation by collisions between the plate and gas molecules.

The number of molecules in a unit volume with velocity in the range of v_x to $v_x + \Delta v_x$ is $dn = n f(v_x)dv_x$, where $f(v_x)$ is Maxwellian distribution function

$$f(v_x) = \sqrt{\frac{m}{2\pi kT}} e^{-\frac{mv_x^2}{2kT}} \tag{3.5.4}$$

where k is the Boltzmann constant (k=1.38×10^{-23}J/K, or, k=8.62×10^{-5}eV/K). If the speed of the plate is $\dot{x}$, due to the movement of the plate, the collision number in a unit time, on a unit area of the plate surface for those molecules in the velocity range v_x to $v_x + \Delta v_x$ is $(v_x + \dot{x})dn \cdot 1 \cdot 1 = n \cdot (v_x + \dot{x})f(v_x)dv_x$.

As the change of momentum for each collision molecule is $2m(v_x + \dot{x})$, the pressure caused by the collisions on the front side of the plate can be found by an integration to take account all the velocity of the molecules

$$P_f = 2mn \int_{-\dot{x}}^{\infty} (v_x + \dot{x})^2 f(v_x)dv_x \tag{3.5.5}$$

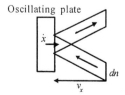

Fig.3.5.1. Collisions of oscillating plate with head-on molecules

Similarly, the pressure caused by the collisions on the back of the plate is

$$P_b = 2mn \int_{\dot{x}}^{\infty} (v_x - \dot{x})^2 f(v_x) dv_x \qquad (3.5.6)$$

The net damping force caused by collisions is $F_r = A(P_f - P_b)$, where A is the area of the plate. If the velocity of the plate, $\dot{x}$, is much smaller than that of the majority of the gas molecules, according to Eqs. (3.5.4), (3.5.5) and (3.5.6), we have

$$F_r \cong 8mnA \int_0^{\infty} v_x \cdot \dot{x} F(v_x) dv_x = 8mnA \sqrt{\frac{kT}{2\pi m}} \dot{x} \qquad (3.5.7)$$

For the gas in the standard condition ($P_0 = 1$ *atm.*, $T_0 = 273$ K), the molecule density of gas is $n_0 = N_A/V_o$, where $N_A = 6.023 \times 10^{23}$ and $V_0 = 0.0224 \text{m}^3$. The molecule density at pressure P and temperature T is

$$n = n_o \frac{PT_o}{P_o T} = \frac{N_A P}{RT} \qquad (3.5.8)$$

where $R = 8.31$ kg·m^2/sec^2/°K is the universal molar gas constant. Therefore, we obtain

$$F_r = 4\sqrt{\frac{2}{\pi}} \sqrt{\frac{M_m}{RT}} PA\dot{x} \qquad (3.5.9)$$

where M_m is Molar mass of the gas. The coefficient of damping force in rarefied air by the free molecule model is

$$c_r = 4\sqrt{\frac{2}{\pi}} \sqrt{\frac{M_m}{RT}} PA \qquad (3.5.10)$$

Eq. (3.5.9) shows that the damping effect in rarefied air decreases linearly with decreasing pressure. Therefore, the Q factor of the system in low vacuum is

$$Q = \frac{M_p \omega}{c_r} \qquad (3.5.11)$$

where M_P is the mass of the plate and ω the natural frequency of the system. As $M_p = Ah\rho$, the quality factor given by Christian model is

$$Q_{Chr} = \frac{H\rho\omega}{4} \cdot \sqrt{\frac{\pi}{2}} \cdot \sqrt{\frac{RT}{M_m}} \frac{1}{P} \qquad (3.5.12)$$

where H is the thickness of the plate and ρ is the mass density of the plate.

Eq. (3.5.12) has been compared with experimental data. The results agree to within an order of magnitude. Though the quality factor is indeed inversely proportional to pressure P, the value of the Q factor is overestimated quantitatively, i.e., the damping force in rarefied air is underestimated by Christian's model.

§3.5.3. Energy Transfer Model for Squeeze-film Damping
Though Christian's model can qualitatively explain the damping effect in low vacuum, the damping effect is underestimated by about an order of magnitude when the calculated quality factors are compared with experimental data by J. Zook et al [22].

Another problem for Christian's model is that it is, in fact, only a damping model for an isolated plate as the effect of any nearby walls is not considered. In most practical cases, vibrating plates are close to driving electrodes or encapsulation walls. Especially, when the plate is electro-statically driven, the driving electrodes are usually very close to the oscillating plate for high driving forces.

The third problem for Christian's model is that the dependence of quality factor on the dimensions of the oscillating plate cannot be considered. Obviously, this is not reasonable.

To overcome these problems, Bao et al [23] proposed another free molecular model. With this new model, the above problems can be solved and the same result as that from Christian's model for isolated vibration can be obtained. As the model calculates the energy losses through the energy transfer effect, this model is called Energy Transfer model. The model is introduced as follows.

(1) Velocity Change of Gas Molecules by Collisions
To calculate the energy transferred from the plate to the surrounding air, let us first consider the velocity change of a gas molecule due to its collision with a moving plate.

Suppose that a gas molecule is moving with a speed of v_2 towards a moving plate with a head on movement of speed v_1 to the molecule as shown in Fig. 3.5.2(a). As the mass of the gas molecule m is much smaller than the mass of the plate M, the gas molecule is bounced back after the collision but the plate moves on with little speed change as shown in Fig. 3.5.2(b). If the velocities of the plate and the molecule are $v_1{}'$ and $v_2{}'$ after the collision, respectively, the conservation of kinetic momentum requires

$$Mv_1 - mv_2 = Mv_1{}' + mv_2{}'$$

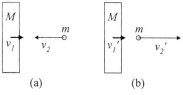

(a)   (b)

Fig. 3.5.2. The velocities of the plate and a molecule (a) before a collision; (b) after a collision

With the notation of $\gamma = m/M$, we have

$$v_1 - \gamma v_2 = v_1{}' + \gamma v_2{}' \tag{3.5.13}$$

By the conservation of kinetic energy, we have

$$v_1^2 + \gamma v_2^2 = v_1'^2 + \gamma v_2'^2 \qquad (3.5.14)$$

From equations (3.5.13) and (3.5.14), for γ much smaller than unity, we have

$$v_2' = v_2 + 2v_1 \qquad (3.5.15)$$

It means that the gas molecule has a speed increment of $2v_1$ due to the head-on collision with the plate with a speed v_1. Similarly, if the plate moves in the same direction with the molecule, the resulting speed of the molecule would be

$$v_2' = v_2 - 2v_1 \qquad (3.5.16)$$

(2) Air Damping of an Isolated Plate

To have a comparison between the Energy Transfer model and Christian model, the quality factor of an isolated oscillating plate in low vacuum is considered using energy transfer model first.

According to Boltzmann statistics, the number of molecules with velocity in the range between v_x and $v_x + dv_x$ colliding with the surface of a static plate normal to the x-axis in a time interval Δt is

$$\Delta n = A n f(v_x) v_x dv_x \Delta t$$

where A is the area of the plate.

If the plate is moving in its normal direction (the x-direction), the number of molecules (with velocity in between v_x to $v_x + \Delta v_x$ and in time interval Δt) colliding with the front of the plate is

$$\Delta n_{front} = A n f(v_x)(v_x + \dot{x}) dv_x \Delta t \qquad (3.5.17)$$

where $\dot{x}$ is the moving speed of the plate. Similarly, the collision number on the back is

$$\Delta n_{back} = A n f(v_x)(v_x - \dot{x}) dv_x \Delta t \qquad (3.5.18)$$

Therefore, the energy transferred from the oscillating plate to the molecules on the front side of the plate in time interval Δt is

$$\Delta E_{front} = A \int_{-\dot{x}}^{\infty} \left[\frac{1}{2} m(v_x + 2\dot{x})^2 - \frac{1}{2} mv_x^2 \right] (v_x + \dot{x}) n f(v_x) dv_x \Delta t$$

Similarly, the energy transferred from the oscillating plate to the molecules on the back of the plate in time interval Δt is

$$\Delta E_{back} = A \int_{\dot{x}}^{\infty} \left[\frac{1}{2} m(v_x - 2\dot{x})^2 - \frac{1}{2} mv_x^{\,2} \right] (v_x - \dot{x}) nf(v_x) dv_x \Delta t$$

The net energy transformed from the oscillating plate to the surrounding molecules ($\Delta E = \Delta E_{front} + \Delta E_{back}$) in time interval Δt is

$$\Delta E = A \int_{-\dot{x}}^{\infty} \frac{1}{2} m(4\dot{x}v_x + 4\dot{x}^2)(v_x + \dot{x}) nf(v_x) dv_x \Delta t + A \int_{\dot{x}}^{\infty} \frac{1}{2} m(-4\dot{x}v_x + 4\dot{x}^2)(v_x - \dot{x}) nf(v_x) dv_x \Delta t$$

If $\dot{x}$ is much smaller than v_x, the equation can be approximated as

$$\Delta E = Amn \int_0^{\infty} 8v_x \dot{x}^2 f(v_x) dv_x \Delta t = 4Amn \sqrt{\frac{2kT}{\pi m}} \cdot \dot{x}^2 \Delta t \qquad (3.5.19)$$

For a sinusoidal oscillation, i.e., $x = a_0 \sin \omega t$, the energy lose in one oscillation cycle is

$$\Delta E_{cycle} = 4Amn \sqrt{\frac{2kT}{\pi m}} \int_0^{2\pi} a_0^{\,2} \omega^2 \cos^2 \omega t \cdot \frac{1}{\omega} d(\omega t) = 4Amn \sqrt{\frac{2kT}{\pi m}} a_0^{\,2} \omega \pi \qquad (3.5.20)$$

According to the definition of quality factor in chapter 2, the quality factor is

$$Q_{E,Iso} = \frac{2\pi E_P}{\Delta E_{cycle}} = \frac{\rho H \omega}{4} \sqrt{\frac{\pi}{2}} \sqrt{\frac{RT}{M_m}} \frac{1}{P} \qquad (3.5.21)$$

When Eq. (3.5.21) is compared with Eq. (3.5.12), we have $Q_{E,Iso} = Q_{Chr}$, i.e., the quality factor by the Energy Transfer model is the same as that by Christian's Momentum Transfer model. Now let us proceed to the discussion on the squeeze-film air damping of plate in low vacuum, which is beyond the ability of Christian's model.

(3) Squeeze-film Air Damping

Let us now consider the "squeeze-film" air damping for an oscillating plate as schematically shown in Fig. 3.5.3 [23]. Suppose that the original gap distance between the plate and the substrate is d_o and the displacement of the oscillating plate is $x = a_o \sin \omega t$.

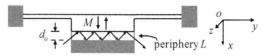

Fig.3.5.3. The squeeze-film air damping in rarefied air

Due to the displacement of the plate, the gap distance is $d = d_o - x$. If the peripheral length of the gap is L, the boundary area of the gap is $L(d_o - x)$. The number of molecules moving across the boundary into the gap per unit time is

$$\frac{1}{4}n\bar{v}L(d_o - x)$$ (3.5.22)

where n is the concentration of the molecules and $\bar{v}$ the average velocity of the molecules

$$\bar{v} = \sqrt{\frac{8kT}{\pi m}}$$

If a molecule entering into the gap has the velocity components of v_{yzo} in the y-z plane and v_{xo} in the x-direction, it gains velocity in the x-direction due to the collisions with the plate when traveling in the gap. If the lateral traveling distance in the gap is l, the time the molecule stays in the gap is $\Delta t = l/v_{yzo}$. As v_{yzo} is on the order of several hundreds meter per second, Δt is much smaller than an oscillating cycle of the plate. The times of collision in the time period of Δt is

$$\Delta N = \frac{\Delta t \times v_{xo}}{2(d_o - x)} = \frac{l v_{xo}}{2(d_o - x)v_{yzo}}$$

As the molecule gains a speed increment of $2\dot{x}$ each time it collides with the plate, the velocity in the x-direction at the end of the traveling in the gap is

$$v_x = v_{xo} + \Delta N \times 2\dot{x} = v_{xo} + \frac{l v_{xo}}{(d_o - x)v_{yzo}}\dot{x}$$

The kinetic energy of the molecule entering the gap is

$$e_{k,in} = \frac{1}{2}m\left[v_{yzo}{}^2 + v_{xo}{}^2\right]$$

while the energy of the molecule leaving the gap is

$$e_{k,out} = \frac{1}{2}m\left[v_{yzo}{}^2 + v_{xo}{}^2 + \frac{2l v_{xo}{}^2}{(d_o - x)v_{yzo}}\dot{x} + \frac{l^2 v_{xo}{}^2}{(d_o - x)^2 v_{yzo}{}^2}\dot{x}^2\right]$$

The extra energy gained by the molecule via collisions with the plate is

$$\Delta e_k = \frac{1}{2}m\left[\frac{2l v_{xo}{}^2}{(d_o - x)v_{yzo}}\dot{x} + \frac{l^2 v_{xo}{}^2}{(d_o - x)^2 v_{yzo}{}^2}\dot{x}^2\right]$$ (3.5.23)

As the average of the first term on the right-handed side is zero in a cycle, only the second term accounts for the energy loss of the plate. According to Eqs. (3.5.22) and (3.5.23), the average energy loss of the plate in one vibration cycle is

$$\Delta E_{cycle} = \frac{1}{4}n\overline{v}L\frac{1}{\omega}\int_0^{2\pi}\frac{m \cdot l^2 \cdot v_{xo}^2}{2(d_o-x)v_{yzo}^2}a_o^2\omega^2\cos^2\omega t d(\omega t)$$

For simplicity, l^2, v_{xo}^2 and v_{yzo}^2 are approximated as $\overline{l^2}$, $\overline{v_{xo}^2}$ and $\overline{v_{yzo}^2}$ respectively. As $\overline{v_{yzo}^2} = 2\overline{v_{xo}^2}$, the energy loss of the plate in one cycle is

$$\Delta E_{cycle} = \frac{1}{4}\rho_o\overline{v}L\frac{1}{\omega}\int_0^{2\pi}\frac{\overline{l^2}a_o^2\omega^2}{4(d_o-x)}\cos^2\omega t \cdot d(\omega t)$$

where $\rho_o=mn$ is the mass density of the gas. If the vibration amplitude a_o is much smaller than d_o, we have

$$\Delta E_{cycle} \cong \frac{1}{4}\rho_o\overline{v}L\frac{1}{\omega}\frac{\pi\overline{l^2}a_o^2\omega^2}{4d_o} = \frac{\pi\overline{l^2}a_o^2\omega}{16}\rho_o\overline{v}\frac{L}{d_o} \tag{3.5.24}$$

According to the definition of Q (see §2.5.4), the quality factor for squeeze-film air damping by the Energy Transfer model in low vacuum is

$$Q_{E,Sq} = \frac{2\pi E_P}{\Delta E_{cycle}} = \frac{16M_p d_o \omega}{\overline{l^2}\rho_o\overline{v}L} \tag{3.5.25}$$

where E_P is the energy of the plate and M_P the mass of the plate. If the mass density of gas at an atmospheric pressure is denoted as ρ_{atm} and the atmospheric pressure is denoted as p_{atm}, the Q factor for a pressure p is

$$Q_{E,Sq} = \frac{16M_p d_o \omega p_{atm}}{\rho_{atm}\overline{v}L\overline{l^2}}\frac{1}{p} \tag{3.5.26}$$

For a rectangular plate with edge lengths of a and b, a thickness of H and a mass density of ρ, Eq. (3.5.26) can be written as

$$Q_{E,Sq} = \frac{16\rho H\omega p_{atm}}{\beta\rho_{atm}\overline{v}}\left(\frac{d_o}{L}\right)\frac{1}{p} \tag{3.5.27}$$

where $\beta = \frac{\overline{l^2}}{ab}$. As $\overline{v} \cong \sqrt{\frac{8kT}{\pi m}} = \sqrt{\frac{8RT}{\pi M_m}}$ and $\frac{p_{atm}}{\rho_{atm}} = \frac{RT}{M_m}$, Eq. (3.5.27) can be written as

$$Q_{E,Sq} = \frac{8\sqrt{\pi}\rho H\omega}{\beta\sqrt{2}}\left(\frac{d_o}{L}\right)\sqrt{\frac{RT}{M_m}}\frac{1}{p} \tag{3.5.28}$$

With a little more approximation, β is found to be 2/π [23]. Therefore, the Q factor for squeeze-film air damping in low vacuum is

$$Q_{E,Sq} = (2\pi)^{\frac{3}{2}} \rho H \omega \left(\frac{d_o}{L}\right) \sqrt{\frac{RT}{M_m}} \frac{1}{p} \qquad (3.5.29)$$

or, the coefficient of damping force

$$c_r = \frac{M_p \omega}{Q_{E,Sq}} = 0.0635 \left(\frac{L}{d_o}\right) \sqrt{\frac{M_m}{RT}} \cdot PA \qquad (3.5.30)$$

According to the discussion given in §3.1, this relation is valid under the condition that the Knudsen number K_n is very large so that the collisions among gas molecules can be neglected. If K_n is not large enough, the actual quality factor might be smaller than that given in Eq. (3.5.29) due to the longer stay time under the plate for the incident molecules.

By comparison of Eq. (3.5.29) with Eq. (3.5.12), the relation between the quality factor of squeeze-film damping by Energy Transfer model, $Q_{E,Sq}$, and that by Christian's Model, $Q_{chr.}$, is

$$Q_{E,Sq} = 16\pi \left(\frac{d_o}{L}\right) Q_{Chr.} \qquad (3.5.31)$$

As the actual value of β might be a little under-estimated, the value of the quality factor might be a little overestimated.

(4) Comparisons With Experiments

Zook et al [22] measured the quality factors of polysilicon micro beams in vacuum. The experimental results (curve I in Fig. 3.5.4) are compared with Christian's model (curve II in Fig. 3.5.4). The results of Christian's model are about an order of magnitude larger.

According to [22], the beam is 200μm long and 40μm wide. With reference to [24], the gap distance is 1.1μm. According to these data and Eq. (3.5.31), we find

$$Q_{E,Sq.} = \frac{1}{8.7} Q_{Chr} \qquad (3.5.32)$$

The result of the Energy Transfer model (curve III in Fig. 3.5.4) matches the experimental data much better than that of Christian's model. As the dimensions of the bridge are on the order of 10^2μm and the free molecular path in atmospheric pressure is on the order of 10^{-1}μm, the condition of $K_n \gg 1$ requires $p \leq 100$Pa.

Based on a comprehensive review of varies existing models and following the basic premise of Bao's Energy Transfer model but with some of its approximations for analytical derivation lifted, S. Hutcherson and W. Ye [25] developed a molecular dynamic simulation code recently. Their simulation results show even better agreement with experimental data than Bao's model.

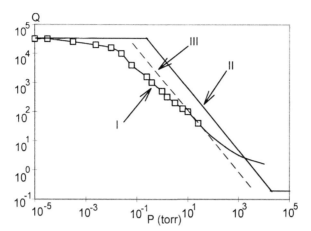

Fig.3.5.4. Comparison with experimental results of a beam (I: experimental data; II: Christian's model; III: the Energy Transfer model)

§3.5.4. Damping in a High Vacuum

According to the free molecule model, the damping force in rarefied air is proportional to the air pressure. Therefore, the air damping force goes down, or, the Q factor of the system goes up, with the decrease in air pressure. However, experimental results show that the Q factor levels off when the vacuum is high enough, i.e., when the effects of internal friction and support losses become the dominant mechanisms of energy dissipation [21]. As the internal friction and the support losses are very hard to predict theoretically, they are evaluated by the quality factor in a high vacuum condition. For micro structures made of silicon, the quality factors at high vacuum, Q_o, range from $10^4 \sim 10^5$. Once Q_o is found through experimental measurements the coefficient of damping force caused by internal friction and support losses can be found using

$$c_o = \frac{M_p \omega}{Q_o} \tag{3.5.33}$$

where M_p is the mass of the oscillating plate. Therefore, the differential equation for vibration in rare air can be modified to

$$m\ddot{x} + (c_r + c_o)\dot{x} + kx = F \tag{3.5.34}$$

where c_r is given in Eq. (3.5.10) or Eq. (3.5.30), depending which model is used.

Assume that Q_o of the oscillating plate is found to be 5×10^4, the plate is made of silicon with a thickness of 200 μm and the natural frequency of the structure is 1 kHz. The critical pressure, P_c, where the Q factor starts to level off with decreasing pressure can be found by equating Eq. (3.5.13) to Eq. (3.5.10) or Eq. (3.5.32) for Christian model or Energy Transfer model. The critical pressures thus found are on the order of 1Pa.

As a summary, the dependence of the Q factor on the air pressure from atmospheric pressure to high vacuum is schematically shown by the curve in Fig. 3.5.5. The Q factor at high pressure is independent of pressure as shown by sector A of the curve. The Q factor in this regime is determined by the geometries and the moving directions of the moving structure. The damping mechanism could be the squeeze-film damping, the slide-film damping, the drag force damping or a combination of these mechanisms.

When the pressure is pumped down to a certain extent ($10^2 \sim 10^3$ Pa), the Q factor starts to rise when the effect of rarefaction starts to play an important role. The Q factor is inversely proportional to the air pressure in this region where the rarefied air damping plays a major role, as shown by sector B of the curve in Fig. 3.5.5. The transition pressure, P_t, from sector A to sector B is usually in the range of several hundreds Pa. The exact value is dependent on the geometries and vibration mode of the structure.

At high vacuum, when the air damping is very small, the effects of internal friction and energy losses via the structure supports have to be considered. The Q factor is determined by air damping as well as the internal friction and the support losses of the structure. Obviously, the Q factor will be mainly determined by the internal friction and the support loss if the vacuum is high enough and the Q factor becomes a constant as shown by sector C of the curve in Fig. 3.5.5. The Q factors of silicon microstructure in high vacuum are usually in the order of 10^5. The exact value is dependent on the geometry design of the structure.

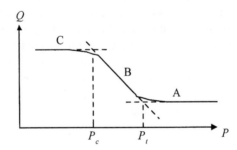

Fig. 3.5.5. The dependence of Q factor on air pressure

§3.6. Problems

Problems on Viscosity Forces

1. If the effective diameter of hydrogen molecule is d=2.25×10^{-4}µm, calculate the coefficient of viscosity of hydrogen at 0°C according to Eq. (3.1.3) in §3.1.

2. A capillary has a diameter of d=80µm and a length of L=10mm. If the pressure difference applied between the two ends of the capillary is 1kPa, find the flow rates for water and for air at 20°C.

3. A capillary has a diameter of d=80µm and a length of L=10mm. If the pressure difference applied between the two ends of the capillary is 1kPa, find the Reynolds' numbers for water and for air at 20°C.

4. For a water droplet of a radius $r = 10\mu m$ falling in air under the gravity of earth, find the balanced velocity of the droplet.

Problems on Squeeze-film Air Damping

5. A silicon structure is shown in the figure below, where $a=500\mu m$, $b=50\mu m$, $h=20\mu m$, $A=4mm$, $B=2mm$, $H=300\mu m$ and $d=20\mu m$. If the mass of the beam and the bending of the mass are negligible, find: (1) the coefficient of damping force, c, for the vibration normal to the mass plane; (2) the damping ratio and the quality factor of the vibration; (3) the resonant frequency. (For silicon $E=1.7\times10^{11}$ Pa and $\rho=2.33gm/cm^3$)

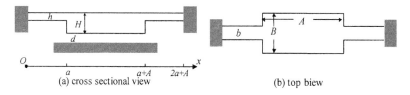

(a) cross sectional view   (b) top biew

6. The width, the thickness and the length of a silicon cantilever beam are b, h and L, respectively, and the gap distance between the beam and the substrate under it is d. Find (1) the expression of damping ratio for basic normal vibration mode, and (2) the quality factor of the basic vibration mode for $b=50\mu m$, $h=5\mu m$, $L=500\mu m$, and $d=10\mu m$.

7. The width, the thickness and the length of a silicon cantilever beam are b, h and L, respectively, and the beam is far from any surrounding object. Find (1) the expression of damping ratio for the basic vibration mode based on the "dish-string" model, and (2) the quality factor of the basic vibration mode for a beam with dimensions of $b=50\mu m$, $h=5\mu m$, and $L=500\mu m$.

8、 A silicon beam-mass structure is shown in the figure below, where $A=2mm$ and $H=200\mu m$. If the free vibration frequency of the mass in its normal direction is $f_0=1kHz$, find the gap distance d for critical damping of the vibration.

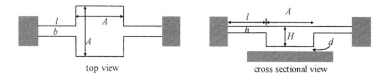

top view   cross sectional view

9. The resonant frequency of a silicon capacitive accelerometer is $f_{res}=1051Hz$ and the quality factor is $Q=2.3$ by measurement. It is found by reverse engineering that (a) the structure of the accelerometer is shown the figure in problem 8; (b) the edge length of the square mass is $A=1.5$ mm and the thickness of the mass is $H=300\mu m$; and (c) the width and length of the two beams supporting the mass are $b=20\mu m$ and $l=200\mu m$, respectively, but it fails to find the thickness of the beams. Based on the available data, find: (1) the thickness of the beams; (2) the distance between the mass and the electrode underneath.

Problems on Squeeze-film Air Damping of Thick Hole-plate

10. Two long rectangular hole-plates have the same width $2a$=2mm, the diameter of holes d_o=$2r_o$=5μm, the density of holes is n=800/mm^2 and the gap distance to the substrate h=2μm. However, the thickness for plate 1 is H_1=2μm and the thickness for plate 2 is H_2=20μm. Find the ratio of the coefficients of damping force for squeeze-film air damping of the two hole-plates.

11. The problem is similar to problem 10, but the diameter of the holes is d_o=20μm and the density of the hole is 50/mm^2. Find the ratio of the coefficients of damping force for squeeze-film air damping of the two hole-plates.

12. The silicon structure shown in the figure below has parameters as follows: a=100μm, b=40μm, h=2μm, H= 2μm, d=2μm, A=B=400 μm, the diameter of the holes $2r_o$=10μm and the hole density n=800/mm^2. If the mass of the beam and the bending of the plate are negligible, find the damping ratio for the normal vibration mode of the structure. (Using E=1.7×10^{11} Pa, ρ=2330kg/m^3)

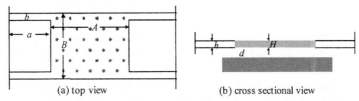

(a) top view   (b) cross sectional view

13. The silicon structure shown in the figure below has the same parameters as problem 12 except that the thickness of the hole-plate is now increased to H=10μm. Find the damping ratio for the basic vibration mode of the structure.

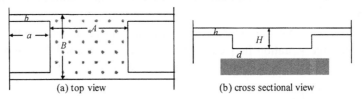

(a) top view   (b) cross sectional view

Problems on Slide-film Air Damping

14. A beam-mass structure is shown in the figure below, where A=B=3mm, b=40μm, h=20μm, H=50μm, d=5μm. If the natural vibration frequency of the mass in normal direction is f_o=2 kHz, find the quality factor of the lateral vibration. (Assuming that the damping under the mass follows the Couette flow model and the damping above the mass follows 1-D Stokes flow model.)

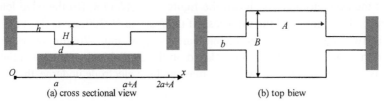

(a) cross sectional view   (b) top biew

15. For the same problem as above but $d=20\mu m$, find the quality factor for the lateral vibration by using 1-D stokes flow model.

References

[1] P. M. Whelan, M. J. Hodgon, Essential Principles of Physics, J.W. Arrowsmith Ltd. Bristol, 1978

[2] F. M. White, "Viscous Fluid Flow", McGraw-Hill Book Company, 1974

[3] L. D. Landau and E. M. Lifshitz, "Fluid Mechanics", Second Edition, New York: Pergamon, 1989

[4] P. M. Morse, "Vibration and Sound", 2nd Edition, 1948, McGraw-Hill Book Company, New York, p.333

[5] W. E. Langlois, Isothermal squeeze films, Quartly of Applied Mathematics, Vol. XX, No. 2, (1962) 131-150

[6] J. Starr, Squeeze film damping in solid-state accelerometers, IEEE Workshop on Solid-state Sensor and Actuator, Hilton Head Island, SC, USA (1990) 44-47

[7] T. Gabrielson, Mechanical-thermal noise in micromechined acoustic and vibration sensors, IEEE Trans. on Electron Devices, Vol.40 (1993) 903-909

[8] B. Davies, S. Montague, J. Smith, M. Lemkin, Micromechanical structures and microelectronics for accelerometer sensing, Proc. SPIE, Vol. 3223 (1997) 237-244

[9] H. Hosaka, K. Itao, S. Kuroda, Damping characteristics of beam-shaped microstructures, Sensors and Actuators A49 (1995) 87-95

[10] M. Sadd, A. Stiffer, Squeeze Film dampers: amplitude effects at low squeeze numbers, J. of Engineering for Industry, Transactions of the ASME, Vol.97, (1975) 1366-1370

[11] M. Bao, H. Yang, Y. Sun, P. French, Modified Reynolds' equation and analytical analysis of squeeze-film air damping of perforated structures, Journal of Micromechanics and Microengineering, Vol. 13 (2003) 795-800

[12] F. Sharipov, V. Seleznev, Data on internal rarefied gas flows, Journal of Physical and Chemical Reference Data, Vol.27, No.3 (1997) 657-706

[13] Y. Cho, A. Pisano, R. Howe, Viscous damping model for laterally oscillating microstructures, J. Microelectromechanical Systems, Vol. 3, No.2 (1994) 81-86

[14] S. Kobayashi, K. Hori, Y. Konada, K. Ohwada, Mechanical quality factor of cantilever micro-resonator operated in air, Technical Digest of the 15th Sensor Symposium of Japan (1997) 25-30

[15] W. Ye, X. Wang, W. Hemmert, D. Freeman, J. White, Air damping in laterally oscillating micro resonators: a numerical and experimental study, J. of Microelectromechancial Systems, Vol. 12 (2003) 557-566

[16] S. Fukui, R. Kaneko, Analysis of ultra-thin gas film lubrication based on linearized Boltzmann equation: first report — derivation of a generalized lubrication equation including thermal creep flow, J. Tribol. Trans. ASME, Vol. 110 (1988) 253-262

[17] T. Veijola, H. Kuisma, J. Lahdenpera, T. RyhanenT, Equivalent-circuit model of the squeezed gas film in a silicon accelerometer, Sensors and Actuators, Vol. A48 (1995) 239-248

[18] M Andrews, I Harres, G. Turner, A comparison of squeeze-film theory with measurements using a microstructure, Sensors and Actuators A, Vol. 36 (1993) 79-87

[19] G. Li, H. Hughes, Review of viscosity damping in micro-machined structures, Proc. of SPIE, Vol. 4176 (2000) 30-46

[20] R. Christian, The theory of oscillating-vane vacuum gauges, Vacuum, Vol. 16 (1966) 175-178

[21] Y. Kawamura, K. Sato, T. Terasawa, S. Tanaka, Si cantilever-oscillator as a vacuum sensor, Digest of Technical Papers, The 4th International Conference on Solid-State Sensors and Actuators, Tokyo, Japan, June 3-5, 1987 (Transducers' 87) 283-286

[22] J. Zook, D. Burns, H.Guckel, J. Sniegowski, R. Engelstad, Z. Feng, Characteristics of polysilicon resonant microbeams, Sensors and Actuators, Vol. A35 (1992) 51-59

[23] M. Bao, H. Yang, H. Yin, Y. Sun, Energy transfer model for squeeze-film air damping in low vacuum, Journal of Micromechanics and Microengineering, 12 (2002) 341-346

[24] H. Guckel, J. Sniegowski, R. Christenson, F. Rossi, The application of fine-grained, tensile polysilicon to mechanically resonant transducers, Sensors and Actuators A, Vol.21 (1990) 346-351

[25] S. Hutcherson, W. Ye, On the squeeze-film damping of micro-resonators in the free-molecular regime, Journal of Micromechanics and Microengineering, 14 (2004) 1726-1733

Chapter 4

Electrostatic Actuation

Conventional mechanical actuators are rarely driven by an electrostatic force because the force is usually too small to displace or lift mechanical parts unless the voltage used is extremely high. With the miniaturization of mechanical structures, the electrostatic force becomes relatively large. Therefore, electrostatic driving has found wide applications in micro mechanical actuators.

The theory governing the electrostatic force is no more than the well-established electrostatics. However, the electrostatic force in micro mechanical systems has the following features, which are very important for the analysis and design of the devices, or, in some cases, they are explored for novel applications:

(a) For microstructures, the electrostatic force is comparable with the elastic force of the mechanical structure and the damping force of the surrounding air. Therefore, all the forces must be considered simultaneously in many cases.

(b) The electrostatic force is nonlinear with the distance. The joint action of the electrostatic force and the elastic force could cause severe nonlinearity or instability problems.

(c) As the distances between mechanical parts and the dimensions of the mechanical structures are quite close, the fringe effects of the electrostatic force have to be considered in many cases.

These effects will manifest themselves in the sections of this chapter.

§4.1. Electrostatic Forces

§4.1.1. Normal Force

Consider an electrostatic actuator consisting of a battery and a capacitor of parallel plate as shown in Fig. 4.1.1(a). One plate of the parallel-plate capacitor (the dark-colored one) is fixed and the other is movable in its normal direction. The capacitance of the parallel plate actuator is

$$C(x) = \frac{A \varepsilon \varepsilon_o}{x} \tag{4.1.1}$$

where A is the overlapping area of the electrodes, x the distance between the two plates, ε_o ($=8.854 \times 10^{-12}$ F/m) the permittivity of a vacuum, ε the relative permittivity of the medium between two electrodes, which is approximately equal to unity for air. For convenience, we assume that the energy of the battery in the original state is E_o and the energy of the capacitor is zero.

Then the capacitor is connected to the battery in parallel as shown in Fig. 4.1.1(b). If the electromotive force of the battery is V, the electric charge stored in the capacitor is

$$Q_C = C(x)V \qquad (4.1.2)$$

The energy stored in the capacitor is

$$E_C = \frac{1}{2}C(x)V^2 \qquad (4.1.3)$$

Due to the charging to the capacitor, the energy of the battery is reduced to

$$E_B = E_o - Q_C V = E_o - C(x)V^2$$

Therefore, the energy of the capacitor-battery system becomes

$$E(x) = E_B + E_C = E_o - \frac{1}{2}C(x)V^2 = E_o - \frac{A\varepsilon\varepsilon_o}{2x}V^2 \qquad (4.1.4)$$

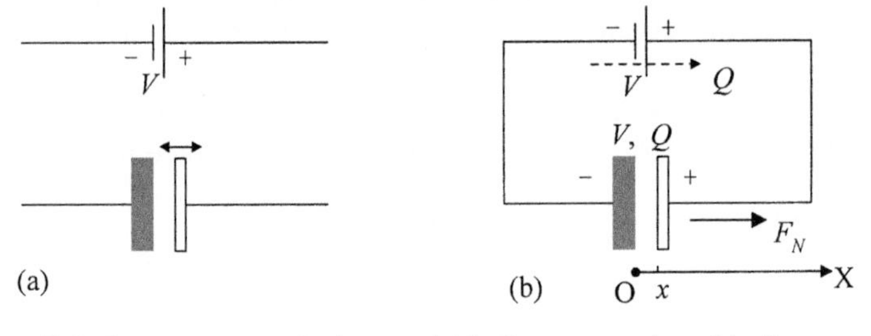

Fig. 4.1.1. A parallel-plate actuator and a battery (a) before connection; (b) after connection

From Eq. (4.1.4), the normal force applied on the movable plate of the capacitor is

$$F_N = -\frac{\partial E(x)}{\partial x} = -\frac{A\varepsilon\varepsilon_o}{2x^2}V^2 \qquad (4.1.5)$$

The negative sign of the force indicates that the force is attractive.

According to Eq. (4.1.5), the attractive force between the two parallel plates keeps constant when the dimensions of the plates and the gap distance are scaled down with the same factor. This is one of the favorable features of electrostatic force in micromechanical applications.

§4.1.2. Tangential Force

Consider a system consisting of a battery and a parallel-plate actuator. But now the two parallel plates are separated by a constant gap distance d and one of the plates (the top one) is movable in its plane as shown in Fig. 4.1.2(a). Suppose that the overlapping distance, y, is much larger than the gap distance, d. The capacitance between the two plates is

$$C(y) = \frac{by\varepsilon\varepsilon_o}{d} \tag{4.1.6}$$

where b is the width of the plates.

Again, the energy of the battery is supposed to be E_o before the capacitor is connected to the battery. Once the capacitor is connected to the battery as shown in Fig. 4.1.2(b), the electric charge stored in the capacitor is

$$Q_C = C(y)V$$

and the energy stored in the capacitor is

$$E_C = \frac{by\varepsilon\varepsilon_o}{2d}V^2 \tag{4.1.7}$$

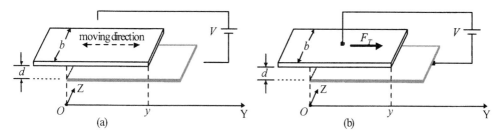

Fig. 4.1.2. The electrostatic force tangential on the plate (a) before connection; (b) after connection

Due to the charging to the capacitor, the energy of the battery is reduced to

$$E_B = E_o - Q_C V = E_o - C(y)V^2$$

Therefore, the energy of the capacitor-battery system becomes

$$E(y) = E_B + E_C = E_o - \frac{1}{2}C(y)V^2 = E_o - \frac{by\varepsilon\varepsilon_o}{2d}V^2 \tag{4.1.8}$$

From Eq. (4.1.8), the force applied on the movable plate of the capacitor is

$$F_T = -\frac{\partial E(y)}{\partial y} = \frac{b\varepsilon\varepsilon_o}{2d}V^2 \tag{4.1.9}$$

The positive sign in Eq. (4.1.9) indicates that the tangential force applied on the movable plate pulls the plate for more overlapping area with the stationary plate. Also we can see from Eq. (4.1.9) that the force is independent of the overlapping distance, y, so that the force is constant with the movement of the plate, and the tangential electrostatic force keeps constant when the width, b, of the plates is scaled down with the distance, d.

Now let us compare the tangential force F_T and the normal force F_N under the conditions of $x=d$ and $A=by$. From Eqs. (4.1.5) and (4.1.9), we have $|F_N|/|F_T| = y/d$. Typically, F_N is larger than F_T, as y is usually larger than d.

§4.1.3. Fringe Effects

(1) Capacitance
In §4.1.1 and §4.1.2, the electrode plates of a mechanical capacitor are considered to be parallel and the dimensions of the plates are much larger than the distance between them. Therefore, the capacitor is approximated as a parallel-plate capacitor and the capacitance can be expressed by Eq. (4.1.1) or (4.1.6). In practical situations for micro sensors and actuators, the dimensions of the mechanical electrodes are often comparable with the distance between them. Some typical situations are shown in Fig. 4.1.3. Therefore, the capacitance between two parallel electrodes cannot be approximated by Eq. (4.1.1) or (4.1.6) with high accuracy. The capacitance caused by the sidewalls and even the back may play an important role. The effect is often referred to as the fringe effect.

| (a) | (b) | (c) | (d) |

Fig. 4.1.3. Cross sections of some typical capacitive structures

Due to the fringe effect, the capacitance of a mechanical structure is larger than that calculated by Eq. (4.1.1) or (4.1.6). Consider a structure of two parallel bars with its cross section as shown in Fig. 4.1.4, where the top plate is movable in the z-direction. The capacitance calculated by parallel-plate approximation is

$$C_o(z) = \frac{2al\varepsilon\varepsilon_o}{z} \tag{4.1.10}$$

where l is the length of the bars that is much larger than a, h and z. Due to the fringe effect, the capacitance between the bars, $C(z)$, is always larger than $C_0(z)$.

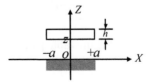

Fig. 4.1.4. The cross section of a parallel-bar capacitor

Generally speaking, the exact value of the capacitance of a micromechanical capacitor cannot be found in a closed form and can only be calculated by numerical methods based on the Poisson equation (i.e., $\nabla^2 V = 4\pi\rho$, where ρ is the charge density) and appropriate boundary conditions. However, for some extreme conditions, approximate relations can be found.

For the structure shown in Fig. 4.1.4 with a small distance between the plates ($z \ll a$), the capacitance can be approximated by [1]

$$C(z) \doteq C_o(z)\left\{1 + \frac{z}{2\pi a}\ln\frac{2\pi a}{z} + \frac{z}{2\pi a}\ln\left[1 + \frac{2h}{z} + 2\sqrt{\frac{h}{z} + \frac{h^2}{z^2}}\right]\right\} \tag{4.1.11}$$

where $C_o(z)$ is the capacitance by a parallel-plate approximation given by Eq. (4.1.10).

Eq. (4.1.11) can be written as $C(z) \equiv \beta C_o(z)$, where β is a correction factor of capacitance for the fringe effect, which is dependent on the dimensions of the structure

$$\beta = 1 + \frac{z}{2\pi a}\ln\frac{2\pi a}{z} + \frac{z}{2\pi a}\ln\left[1 + \frac{2h}{z} + 2\sqrt{\frac{h}{z} + \frac{h^2}{z^2}}\right] \tag{4.1.12}$$

Based on Eq. (4.1.12), the dependence of β on the z, h and a is shown by the curves in Fig. 4.1.5. It is found that β can be larger than 1 appreciably. For example, for $z=0.5a$ and $h=0.2a$, we have $\beta=1.30$.

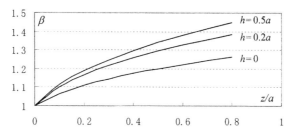

Fig. 4.1.5. Dependence of β on the dimensions of a structure

On the other hand, if the distance between the two bars is quite large ($z \gg a$), the structure can be approximated as two parallel conductive filaments. If the filaments have a circular cross section of radius a and the distance between the two surfaces of the filaments is z, the electrostatic theory [2] gives

$$C(z) \doteq C_o(z)\frac{\pi z}{2a\ln(z/a+2)} \tag{4.1.13}$$

Therefore, $C(z)$ is much larger than $C_o(z)$ for a large distance, but $C_o(z)$ and $C(z)$ are both very small so that the differential capacitance between them might not be significant.

For the same structure shown in Fig. 4.1.4 and assuming that the top plate can only move laterally with a constant normal distance, d, between the two bars, the capacitance by the parallel-plate approximation is

$$C_o(x) = \frac{l(2a - |x|)\varepsilon\varepsilon_o}{d} \qquad (-2a < x < 2a)$$
$$C_o(x) = 0 \qquad\qquad (x < -2a, x > 2a) \tag{4.1.14}$$

where l is the length of the bar and x is the center position of the moving bar. The $C_o(x) \sim x$ relation is shown in Fig. 4.1.6 by the dashed lines. When fringe effects are considered, the capacitance $C(x)$ is generally larger than $C_o(x)$ as schematically shown by the solid curve in Fig. 4.1.6.

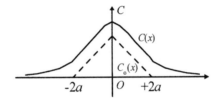

Fig. 4.1.6. Dependence of capacitance on lateral displacement

(2) Electrostatic Force
For a pair of parallel bars as shown in Fig. 4.1.4, from Eq. (4.1.11), the normal force is

$$F_N = \frac{1}{2}\frac{\partial C(z)}{\partial z}V^2 = \frac{1}{2}\left(\beta(z)\frac{\partial C_o(z)}{\partial z} + \frac{\partial \beta(z)}{\partial z}C_o(z)\right)V^2 \tag{4.1.15}$$

where $C_o = A\varepsilon\varepsilon_o/z$. As $\beta(z) \geq 1$ and $\partial\beta/\partial z > 0$, the force is always larger than that by the parallel-plate approximation. If $h \ll z \ll a$, Eq. (4.1.12) can be approximated as

$$\beta = 1 + \frac{h}{\pi a} + \frac{z}{2\pi a}\ln\frac{2\pi a}{z} + \frac{1}{\pi a}\sqrt{zh + h^2}$$

Therefore, Eq.(4.1.15) can be simplified as

$$F_N = \left[\left(1 + \frac{z}{2\pi a} + \frac{h}{\pi a}\right) + \frac{zh + 2h^2}{2\pi a\sqrt{zh + h^2}}\right]\frac{1}{2}\cdot\frac{\partial C_o}{\partial z}V^2 \equiv \gamma\cdot\frac{1}{2}\cdot\frac{\partial C_o}{\partial z}V^2 \tag{4.1.16}$$

where γ is a correction factor of force for the fringe effect.

Based on Eq.(4.1.16), the dependence of γ on z and h is shown in Fig. 4.1.7. For example, for $z=0.3a$ and $h=0.1a$, we have $\gamma=1.12$.

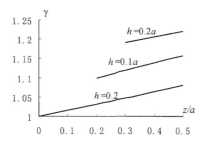

Fig. 4.1.7. The dependence of γ on z and h

For the same structure as shown in Fig. 4.1.4 but with lateral movement of the movable plate, the tangential force is $F_T = \partial C(x)/\partial x \cdot V^2$. With the parallel-plate approximation, the force is uniform in the overlapping region but has opposite signs for left and right sides as shown by the dashed curve in Fig. 4.1.8. The force drops to zero when the top plate moves out of the overlapping region. When fringe effects are considered, the lateral force is related to the slope of the $C(x)$ curve in Fig. 4.1.6. Thus, the force decays gradually when the top plate moves out of the overlapping region, as shown in the solid curve in Fig, 4.1.8.

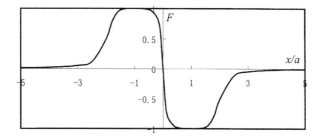

Fig. 4.1.8. The tangential force on the movable plate with the fringe effect

§4.2. Electrostatic Driving of Mechanical Actuators

§4.2.1. Parallel-plate Actuator

(1) Voltage Driving and the Pull-in Effect

Let us consider a parallel-plate actuator with a mass supported by elastic beams (flexures) on both sides so that it can move only in its normal direction as shown in Fig. 4.2.1(a). The mass is used as a movable electrode and the fixed electrode is under the mass with an original gap distance of d. When the electrodes are supplied with a voltage difference, V, an electrostatic force is applied to the mass, pulling it towards the fixed electrode as shown in Fig. 4.2.1(b). Once the mass is displaced, an elastic recovery force by the flexures tends to pull the mass back towards its original position. The balanced position of the mass is simply determined by the condition of force balance. However, the problem is not straightforward due to the nonlinear nature of the electrostatic force, which may cause instability problem in some conditions.

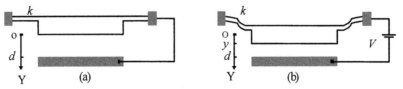

Fig. 4.2.1. A parallel-plate actuator (a) without electrostatic force; (b) with electrostatic force

Suppose that the displacement of the mass is y. Due to the joint action of the electrostatic force and the elastic force, the condition of force balance is

$$F = F_e + F_k = 0$$

where F_e is the electrostatic force and $F_k = -ky$ is the elastic recovery force. The balanced displacement is determined by

$$\frac{A\varepsilon\varepsilon_o V^2}{2(d-y)^2} - ky = 0 \qquad\qquad (4.2.1)$$

First. Eq. (4.2.1) is investigated using a graphic method. The curves of F_e and $|F_k|$ as functions of displacement, y, are shown in Fig. 4.2.2. The curve for F_e is a hyperbola while the curve for F_k is a straight line starting from the origin of the coordinates. If k is large enough, the two curves intersect at points a and b as shown in the figure.

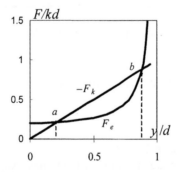

Fig. 4.2.2. The dependence of electrostatic force F_e and elastic force $|F_k|$ on displacement

We can see that the solution corresponding to point b is not a stable state. If a small disturbance moves the mass back a little, the recovery force F_k will be larger than the electrostatic attractive force in quantity and move it back further until it falls to point a [3,4]. On the other hand, if a disturbance moves the mass a little farther from point b, the electrostatic force will always be larger than the elastic recovery force and the mass will move forward continuously until falling into contact with the fixed electrode. However, the solution corresponding to point *a* is a stable one as the mass will always return to the balanced position after a disturbance force moving it away from the point.

From the mathematic condition for a stable state $\partial F/\partial y < 0$, we have

$$\frac{A\varepsilon\varepsilon_o V^2}{(d-y)^3} - k < 0 \qquad (4.2.2)$$

From Eqs. (4.2.1) and (4.2.2), we have

$$y < \frac{1}{3}d \qquad (4.2.3)$$

This means that the balanced displacement is stable when the balanced position of the plate is less than one third of its original distance from the fixed electrode.

For a specific mechanical structure, *k* is a constant. From Fig. 4.2.2, we can see that the curve for F_e moves up with voltage *V*. Therefore, the points *a* and *b* move closer with the increased *V*. It is conceivable that, for a critical voltage V_{po}, points *a* and *b* merge. For any voltage larger than V_{po}, there will be no intersection between the two curves. As, in this case, F_e is always larger than $|F_k|$ for $V > V_{po}$, the mass will always move towards the fixed electrode and falls into contact with the fixed electrode finally. This phenomenon is called the pull-in effect and V_{po} is referred to as a pull-in voltage [4].

The pull-in voltage V_{po} can be found from Eq. (4.2.1) directly. However, for generalization, further discussion will be made using the following dimensionless notations

$$\tilde{y} = \frac{y}{d} \text{ and } p = \frac{F_{eo}}{kd}$$

Thus, Eq. (4.2.1) can be written as

$$\tilde{y}(1 - \tilde{y})^2 = p \qquad (4.2.4)$$

The maximum of $\tilde{y}(1 - \tilde{y})^2$ in the region of 0 to 1 is 4/27 at $\tilde{y} = 1/3$. Therefore, the condition for a stable solution is

$$p \leq \frac{4}{27}$$

The pull-in voltage is thus found to be

$$V_{po} = \sqrt{\frac{8kd^3}{27A\varepsilon\varepsilon_o}} \qquad (4.2.5)$$

For a voltage *V* smaller than V_{po}, the displacement of the mass can be found from Eq. (4.2.4) by iterated calculations. If *p* is small, $\tilde{y}$ increases with *p* quite linearly. However, $\tilde{y}$

increases abruptly with p when p reaches 4/27 (i.e., V reaches the pull-in voltage V_{po}). Once the mass is pulled-in, it would not be released until the voltage is taken away completely (i.e., $V=0$). The dependence of $\tilde{y}$ on p is shown in Fig. 4.2.3.

Now let us consider an example. For a double supported silicon beam-mass structure as shown in Fig. 4.2.1, where the beam width is $b=200\mu m$, the thickness $h=5\mu m$, the length $l=500\mu m$, the area of the mass $A=2mm\times2mm$ and the original distance between the mass and the fixed electrode $d=5\mu m$. Let us find the pull-in voltage and the displacement at $V=5V$.

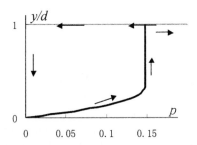

Fig. 4.2.3. The dependence of the normalized displacement $\tilde{y}$ on p

From Eq. (2.2.45), we have $k = 2Ebh^3/l^3 = 68\,N/m$ and $kd = 3.4\times10^{-4}\,N$. From Eq. (4.2.5), we have $V_{po}=8.4V$. For $V=5V$, we have $p=0.0521$. The balanced displacement found from Eq. (4.2.4) by iteration calculation is $\tilde{y}=0.0588$, or $y=0.294\mu m$.

(2) The Elimination of Pull-in Effect by a Series Capacitor

According to the above discussion, the stable displacement of the movable plate in the normal direction by an electrostatic force is limited to one third of the original distance, d, between the two parallel plates. The movable plate will be pulled into contact with the fixed electrode when the voltage reaches a critical value, the pull-in voltage. Even before the pull-in, the displacement increases with p super-linearly. To avoid the pull-in effect and to improve the linearity of the $y\sim p$ relationship, a capacitor, C_S, is inserted in series with the mechanical capacitor [4] as shown in Fig. 4.2.4(a). C_S can be an electric capacitor component of any form.

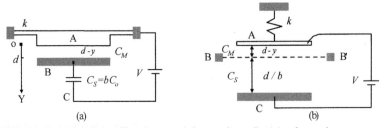

Fig. 4.2.4. Elimination of pull-in effect by a serial capacitor, C_S, (a) schematic structure; (b) effective gap distance mode

If the original distance between the two plates is d, the original capacitance of the parallel-plate capacitor is $C_o = A\varepsilon\varepsilon_o/d$, where A is the area of the plates. With displacement y, the capacitance of the mechanical capacitor between the two plates is

$$C_M = \frac{A\varepsilon\varepsilon_o}{d - y}.$$

If the capacitance of the series capacitor is $C_S = bC_o$, the capacitor (in whatever type of electric components) can be modeled as a parallel-plate capacitor with a plate area of A and a gap distance of d/b. Therefore, capacitor C_M and capacitor C_S can be modeled as a single parallel-plate capacitor with a plate area of A and an effective gap distance of d_{eff}

$$d_{eff} = d + \frac{1}{b}d \tag{4.2.6}$$

However, the movable plate can only move a maximum distance of d (from A to BB' as shown in Fig. 4.2.4), as the space between BB' to C does not physically exist.

Quite often, the substrate is covered with a layer of SiO$_2$. If the capacitor of the SiO$_2$ layer is used for the series capacitor, we have

$$d_{eff} = d + \frac{d_1}{\varepsilon_1}$$

where d_1 is the thickness of the SiO$_2$, ε_1 the relative permittivity of SiO$_2$ and d is the distance between the SiO$_2$ and the movable plate. Or, we have $b = \varepsilon_1 d/d_1$.

According to the effective gap distance model shown in Fig. 4.2.4 (b), the pull-in voltage is

$$V_{po} = \sqrt{\frac{8kd_{eff}^3}{27A\varepsilon\varepsilon_o}} \tag{4.2.7}$$

If the capacitance of C_S is small enough to meet the condition of $d_{eff} > 3d$, "pull-in" will not happen. This requires $b < 1/2$ or, $C_S < C_o/2$.

Thus, a maximum voltage V_m for the movable plate to reach the stationary plate (plate B) is determined by

$$kd = \frac{A\varepsilon\varepsilon_o V_m^2}{2d^2/b^2} \tag{4.2.8}$$

The movement of the mass is stopped by the fixed electrode if the applied voltage V is larger than V_m. From Eq. (4.2.8), voltage V_m is found to be

$$V_m = \frac{1}{b}\sqrt{\frac{2kd^3}{A\varepsilon_o}} \cong \frac{2.6}{b}V_{po} \tag{4.2.9}$$

For example, if $b=1/3$, we have $V_m=7.8V_{po}$. Fig. 4.2.5 shows the $\tilde{y} \sim p$ relationship in the region between $\tilde{y}=0$ to 1 for $b=1/3$. The linearity of $\tilde{y} \sim p$ relation now is much better than that in Fig. 4.2.3. However, the cost for the improvement is the extra voltage drop in capacitor C_S.

If $C_S > C_o/2$, or $b>1/2$, the movable plate is pulled-in at the voltage V_p. Once pulled-in, a minimum voltage is needed to hold the plate in the pull-in state. The voltage is often called the holding voltage and it has the same expression as Eq. (4.2.9).

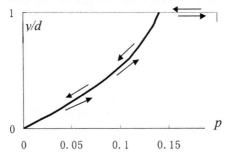

Fig. 4.2.5. The improved linearity of $y \sim p$ relationship (with $b=1/3$)

(3) Charge Driving of Parallel-plate Actuator

As discussed in last section, when a parallel-plate actuator is driven by a constant voltage, the driving voltage has to be restricted to a certain limit so that the balanced displacement does not exceed one third of its original gap distance. Otherwise, the pull-in effect occurs and the system can no longer work properly. To avoid the pull-in effect to occur, some schemes are proposed. One of them is the series capacitor scheme as discussed above. The cost for this scheme is the much higher operation voltage. Now let us discuss the charge driving scheme [3], which may alleviate the high voltage problem significantly.

Suppose that a parallel-plate actuator is charged to a voltage V at its original state in a very short time period and then the power supply is disconnected before the mass can move an appreciable distance as shown in Fig. 4.2.6(a), where C_p is the parasitic capacitance in parallel with the parallel-plate actuator. Then the plate moves under the electrostatic force caused by the charge, as shown in Fig. 4.2.6(b).

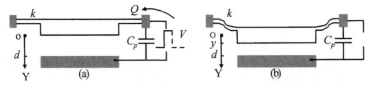

Fig. 4.2.6. Charge driving of a parallel plate actuator

The electric charge stored in the structure is

$$Q = (C_o + C_p)V \tag{4.2.10}$$

where $C_o = A\varepsilon\varepsilon_o/d$. The charge creates an attractive force to make the plate to move. The balanced displacement of the plate y_o should satisfy the following condition

$$ky_o = \frac{A\varepsilon\varepsilon_o Q^2}{2(d - y_o)^2 \left(\dfrac{A\varepsilon\varepsilon_o}{d - y_o} + C_p\right)^2} \tag{4.2.11}$$

If the notations $\tilde{y} = \dfrac{y}{d}$ and $b = \dfrac{C_p}{(C_p + C_o)}$ are used, we have

$$\tilde{y}_o(1 - b\tilde{y}_o)^2 = \frac{A\varepsilon\varepsilon_o Q^2}{2kd^3(C_p + C_o)^2} \tag{4.2.12}$$

If we define $\zeta = b\tilde{y}_o$, the equation for ζ is

$$\zeta(1 - \zeta)^2 = \frac{A\varepsilon\varepsilon_o V^2}{2kd^3} b \tag{4.2.13}$$

We know from Eq. (4.2.13) that the maximum of the left-sided term is 4/27 for $\zeta = 1/3$, or $\tilde{y}_o = 1/3b$. Obviously, if $b < 1/3$ (i.e., $C_p < C_o/2$), we have $\tilde{y}_o > 1$. This means that no pull-in effect occurs in the condition of $C_p < C_o/2$.

In the condition of $b > 1/3$ (or, $C_p > C_o/2$), the plate is pulled-in at a maximum balanced displacement $y_o = \dfrac{d}{3b} = d\dfrac{(C_o + C_p)}{3C_p}$ for a charging voltage of

$$V_{p,q} = \sqrt{\frac{8kd^3}{27A\varepsilon\varepsilon_o} \frac{1}{b}} \tag{4.2.14}$$

As b is always smaller than 1, the maximum balanced displacement for the charge driving scheme is always larger than 1/3. For example, if $C_p = C_o$, we have $y_o = 2d/3$ and the pull-in voltage for the charge driving scheme is $V_{p,q} = \sqrt{2}\, V_{po}$.

It may be mentioned here that though the charge driving scheme has the advantage of lower driving voltage than the series capacitor scheme, the implementation of the scheme is not easy due to the difficulties in charge control and the leakage control.

§4.2.2. Torsion Bar actuator

(1) Angular Displacement

A typical torsion bar actuator consisting of a pair of torsion bars and a rectangular plate is shown in Fig. 4.2.7, where (a) is a top view and (b) a cross sectional view of the structure. An electrode is placed under the plate to the right as shown in the figure. The plate and the torsion bars are usually made of polysilicon about 2 μm thick, suspending over the substrate about 2 μm high. Each torsion bar is attached to the plate on one end and anchored to the substrate on the other end.

If a torque T_e is applied by electrostatic force, the angular displacement of the plate is

$$\varphi = \frac{1}{k_\varphi} T_e \qquad\qquad (4.2.15)$$

where k_φ is the torsion constant of the torsion bars. According to §2.2.5, k_φ is

$$k_\varphi = \frac{2k_1 b h^3 G}{l}$$

where b is the width, h the thickness, l the length of the torsion bar, G the shearing modulus of the material and k_1 the numerical factor related to the ratio of b/h from Table 2.2.1.

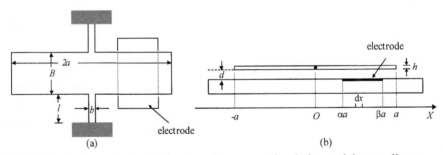

Fig. 4.2.7. A torsion bar actuator (a) top view; (b) cross sectional view and the coordinate

With an angular displacement, a recovery torque $T = -k_\varphi \varphi$ is produced by the torsion bars to balance the applied torque

$$T + T_e = 0 \qquad\qquad (4.2.16)$$

If the voltage applied between the plate and the electrode is V, the torque (for a small φ) caused by the electrostatic force is [5,6]

$$T_e = \int_{\alpha a}^{\beta a} \frac{B\varepsilon\varepsilon_o V^2 x dx}{2(d-\varphi x)^2} = \frac{B\varepsilon\varepsilon_o V^2}{2\varphi^2}\left[\ln(1-\beta a\varphi/d) - \ln(1-\alpha a\varphi/d) + \frac{1}{(1-\beta a\varphi/d)} - \frac{1}{(1-\alpha a\varphi/d)} \right]$$

By using the normalized angular displacement, $\phi = a\varphi / d$, we have

$$T_e = \frac{B\varepsilon\varepsilon_o V^2}{2\phi^2} \frac{a^2}{d^2} \left[\ln\frac{1-\beta\phi}{1-\alpha\phi} + \frac{1}{(1-\beta\phi)} - \frac{1}{(1-\alpha\phi)} \right]$$

(4.2.17)

From Eqs. (4.2.15), (4.2.16) and (4.2.17), the equation for ϕ is

$$\frac{B\varepsilon\varepsilon_o V^2}{2} \frac{a^3}{d^3} \left[\ln\frac{1-\beta\phi}{1-\alpha\phi} + \frac{1}{(1-\beta\phi)} - \frac{1}{(1-\alpha\phi)} \right] - k_\phi \phi^3 = 0$$

(4.2.18)

The relation between V and ϕ is

$$V^2 = \frac{2k_\phi d^3}{B\varepsilon\varepsilon_o a^3} \cdot \frac{\phi^3}{\ln\dfrac{1-\beta\phi}{1-\alpha\phi} + \dfrac{1}{(1-\beta\phi)} - \dfrac{1}{(1-\alpha\phi)}}$$

(4.2.19)

By defining a constant $V_o = \sqrt{2k_\phi d^3 /(B\varepsilon\varepsilon_o a^3)}$, Eq. (4.2.19) can be written as

$$V = V_o \cdot \sqrt{\frac{\phi^3}{\ln\dfrac{1-\beta\phi}{1-\alpha\phi} + \dfrac{1}{(1-\beta\phi)} - \dfrac{1}{(1-\alpha\phi)}}}$$

(4.2.20)

A typical relation between V and ϕ determined by Eq. (4.2.20) is shown by the curve in Fig. 4.2.8. With reference to the figure, the dependence of ϕ on V is discussed as follows.

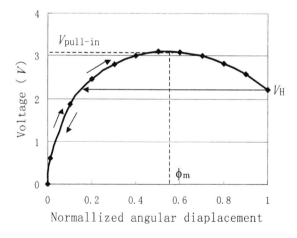

Fig. 4.2.8. Relation between applied voltage V and normalized angular displacement ϕ

For very small voltage, the normalized angular displacement is also very small and it increases with the applied voltage. In this case, Eq. (4.2.20) can be approximated as

$$V = V_o \sqrt{\frac{2\phi}{\beta^2 - \alpha^2}} \qquad (4.2.21)$$

or, in a reverse form,

$$\phi = \frac{\beta^2 - \alpha^2}{2V_o^2} V^2 \qquad (4.2.22)$$

If the voltage is not small, the slope of the curve decreases with the angular displacement, and the voltage reaches a maximum value at a critical angle, ϕ_m. Once the voltage reaches the maximum, the system is no more stable and the plate is pulled into contact with the substrate (i.e., ϕ goes to 1). The pull-in effect of the torsion bar structure is similar to the one we saw in §4.2.1 for a parallel-plate actuator and the maximum voltage is the pull-in voltage, $V_{pull-in}$, for the torsion bar actuator. After the plate is pulled-in, a voltage larger than a critical value V_H is needed to hold the plate in the pull-in state. This value is called the holding voltage. Once the voltage falls below V_H, the plate will be released from the pull-in state as shown by the horizontal arrow in Fig. 4.2.8. By using $\phi=1$ for Eq. (4.2.20), the holding voltage V_H is determined by

$$V_H = V_o \sqrt{\frac{1}{\ln\dfrac{1-\beta}{1-\alpha} + \dfrac{1}{1-\beta} - \dfrac{1}{1-\alpha}}} \qquad (4.2.23)$$

Now let us return to the discussion on the pull-in voltage. From Eq. (4.2.19), and the condition for the maximum voltage, we have

$$3\left(\ln\frac{1-\beta\phi}{1-\alpha\phi} + \frac{1}{1-\beta\phi} - \frac{1}{1-\alpha\phi} \right) - \frac{\beta^2\phi^2}{(1-\beta\phi)^2} + \frac{\alpha^2\phi^2}{(1-\alpha\phi)^2} = 0 \qquad (4.2.24)$$

The solution to the equation is the critical angular displacement for pull-in, ϕ_m. ϕ_m can be found by a numerical calculation. Once ϕ_m is found, the pull-in voltage is known by

$$V_{pull-in} = V_o \cdot \sqrt{\frac{\phi_m^3}{\ln\dfrac{1-\beta\phi_m}{1-\alpha\phi_m} + \dfrac{1}{(1-\beta\phi_m)} - \dfrac{1}{(1-\alpha\phi_m)}}} \qquad (4.2.25)$$

For Simplicity, let us consider the condition of $\alpha=0$, as α is usually small. For $\alpha=0$, the numerical solution to Eq. (4.2.24) is found to be $\beta\phi_m = 0.44$, or, $\phi_m = 0.44/\beta$. As ϕ_m can only fall in between 0 to 1, β should be larger than 0.44 for having a pull-in voltage, i.e., if β>0.44, there exists a real-valued ϕ_m and there is a pull-in voltage. Otherwise, for β<0.44, the pull-in effect can be avoided for the system. Some curves showing the $V{\sim}\phi$ relation for some β values are given in Fig. 4.2.9.

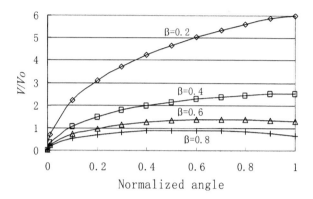

Fig. 4.2.9. The curves of $V \sim \phi$ relation for some β values

By substituting α=0 and $\beta\phi_m = 0.44$ into Eq. (4.2.25), the pull-in voltage is

$$V_{pull-in} = 0.643\beta^{-1.5}V_o \qquad\qquad (4.2.26)$$

This relation is for the condition of α=0, but it is a good approximation when α is small.

(2) Instability Caused by Bias Voltage
    If the electrode on the substrate covers the whole area of the plate of the torsion bar actuator (or the substrate is biased by a voltage), as shown in Fig. 4.2.10, the resultant torque caused by the applied voltage is zero if the plate is at the state of φ=0 due to the symmetry of the structure. Obviously, the voltage should cause no angular displacement. However, further discussion below will show that the state of φ=0 of the plate becomes unstable when the voltage is larger than a critical value.

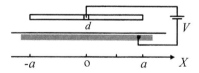

Fig. 4.2.10. A torsion bar structure under a uniform electrostatic force

    To discuss the instability problem, the energy approach is used. The capacitance between the plate and the substrate is a function of the angular displacement of the plate

$$C(\varphi) = \int_{-a}^{a} \frac{B\varepsilon\varepsilon_o dx}{(d - \varphi x)} = \frac{B\varepsilon\varepsilon_o}{\phi} \frac{a}{d} \left[\ln(1+\phi) - \ln(1-\phi) \right] \tag{4.2.27}$$

where B is the width of the plate and ϕ is the normalized angular displacement $\phi = \varphi a/d$. The variation of capacitance is a function of angular displacement

$$\Delta C = C(\varphi) - C_o = C_o \frac{\ln(1+\phi) - \ln(1-\phi) - 2\phi}{2\phi} \tag{4.2.28}$$

where C_o is the capacitance for the state of $\varphi = 0$.

With reference to §4.1 and assuming that the system energy at $\varphi = 0$ is zero, the system energy is

$$E(\varphi) = \frac{1}{2} k_\varphi \varphi^2 - \frac{1}{2} V^2 \Delta C \tag{4.2.29}$$

or,

$$E(\phi) = \frac{d^2}{2a^2} \left(k_\varphi \phi^2 - \frac{a^2}{d^2} V^2 C_o \frac{\ln(1+\phi) - \ln(1-\phi) - 2\phi}{2\phi} \right) \tag{4.2.30}$$

The dependence of $E(\phi)$ on V is schematically shown in Fig. 4.2.11. Some interesting features can be found from the curves.

(i) If $V=0$, according to Eq. (4.2.30), the curve of $E(\phi)$ is simply a parabola with a minimum at $\phi = 0$. It represents a stable balanced state at $\varphi = 0$. This means that the plate will always stay at the $\phi = 0$ state. It will return to this state after any disturbance.

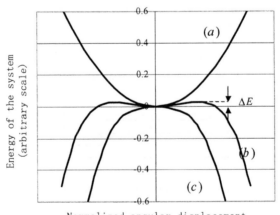

Fig. 4.2.11. The curves of $E(\phi)$: (a) $V=0$; (b) $V_C > V > 0$; (c) $V > V_C$

(ii) For a finite V not very large, the energy of the system still has a minimum at $\varphi=0$. The system energy increases with φ at first, reaches a maximum at a critical angle and then drops continuously when φ exceeds the critical angle. It means that $\varphi=0$ is still a stable balanced position; the plate will return to its balanced position from a disturbance that is not too large. However, the plate will not return to the $\varphi=0$ state if the energy of the disturbance is larger than the energy barrier ΔE. Instead, the plate will continue to turn until it is stopped by the substrate.

(iii) If V is larger than a critical value V_C, the curve of $E(\phi)$ has a maximum at $\varphi=0$. Now the state $\varphi=0$ is not a stable balanced position any more. This means that the plate is pulled-in on one side or the other.

Now let us proceed to some quantitative analyses on the three situations. For a φ close to zero, the variation of capacitance is approximated as

$$\Delta C = C(\varphi) - C_o = C_o\left(\frac{1}{3}\phi^2 + \frac{1}{5}\phi^4 + \cdots\right) \tag{4.2.31}$$

From Eqs. (4.2.29) and (4.2.31), we have

$$E(\varphi) \doteq \left(\frac{d^2}{2a^2}k_\varphi - \frac{1}{6}C_oV^2\right)\phi^2 - \frac{1}{10}C_oV^2\phi^4 \tag{4.2.32}$$

From Eq. (4.2.32) and the condition for balance, $\partial E/\partial\phi = 0$, we have

$$\left(\frac{d^2}{a^2}k_\varphi - \frac{1}{3}C_oV^2\right)\phi - \frac{2}{5}C_oV^2\phi^3 = 0$$

There are three solutions to the above equation

$$\phi_1=0; \quad \phi_{2,3} = \pm\sqrt{\frac{5d^2}{2a^2C_oV^2}\left(k_\varphi - \frac{a^2}{3d^2}C_oV^2\right)} \tag{4.2.33}$$

To examine the stability problem at $\phi=0$, we check the second derivative of $E(\varphi)$

$$\frac{\partial^2 E(\phi)}{\partial\phi^2} = \left(\frac{d^2}{a^2}k_\varphi - \frac{C_oV^2}{3}\right) - \frac{6}{5}C_oV^2\phi^2$$

For solution $\varphi=0$, we have

$$\frac{\partial^2 E(\phi)}{\partial\phi^2} = \frac{d^2}{a^2}\left(k_\varphi - \frac{C_oa^2V^2}{3d^2}\right) \tag{4.2.34}$$

(i) If V is small so that $k_\varphi > C_o a^2 V^2 / 3d^2$, we have $\partial^2 E(\varphi)/\partial\varphi^2 > 0$. $E(\varphi)$ has a minimum at $\varphi = 0$, corresponding to a stable state. Meanwhile, φ_2 and φ_3 correspond to two unstable states.

(ii) If V is large so that $k_\varphi < C_o a^2 V^2 / 3d^2$, i.e., V is large larger than a critical voltage

$$V_{C,\varphi} = \sqrt{\frac{3d^2 k_\varphi}{C_o a^2}} \qquad (4.2.35)$$

we have $\partial^2 E(\varphi)/\partial\varphi^2 < 0$. Thus, $E(\varphi)$ has a maximum at $\varphi = 0$ and the system is not stable.

The above argument shows that if the voltage is larger than the critical value $V_{C,\varphi}$, the state of $\varphi = 0$ is no longer a stable one. Any small disturbance will cause the plate to tilt continuously in one direction or the other until one of its ends lands on the substrate. This instability problem has to be considered for the operation of a torsion bar actuator and has been utilized in the digital mirror devices by TI to maintain the tilted positions of the mirrors when the data in the memory cells are being renewed.

§4.2.3. Comb Drive Actuator

(1) Displacement in Tangential Direction

Comb drive actuators make use of tangential electrostatic forces for driving. As discussed in §4.1.2, a normal electrostatic force always exists along with the tangential force and it is usually much larger than the tangential force. To eliminate the effect of normal force the stationary electrodes are arranged symmetrically on both sides of each movable finger so that the normal forces from both sides cancel out, as shown in Fig. 4.2.12(a).

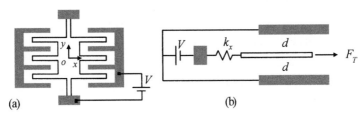

(a)   (b)

Fig. 4.2.12. The operation of a comb-drive actuator (a) schematic structure; (b) simplified model

As the tangential force F_T is independent of displacement, from the model shown in Fig. 4.2.12(b) and Eq. (4.1.9), the equation for the balanced displacement is

$$2\frac{nh\varepsilon\varepsilon_o V^2}{2d} - k_x x = 0 \qquad (4.2.36)$$

where h is the height of the fingers, d the distance between the movable electrode and the stationary electrode, n the number of active fingers and k_x the elastic constant of the flexures in the tangential (x-) direction. Therefore, the balanced displacement of the movable plate is

$$x = \frac{nh\varepsilon\varepsilon_o V^2}{k_x d}$$

(4.2.37)

The displacement of the plate is directly proportional to the square of the applied voltage V.

(2) Instability in Normal Direction

Even though the attractive forces from the stationary electrodes on both sides of a finger are canceled out each other, they may still cause instability in the y-direction. For the instability problem in the y-direction, the model for the structure is modified as shown in Fig. 4.2.13, where A and C are stationary electrodes, B is a movable plate located in between A and C, and k_y is the elastic constant of the flexures in the y-direction. The distance from B to A or C is d. Now we examine the instability problem in the y-direction using the energy approach.

The capacitance between the movable electrode and the two stationary electrodes is

$$C(y) = \frac{A\varepsilon_o}{d+y} + \frac{A\varepsilon_o}{d-y} = \frac{2A\varepsilon_o}{d(1-\widetilde{y})^2}$$

(4.2.38)

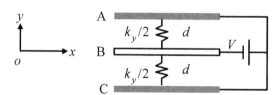

Fig. 4.2.13. The model for the stability problem in the normal direction

By using notations $C_o = 2A\varepsilon_o / d$ and $\widetilde{y} = y / d$, we have

$$C(\widetilde{y}) = \frac{C_o}{1-\widetilde{y}^2}$$

The variation of capacitance on the displacement in the y-direction is

$$\Delta C(\widetilde{y}) = C(\widetilde{y}) - C_o = C_o \frac{\widetilde{y}^2}{1-\widetilde{y}^2}$$

(4.2.39)

With reference to the energy at the original position ($y=0$), the energy of the system is

$$E(\widetilde{y}) = \frac{1}{2}k_y d^2 \widetilde{y}^2 - \frac{1}{2}\Delta C(\widetilde{y})V^2 = \frac{(k_y d^2 - C_o V^2)\widetilde{y}^2 - k_y d^2 \widetilde{y}^4}{2(1-\widetilde{y}^2)}$$

(4.2.40)

The dependence of $E(\tilde{y})$ on V is schematically shown in Fig. 4.2.14 and some interesting features can be found from the curves.

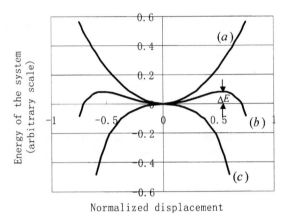

Fig. 4.2.14. The curves of $E(\tilde{y})$ for: (a) $V=0$, (b) $V_C>V>0$, (c) $V>V_C$

    (i) If $V=0$, according to Eq. (4.2.40), the curve of $E(\tilde{y})$ is simply a parabola with a minimum at $\tilde{y}=0$. It represents a stable balanced state at $y=0$. This means that the plate will always return to this state after any disturbance.

    (ii) For a finite V that is not very large, the energy of the system still has a minimum at $y=0$. It increases with y at first, reaches a maximum and then drops continuously when y exceeds a critical displacement. It means that $y=0$ is still a stable balanced position, but the plate will not return to $y=0$ state if the energy of the disturbance is larger than the energy barrier ΔE. Instead, the plate will move and hit the stationary electrode.

    (iii) If V is larger than a critical value V_C, the curve of $E(\tilde{y})$ has a maximum at $y=0$. Now $y=0$ is not a stable position any more. This means that the plate is always pulled-in on one side or the other.

    Now let us proceed to quantitative analyses on the problem. By letting the first order derivative of $E(\tilde{y})$ be zero, we have

$$kd^2\tilde{y} - \frac{\tilde{y}}{\left(1-\tilde{y}^2\right)^2}C_oV^2 = 0 \qquad (4.2.41)$$

    Three solutions to the equation above are

$$\tilde{y}_1 = 0 \text{ and } \tilde{y}_{2,3} = \pm\sqrt{1-\sqrt{\frac{C_oV^2}{kd^2}}} \qquad (4.2.42)$$

As a criterion of instability, we examine the second order derivative of $E(\tilde{y})$

$$\frac{\partial^2 E(\tilde{y})}{\partial \tilde{y}^2} = kd^2 - \frac{1+3\tilde{y}^2}{(1-\tilde{y}^2)^3}C_o V^2$$

For the state of $\tilde{y}_1 = 0$, $\dfrac{\partial^2 E}{\partial \tilde{y}^2} = kd^2 - C_o V^2$. Further discussion can be made as follows.

(i) If $V < \sqrt{kd^2/C_o} = \sqrt{kd^3/2A\varepsilon\varepsilon_o}$, or $\partial^2 E(\tilde{y})/\partial \tilde{y}^2 > 0$, $E(\tilde{y})$ has a minimum at $\tilde{y}_1 = 0$. Thus, $\tilde{y}_1 = 0$ corresponds to a stable state and $\tilde{y}_2$ and $\tilde{y}_3$ are two real-valued solutions corresponding to two unstable balanced positions on both sides.

Take $\tilde{y}_2$ as an example. If plate B is moved a little farther from $\tilde{y}_2 d$, it will move continuously ahead, fall into contact with electrode A and stay there forever. On the contrary, if plate B moves a little back to the center, it will move back to the stable state at $y=0$. The region from where plate B can return to the stable state at $y=0$ is between $\tilde{y}_2 d$ and $-\tilde{y}_2 d$ (i.e., $\tilde{y}_3 d$). Obviously, the larger the voltage V, the smaller the stable region.

(ii) If V is larger than a critical voltage $V_C = \sqrt{kd^2/C_o}$, $E(\tilde{y})$ has a maximum at $\tilde{y}_1 = 0$. Therefore, $\tilde{y}_1 = 0$ corresponds to an unstable state. In this case, $\tilde{y}_2$ and $\tilde{y}_3$ are not real-valued and plate B will always be pulled into contact with one of the stationary electrodes.

Therefore, for stable operation, k_y must be large enough. It does not seem too difficult to design a comb actuator with a large elastic constant, k_y. However, the twist (i.e., rotation) movement in the x-y plane may cause problem, as the elastic constant for rotation movement is relatively small in most comb drive actuator. By referring to Fig. 4.2.12(a), the fingers on one side may move up and the fingers on the other side move down. For the twist movement, the structure can be modeled by Fig. 4.2.15(a) or further simplified as shown in Fig. 4.2.15(b).

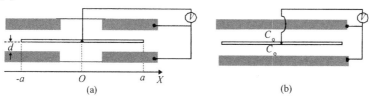

Fig. 4.2.15. Models for the twist movement of a comb drive actuator

With reference to Eq.(4.2.35), the critical voltage that causes instability in the twist movement is

$$V_{C,\varphi} = \sqrt{\frac{3d^2 k_\varphi}{2C_o a^2}} \qquad (4.2.43)$$

where k_φ is the elastic constant for twisting and C_o the nominal capacitance of one side.

§4.3. Step and Alternative Voltage Driving

§4.3.1. Step Voltage Driving

In §4.2.1, the balanced displacement of a movable plate is discussed for a constant driving voltage. The results apply only when the voltage is ramped up to its nominal value slowly or when the structure is heavily damped so that there is no overshooting caused by the electrostatic force.

However, in many micro sensor and actuator applications (such as micro relays and optical switches) movable plates are driven by electrostatic force caused by a step voltage and the damping ratio of the system is much smaller than 1. Therefore, in this section, the dynamic response of a movable plate to a step voltage driving is investigated for the condition of light damping. Here in this section, only parallel plate actuator is considered.

(1) Over Shooting Movement and Pull-in Voltage

Consider a parallel plate actuator as schematically shown in Fig. 4.3.1. When a step voltage V is applied at $t=0$, the electrostatic force applied on the movable plate is

$$F_e(y) = \frac{A\varepsilon_o V^2}{2(d-y)^2}$$

where y is the time-dependent displacement of the movable plate.

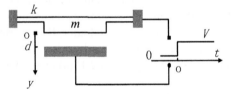

Fig. 4.3.1. A movable plate driven by a step voltage

The balanced positions for the plate are determined by Eq. (4.2.1)

$$\frac{A\varepsilon_o V^2}{2(d-y)^2} - ky = 0$$

If the voltage V is not too large, there are two solutions , y_0 and y_1, to the equation as shown in Fig. 4.3.2, where y_0 is a stable balanced position and y_1 is an unstable balanced position.

For light damping, the movable plate will not settle down at the balanced position, y_0, directly after the voltage is applied. The plate will move toward the balanced position, y_0, pass it with a maximum speed v_m, reach a maximum displacement y_m (assume that y_m does not exceed the unstable balanced position y_1), return and pass the balanced position again, and so forth. With very little damping, the plate oscillates around the balanced position y_0 with diminishing amplitude and settles down at y_0 finally.

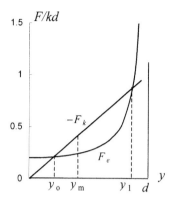

Fig. 4.3.2. Electrostatic and elastic force on the movable plate

Similar to the discussion in §4.2, the plate is pulled-in by the fixed electrode if the step voltage is larger than a critical value, the pull-in voltage of the step voltage driving, which is smaller than the pull-in voltage for the static voltage driving.

In the following, we will discuss the maximum speed of the movable plate when it first passes the balanced position, y_o, the maximum displacement, y_m, the plate can reach and the pull-in voltage for the step voltage driving with light damping. The oscillating frequency around y_o before it settles down there will be addressed in §4.3.2.

(a) The Maximum Speed
If the energy dissipation by damping in a cycle is negligible, we have

$$\frac{1}{2}mv_m^2 = \int_0^{y_o}(F_e + F_k)dy = \frac{A\varepsilon\varepsilon_o V^2}{2(d-y_o)}\frac{y_o}{d} - \frac{1}{2}ky_o^2 \tag{4.3.1}$$

where m is the mass of the plate and v_m is the maximum speed. Thus, we have

$$v_m = \sqrt{\frac{A\varepsilon\varepsilon_o V^2}{m(d-y_o)}\frac{y_o}{d} - \frac{ky_o^2}{m}}$$

Once y_o is calculated from Eq. (4.2.1), v_m can be found easily.

(b) Overshooting Distance
The overshooting distance, y_m, is defined as the maximum displacement of the plate due to overshooting. y_m can be found by

$$\int_0^{y_m}(F_e + F_k)dy = 0 \text{ , or, } y_m(d-y_m) = \frac{A\varepsilon\varepsilon_o V^2}{kd}$$

By using the normalized notations as used in §4.2.1, the equation for $\tilde{y}_m$ is

$$\tilde{y}_m(1-\tilde{y}_m)=2p \qquad (4.3.2)$$

(c) Pull-in Voltage

From Eq. (4.3.2), the maximum value for $\tilde{y}_m(1-\tilde{y}_m)$ is $1/4$ at $\tilde{y}_m=1/2$. Therefore, the voltage for a reasonable solution of y_m is $p<1/8$. This determines that, for a step driving voltage, the pull-in voltage, V_{ps}, is

$$V_{ps}=\sqrt{\frac{kd^3}{4A\varepsilon_o}}=\sqrt{\frac{27}{32}}V_{po}=0.92V_{po} \qquad (4.3.3)$$

where V_{po} is the pull-in voltage for the quasi-static driving discussed in §4.2.1.

(2) Pull-in Time and Release Time

If the voltage of the step signal is larger than the pull-in voltage, V_{ps}, the plate is pulled-in by the electrostatic force. The time needed to pull the plate into contact with the fixed electrode is related to the voltage. According to the principle of energy conservation, the speed of the plate is determined by

$$\frac{1}{2}m\dot{y}^2=\int_0^y F(y)dy=\frac{A\varepsilon_o V^2 y}{2d(d-y)}-\frac{1}{2}ky^2$$

By using the normalized notations, we have

$$\dot{\tilde{y}}=\omega_o\sqrt{\frac{2p\tilde{y}-\tilde{y}^2+\tilde{y}^3}{1-\tilde{y}}}$$

The time needed to pull the plate into contact with the fixed electrode (the "on" time) is

$$t_{on}=\int_0^1\frac{d\tilde{y}}{\dot{\tilde{y}}}=\int_0^1\frac{\sqrt{1-\tilde{y}}\cdot d\tilde{y}}{\omega_o\sqrt{\tilde{y}}\cdot\sqrt{2p-\tilde{y}(1-\tilde{y})}}$$

By using $\zeta=\sqrt{\tilde{y}}$, the turn-on time is

$$t_{on}=\frac{2}{\omega_o}\int_0^1\frac{\sqrt{1-\zeta^2}}{\sqrt{2p-\zeta^2(1-\zeta^2)}}d\zeta\equiv\frac{2}{\omega_o}f(p) \qquad (4.3.4)$$

In the above expression, $f(p)$ is a definite integration. As the maximum value of $\zeta^2(1-\zeta^2)$ is $1/4$, $f(p)$ is real-valued only if $p>1/8$. The dependence of $f(p)$ on p is given in Fig. 4.3.3.

If the applied voltage is large so that $p >> 1/8$, we have the approximation

$$f(p) = \frac{1}{\sqrt{2p}} \int_0^1 \sqrt{1-\zeta^2}\, d\zeta \equiv \frac{\pi}{4\sqrt{2p}}$$

In this case, the turn-on time is inversely proportional to the voltage V

$$t_{on} \doteq \frac{\pi}{\omega_o \sqrt{2p}} = \frac{\pi}{2V} \frac{\sqrt{md^3}}{\sqrt{A\varepsilon_o}} \tag{4.3.5}$$

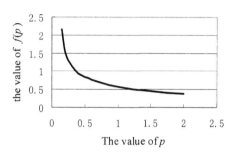

Fig. 4.3.3. The dependence of $f(p)$ on p

Now let us discuss the release time (or, the "off" time), the time needed for the plate to return from the pull-in state to its original position after the voltage is removed. As the initial conditions for the plate are $y = d$, $\dot{y} = 0$ and the inertial potential energy of the plate is $E_p = kd^2/2$ for $t=0$, according to the principle of energy conservation, we have

$$\frac{1}{2}m\dot{y}^2 = \frac{1}{2}kd^2 - \frac{1}{2}ky^2$$

The velocity of the plate as a function of y is

$$\dot{y} = -\sqrt{\frac{k}{m}d^2\left(1-\tilde{y}^2\right)}$$

The expression for the release time (or, the "off" time) is

$$t_{off} = \int_d^0 \frac{dy}{-\sqrt{\dfrac{k}{m}d^2\left(1-\tilde{y}^2\right)}}$$

By letting $\tilde{y} = \sin x$, the definite integration is

$$\int_0^1 \frac{d\tilde{y}}{\sqrt{1-\tilde{y}^2}} = \int_0^{\pi/2} dx = \frac{\pi}{2}$$

Thus, the release time is found to be

$$t_{off} = \frac{\pi}{2}\sqrt{\frac{m}{k}} = \frac{\pi}{2\omega_o} = \frac{1}{4f_o} \qquad (4.3.6)$$

Note that the release time, t_{off}, is only dependent on the natural frequency of the structure.

Now let us examine an example to give the order of magnitude of t_{on} and t_{off}. For a double supported beam-mass microstructure as shown in Fig. 4.3.1, the beam width is 200 μm, the thickness 5 μm, the beam length 500 μm, the area of the mass 4 mm^2 and the thickness 200 μm. The natural vibration frequency of the structure is found to be ω_o=6040/s. If the original gap distance between the mass and the fixed electrode is 5 μm, the pull-in voltage is V_{po}=8.43 V. For a step voltage of 15 V, we find the pull-in time t_{on}=0.27 ms and the release time t_{off}=0.26 ms.

§4.3.2. Negative Spring Effect and Vibration Frequency

(1) Parallel-plate Actuator

As described in §4.3.1, with a step voltage, the movable plate of a parallel-plate actuator oscillates around a new balanced position y_o before it settles down there. With the voltage, the oscillating frequency is also shifted away from its natural vibration frequency. Now let us discuss the dependence of the vibration frequency on the applied voltage.

If the voltage is V, the resultant force on the mass is

$$F = \frac{A\varepsilon\varepsilon_o V^2}{2(d-y)^2} - ky$$

The effective elastic constant (or, the effective spring constant) is

$$k_{eff} = -\frac{\partial F}{\partial y}\bigg|_{y=y_o} = k - \frac{A\varepsilon\varepsilon_o V^2}{(d-y_o)^3} \qquad (4.3.7)$$

The first term on right side of the equation is the mechanical elastic constant and the second term is equivalent to an elastic constant caused by the electrostatic force

$$k_e = -\frac{A\varepsilon\varepsilon_o V^2}{(d-y_o)^3} \qquad (4.3.8)$$

k_e in the above equation is referred to as an "electrical spring constant" or an "electrical elastic constant". As an electric voltage causes an attractive force on the movable plate and the electric force is inversely proportional to the distance, the electrical elastic constant is a negative one. Thus, this effect is also referred to as a "negative spring effect". Due to the negative elastic effect the vibration frequency of the plate is reduced.

From Eqs. (4.3.7) and (4.2.1), the effective spring constant can be found

$$k_{eff} = k + k_e = k\frac{d - 3y_o}{d - y_o}$$

Therefore, the oscillation frequency around the balanced position y_o is

$$\omega'_o = \sqrt{\frac{k_{eff}}{m}} = \omega_o\sqrt{\frac{d - 3y_o}{d - y_o}} \tag{4.3.9}$$

where $\omega_o = \sqrt{k/m}$ is the natural vibration frequency of the plate without an electrostatic force. When the voltage approaches V_{po} so that y_o is close to $d/3$, ω_o' is reduced to zero.

The negative spring effect appears also in other micro structures and may play an important role when k_e is comparable with k. In some devices, negative spring effect is explored for special functions. For example, it has been used for fine-tuning the vibration frequency of a vibratory gyroscope.

(2) Torsion Bar Actuator

(a) Vibration Frequency on Driving Voltage

For the torsion bar structure as shown in Fig 4.2.7, the electrical elastic constant for angular displacement can be found by

$$k_{\varphi,e} = -\left.\frac{\partial T_e}{\partial \varphi}\right|_{\varphi=\varphi_o}$$

where T_e is the torque given by Eq. (4.2.17) and φ_o the balanced angular displacement determined by Eq. (4.2.18). Therefore, the effective elastic constant for angular displacement is $k_{\varphi,eff} = k_\varphi + k_{\varphi,e}$. The vibration frequency of the plate at the balanced position is

$$\omega' = \sqrt{\frac{k_{\varphi,eff}}{I_\varphi}} = \omega_o\sqrt{\frac{k_{\varphi,eff}}{k_\varphi}}$$

Generally, the frequency has to be found by a numerical calculation. Due to the negative value of the electrical spring effect, the vibration frequency decreases with the voltage and the vibration frequency is reduced to zero at pull-in voltage.

(b) Vibration Frequency on Bias Voltage

For a torsion bar structure with a substrate bias as shown in Fig. 4.2.10, the plate stays at its balanced position φ=0, if the bias voltage does not exceed a critical value. However, the vibration frequency at the balanced position φ=0 is still affected by the bias voltage. To discuss the dependence of the vibration frequency on bias voltage V, the results of energy analysis is useful.

According to Eq. (4.2.34), the substrate bias creates a negative electrical elastic constant $k_{\varphi,e} = -C_o a^2 V^2 / 3d^2$ and the effective elastic constant for the torsion bar is

$$k_{\varphi,eff} = k_\varphi - \frac{C_o a^2 V^2}{3d^2} \tag{4.3.10}$$

Therefore, the vibration frequency at the balanced position is

$$\omega' = \sqrt{\frac{k_{\varphi,eff}}{I_\varphi}} = \omega_o \sqrt{1 - \frac{C_o a^2 V^2}{3k_\varphi d^2}} \tag{4.3.11}$$

where $\omega_o = \sqrt{k_\varphi / I_\varphi}$. Obviously, ω' decreases with the bias voltage V and becomes zero when V reaches a critical voltage $V_C = \sqrt{3k_\varphi d^2 / C_o a^2}$. Then, the system is no more stable.

§4.3.3. Alternative Voltage Driving

A variety of micro mechanical sensors and actuators work in an oscillating state, such as resonant pressure sensors, resonant accelerometers, angular rate sensors (gyroscopes), comb resonators, etc. For these devices to work, some method of alternative driving is needed to excite the mechanical structure into a vibration or a resonant state.

Alternative driving is conducted by applying an alternating voltage between the movable plate and the fixed electrodes. The most typical driving schemes are as follows:

(1) Simple Alternative Voltage

If an alternating voltage, $V_1 \sin \omega t$, is applied between a movable plate and a fixed electrode as shown in Fig. 4.3.4, the force applied on the plate is

$$F_e = \frac{A\varepsilon_o}{2d^2} (V_1 \sin \omega t)^2 \tag{4.3.12}$$

where d is the distance between the movable plate and the fixed electrode and A the area of the plates. Eq. (4.3.12) can be written as

$$F_e = \frac{A\varepsilon_o V_1^2}{4d^2} (1 - \cos 2\omega t) \equiv F_o + F_2$$

where F_o is an attractive dc force and F_2 an alternating force component with a radial frequency of 2ω. As the frequency of the driving force is not the same as the frequency of the driving voltage, further control through feedback signal might be difficult. Therefore, this simple driving scheme is rarely employed.

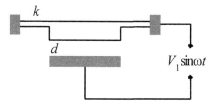

Fig. 4.3.4. Simple alternative voltage driving

(2) Alternative Voltage With a DC Bias

To create an alternative force component with the same frequency as the alternative driving voltage, an alternative voltage with a dc bias can be used as shown in Fig. 4.3.5

$$V = V_o + V_1 \sin \omega t$$

Thus, the electrostatic force is

$$F_e = \frac{A\varepsilon_o}{2d^2} \left(V_o + V_1 \sin \omega t \right)^2 = \frac{A\varepsilon_o}{2d^2} \left[\left(V_o^2 + \frac{V_1^2}{2} \right) + 2V_oV_1 \sin \omega t - \frac{V_1^2}{2} \cos 2\omega t \right]$$

$$\equiv F_o + F_1 + F_2$$

where F_o is a constant force component, F_1 an alternative force component with the same frequency as the driving voltage and F_2 the force component with a double frequency.

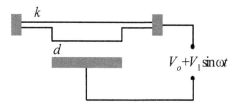

Fig. 4.3.5. Driven by an alternative voltage with a dc bias

If ω is close to the resonant frequency of the structure, ω_o, the effect of F_1 will be much larger than that of F_2 due to the mechanical resonance. As the frequencies of F_2 and F_1 are far apart, the effects of F_2 are small and can be further reduced by an electronic means.

To reduce the effect of the constant force component, the ratio between V_1 and V_o is adjusted for a high ratio of F_1 to F_o. Suppose that $V_1 = \alpha V_o$, the ratio between F_1 and F_o is

$$\frac{F_1}{F_o} = \frac{2V_oV_1}{V_o^2 + V_1^2/2} = \frac{2\alpha}{1 + \alpha^2/2}$$

For a maximum ratio of F_1 to F_o, α has to be $\sqrt{2}$. In this case, we have $V_1 = \sqrt{2}V_o$ and $F_1 = \sqrt{2}F_o = 2\sqrt{2}F_2$. If the voltage driver has a single voltage supply, V_1 is not allowed to exceed V_o for a good sinusoidal waveform. In this case, the condition of $V_1 \cong V_o$ is often used and $F_1 = 1.33F_o = 4F_2$.

(3) Push-pull Driving

For a structure forces can be applied on the movable plate from both sides, a push-pull driving scheme is often considered as the best solution for driving. A comb resonator is a typical structure suitable for a push-pull driving. The scheme is shown in Fig. 4.3.6. The driving voltages on the left side, V_L, and the right side, V_R, are given below

$$V_L = V_o + V_1 \sin\omega t$$
$$V_R = V_o - V_1 \sin\omega t \tag{4.3.13}$$

The resulting driving force on the comb resonator is

$$F = \frac{nh\varepsilon\varepsilon_o}{d}\left(V_L^2 - V_R^2\right) \tag{4.3.14}$$

where n is the number of active fingers, d the lateral distance between the movable fingers and the fixed fingers, and h the height (the thickness) of the fingers.

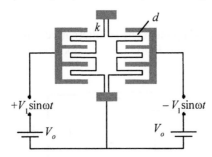

Fig. 4.3.6. A push-pull driving scheme for a comb drive resonator

According to Eqs. (4.3.13) and (4.3.14), the electric driving force is

$$F_e = \frac{4nh\varepsilon\varepsilon_o}{d}V_oV_1 \sin\omega t \tag{4.3.15}$$

The driving force has a single frequency component and the driving force can be adjusted by changing either V_o or/and V_1.

§4.4. Problems

(Note: for all the problems in this chapter, the fringe effect is not considered.)

Problems on Electrostatic Forces

1. A silicon parallel-plate actuator is shown in the figure below. The movable plate is charged to a voltage V quickly at its original position, and soon the voltage supply is disconnected before the plate can move an appreciable distance (Fig.a). Find (1) the expression of the force on the movable plate for a displacement y (Fig.b); (2) the force for $y=2\mu m$. (The nominal gap distance is $d=5\mu m$, the area of plate $A=4mm^2$ and $V=10V$).

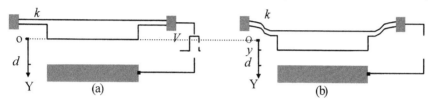

        (a)   (b)

2. For a comb drive actuator, the movable fingers are charged quickly to a voltage V at the original state, and soon the supply voltage is disconnected before the fingers can move an appreciable distance (Fig.a). Find (1) the expression of the tangential force on the movable fingers when the overlapping distance is y_1, as shown in Fig. b; (2) the force for $d=5\mu m$, $y_o=50\mu m$, $y_1=75\mu m$, $H=20\mu m$ (the thickness of the fingers), $n=3$ (the number of the fingers) and $V=10V$.

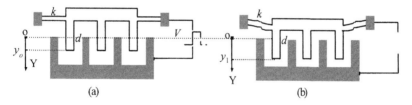

        (a)   (b)

3. An electrostatic levitation system is shown in the figure below. The area of the silicon plate is A and the area of each electrode on top of the silicon plate is $A/2$. (Assuming that the area of the slot between the two electrodes is neglected.) If the silicon plate is kept in parallel with the electrodes by some mechanism, (1) find the balanced distance d between the silicon plate and the electrodes; (2) calculate the value of the balanced distance for $A=4mm^2$, $H=300\mu m$, $\rho=2330kg/m^3$ and $V=100V$; and (3) determine if the balance is stable.

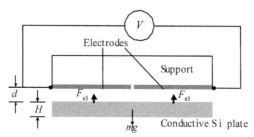

4. If the movable plate of a parallel plate actuator is charged to a voltage V and soon the supply voltage is disconnected before the plate can move an appreciable distance (Fig.a). Due to the electrostatic force caused by the charge on the plate, the plate is displaced as shown in Fig. b. Find the balanced displacement of the plate for d=5µm, V=5V, k=10N/m and the area of the plate A=4mm².

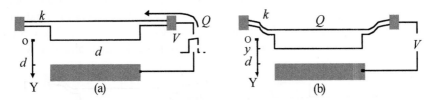

(a)   (b)

5. A comb drive actuator is shown in the figure below. The movable fingers are charged to a voltage V quickly at its original state as shown in Fig. a and soon the voltage supply is disconnected before the fingers can move an appreciable distance. Then the structure goes to a balanced state as shown in Fig. b. Find: (1) the expression of the balanced displacement y_1; (2) the value of y_1 for H=20µm (the thickness of the fingers), n=3 (the number of the fingers), d=2µm, V=10V, k=0.001N/m and y_o=20µm.

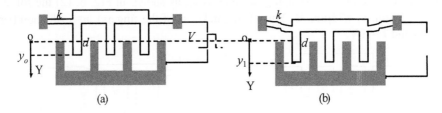

(a)   (b)

Problems on Electrostatic Driving

6. A silicon beam-mass structure is shown in the figure below, where l=500µm, b=100µm, h=10µm, A=B=2mm, d=4µm. Find (1) the maximum balanced displacement y_{max}; (2) the pull-in voltage V_{po}. (E=1.7×10¹¹ Pa)

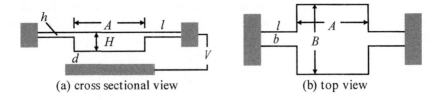

(a) cross sectional view   (b) top view

7. A silicon beam-mass structure is shown in the figure below, where A =2.4mm, B = 1.2mm, b_1 =50µm, b_2 =100µm, l_1 =100µm, l_2 =150µm, h =5µm, d =3µm, H =250µm. Find the pull-in voltage V_{po}.

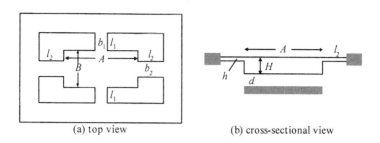

(a) top view   (b) cross-sectional view

8. A silicon beam-mass structure as shown in the figure below, where l=500μm, b=100μm, h=10μm, A=B=2mm, d_o=4μm and d_1=1μm (SiO$_2$). Find (1) the maximum balanced displacement y_{max}; (2) the pull-in voltage V_{po}; (3) the holding voltage V_H. (E=1.7×10^{11} Pa, ρ=2330kg/m^3 and, for SiO$_2$, ε_1=3.8.)

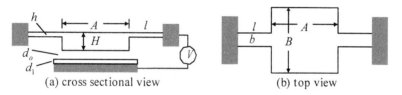

(a) cross sectional view   (b) top view

9. A silicon beam-mass structure is shown in the figure below, where l=500μm, b=50μm, h=10μm, A=B=2mm and d=3μm. If C_S=25pF, find (1) the maximum stable balanced displacement y_{max}; (2) the pull-in voltage V_{po}; and (3) the holding voltage V_H.

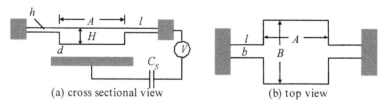

(a) cross sectional view   (b) top view

10. A silicon structure is shown in the figure below, where b=50μm, h=10μm, l=500μm, A=B=2mm, d_o=3μm and d_1=1μm. If ε_1=3.8 and C_S =100pf, find: (1) the maximum balanced displacement y_{max}; (2) the pull-in voltage V_{po}; and (3) the holding voltage V_H.

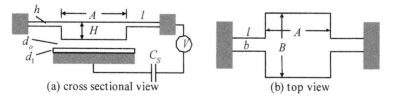

(a) cross sectional view   (b) top view

11. A silicon torsion bar structure is shown in the figure below, where b= 6μm, h=2μm, l=10μm, B=400μm, $2a$=800μm, α=0, β=0.8 and d=2μm. Find (1) the angular displacement

of the plate if the voltage applied is $V=2.5$V, and (2) the maximum stable angular displacement and the corresponding voltage (the pull-in voltage).

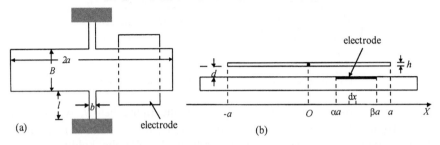

(a)   electrode   (b)

12. A silicon torsion bar structure is shown in the figure below, where $b=10\mu$m, $h= 2\mu$m, $l=20\mu$m, $B=400\mu$m, $2a=800\mu$m and $d=2\mu$m. Find the critical voltage that causes the instability of the plate. ($G=0.65\times10^{11}$ Pa)

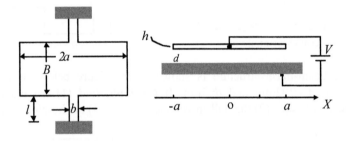

13. A silicon torsion bar structure is shown in the figure below, where $b=10\mu$m, $h= 2\mu$m, $l=20\mu$m, $B=400\mu$m, $2a=800\mu$m, $d_o=2\mu$m and $d_1= 0.7\mu$m. Find the critical voltage that causes the instability of the plate. ($G=0.65\times10^{11}$ Pa and $\varepsilon_1=3.8$)

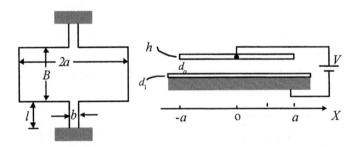

Problems on Step and Alternative Driving

14、A silicon beam-mass structure is shown in the figure below, where $l=500\mu$m, $b=100\mu$m, $h=10\mu$m, $A=B=2$mm, $H=300\mu$m and $d=4\mu$m. If the bias voltage is 5V, find the free vibration frequency at its balanced position.

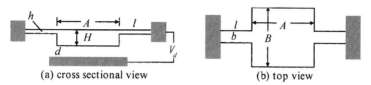

(a) cross sectional view   (b) top view

15、A silicon beam-mass structure is shown in the figure below, where l=500μm, b=100μm, h=10μm, A=B=2mm, H=300μm, d_o=4μm and d_1=1μm。If ε_1=3.8 and V= 6V, find the free vibration frequency at its balanced position.

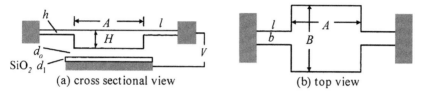

(a) cross sectional view   (b) top view

16. A silicon beam-mass structure is shown in the figure below, where b=50μm, h=10μm, l=500μm, A=B=2mm, H=200μm and d=3μm. If C_S=20pF and V=5V, find the vibration frequency of the mass at its balanced position.

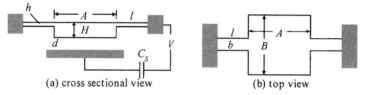

(a) cross sectional view   (b) top view

17. A beam-mass structure is shown in the figure below. The area of the mass is A, the mass of the central plate is m and the elastic constant of the beams is k and the voltage applied is V. Find the expression for the vibration frequency of the mass.

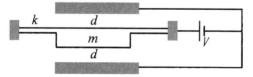

18. A silicon torsion bar structure is shown in the figure below, where h=1.2μm, l=20μm, b=6μm, B=200μm, $2a$=400μm, d=2μm and V=2V. Find the frequencies of angular vibration of the plate at its balanced position.

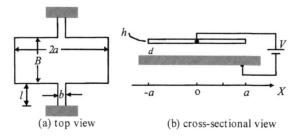

(a) top view   (b) cross-sectional view

19. A silicon torsion bar structure is shown in the figure below, where h=1.2μm, l=20μm, b=6μm, B=200μm, $2a$=400μm, d_o=2μm, d_1=1.0μm (SiO$_2$) and V=2V. Find the frequency of angular vibration of the plate at its balanced position. (ε_1=3.8 for SiO$_2$ and G=0.65×10^{11} Pa for Si)

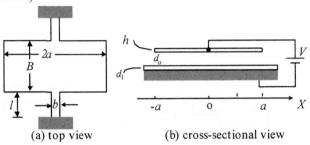

    (a) top view   (b) cross-sectional view

References

[1] H. Yang, "Microgyroscope and Microdynamics", (Ph. D. Dissertation, December, 2000)

[2] D. Halliday, R. Reswick, "Physics (Part II)", John Wiley, 1966

[3] R. Puers, D. Lapadatu, Electrostatic forces and their effects on capacitive mechanical sensors, Sensors and Actuators A56 (1996) 203-210

[4] J. Seeger, S. Crary, Stabilization of electrostatically actuated mechanical devices, Digest of Technical Papers, The 9th Intl. Conf. on Solid-State Sensors and Actuators, Chicago, IL, USA, June 16-19, 1997 (Transducers'97) 1133-1136

[5] O. Degani, E. Socher, A. Lipson, T. Leitner, D. J. Setter, S. Kaldor, Y. Nemirovsky, Pull-in study of an electrostatic torsion microactuator, Journal of Microelectromechanical Systems, Vol. 7 (1998) 373-379

[6] X. M. Zhang, F.S. Chau, C. Quan, Y.L. Lam, A.Q. Liu, A study of the static characteristics of a torsional micromirror, Sensors and Actuators, A 90 (2001) 73-81

Chapter 5

Capacitive Sensing and Effects of Electrical Excitation

Though capacitive sensing has some successful applications in conventional transducers for industry, it suffers from high electromagnetic interference and the increased complexity of the measurement electronics for miniaturized structures. However, capacitive sensing has many attractive features for MEMS: in most micro machining technologies, minimal additional process is needed, capacitors operate both as sensors and actuators, excellent sensitivity has been demonstrated, the transduction mechanism is intrinsically insensitive to temperature, the sensitivity of the sensor keeps constant with the scaled-down of the structure, etc [1,2]. Thanks to the rapid progress in sensing techniques and the integration of micro mechanical structures with microelectronics, capacitive sensing has gained its dominating position in MEMS.

The theory governing capacitive sensing is no more than the well-established electrostatics. However, special considerations have to be made for its applications:

(i) As the capacitance of a MEMS capacitor is usually small, the effects of stray capacitance and parasitic capacitance are relatively large. Therefore, special attention has to be pay to these effects in the design of sensing structure and circuitry.

(ii) As micro mechanical capacitors operate both as sensors and actuators, the electrical excitation signal for capacitive sensing changes the capacitance being measured. The effect interferes with the measurement and reduces the signal level that lead to the "pull-in" failure of the capacitive sensors.

In §5.1 of the chapter, some basic methods of capacitive sensing in MEMS are studied first. In §5.2, the effects of excitation signal on capacitive sensing are analyzed for a quasi-static signal condition. Then, in §5.3 and §5.4, the effects of excitation signal on capacitive sensing are analyzed for dynamic signals (the step signal and the pulse signal). As dynamic signals for a sensing system with light damping represent the worst working condition, the results in §5.3 and §5.4 are of practical importance and preferred for design consideration of MEMS devices.

§5.1. Capacitive Sensing Schemes

§5.1.1. DC Bias Sensing

The simplest method for capacitor sensing is the dc bias sensing scheme as illustrated in Fig. 5.1.1, where V_B is the dc bias voltage, C_{ac} a capacitor of a varying capacitance and R_B a bias resistor of large resistance. As this kind of sensing method can only sense the variation of capacitance, it has the applications in silicon capacitive microphones [3,4]. The working principle is described as follows.

According to §1.3.3, a capacitive microphone is essentially a parallel-plate capacitor with one plate supported by flexures. The plate can vibrate with the sound pressure. The ratio between the displacement Δx of the plate and the sound pressure Δp on the plate is defined as the mechanical sensitivity of the microphone

$$S_m = \frac{\Delta x}{\Delta p} = \frac{A}{k}$$

where k is the effective elastic constant of the flexures and A the effective area of the movable plate. For a nominal gap distance of d between the plates, the nominal capacitance is $C_o = A\varepsilon\varepsilon_o/d$. If the plate oscillates with amplitude δ and a radial frequency ω, the capacitance of the mechanical capacitor is

$$C_{ac} = \frac{A\varepsilon\varepsilon_o}{d + \delta\sin\omega t} \doteq C_o(1 - \frac{\delta}{d}\sin\omega t)$$

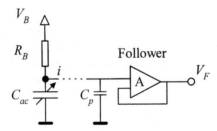

Fig. 5.1.1. DC bias sensing scheme for a microphone

If the time constant $R_B C_o$ is much larger than $1/\omega$, the electric charge stored in the capacitor keeps constant. The voltage V_B' at point i on the capacitor is determined by

$$C_o V_B = (C_o - \frac{\delta}{d} C_o \sin\omega t) V_B'$$

Therefore, we have

$$V_B' \cong V_B(1 + \frac{\delta}{d}\sin\omega t) = V_B + v_{ac}$$

where $v_{ac} = (V_B/d)\cdot\delta\sin\omega t \equiv v_{ac,i}\sin\omega t$ is the alternative component of the voltage signal on the microphone (at point i), representing the sound pressure. Thus, $S_e = V_B/d$ is referred to as the open-circuit electrical sensitivity of the microphone. Thus, the overall open circuit sensitivity of the microphone is

$$S_{open} = S_e S_m = \frac{AV_B}{kd} = \frac{v_{ac,i}}{\Delta p}$$

Signal v_{ac} is usually converted to a low impedance signal via a follower as shown in Fig. 5.1.1 before it is further amplified. In the process, the signal will be attenuated due to the existence of the parasitic capacitance C_p and the input capacitance C_i of the follower. The sensitivity at the output of the follower is

$$S_F = \frac{C_o}{C_o + C_p + C_i} S_{open} \tag{5.1.1}$$

It may be mentioned here that the open-circuit electrical sensitivity $S_e = V_B/d$ is directly proportional to the bias voltage V_B. A higher sensitivity can be achieved by a higher bias voltage. However, the value of the bias voltage is restricted by the pull-in voltage of the parallel-plate structure of the microphone. For reliable operation, V_B should be smaller than the pull-in voltage $V_{pull\text{-}in}$ by a certain ratio. If $V_B = \alpha V_{pull\text{-}in}$, the open circuit sensitivity is

$$S_{open} = \frac{\alpha A V_{pull-in}}{kd} = \alpha \sqrt{\frac{8Ad}{27k\varepsilon\varepsilon_o}}$$

This equation is useful in designing a micro mechanical capacitive microphone.

§5.1.2. Diode-quad Sensing

Suppose that the capacitance of a sensing capacitor is C_S. Usually, the variation of the capacitance is a small fraction of its nominal value. Therefore, for sensing purpose, C_S is compared with a reference capacitor C_R, which is close to C_S. The diode-quad sensing scheme [5,6] for comparing C_S with C_R is shown in Fig. 5.1.2, where $D_1 \sim D_4$ are diodes with identical characteristics, and C_C is coupling capacitors, which is much larger than C_S or C_R. The excitation signal is a square wave with amplitude of $\pm V_P$ and a frequency much higher than natural frequency of the mechanical structure.

(1) Capacitance to Voltage Conversion

The alternative component of the excitation signal enters point A and B via C_C with little attenuation, as C_C is much larger than C_R and C_S. When the excitation voltage is $+V_P$, capacitor C_S is charged from point B to point C via diode D_2 while capacitor C_R is charged from point A to point D via diode D_3. When the excitation voltage is $-V_P$, capacitor C_S discharges from C to A via diode D_1 and capacitor C_R discharges from point D to point B via diode D_4. Due to the rectifying effect of the diodes, in a cycle of excitation signal, a certain amount of charge is transferred from point B to point A via the route of B-C-A and, in the meantime, a certain amount of charge is transferred from A to B via the route of A-D-B. There would be no net charge transfer between points B and A if C_S is exactly equal to

C_R. Since C_S is different from C_R, there is a net charge transfer between points A and B, and a voltage difference between points A and B develops.

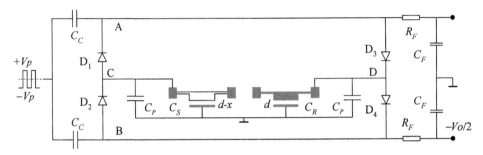

Fig. 5.1.2. Capacitive sensing by a diode-quad rectifier

Suppose that the final voltage at A is $V_o/2$ and the voltage at B is $-V_o/2$ after many cycles of driving. The charge transferred from point B to A in one cycle is

$$\Delta Q_{BA} = 2\left(V_P - \frac{1}{2}V_o - V_F\right)\cdot(C_S + C_P)$$

where V_F is the on-state voltage drop of the diode and C_P is the parasitic capacitance in parallel with C_S. Similarly, the charge transferred from A to B in one cycle is

$$\Delta Q_{AB} = 2\left(V_P + \frac{1}{2}V_o - V_F\right)\cdot(C_R + C_P)$$

At balance, we have $\Delta Q_{BA}=\Delta Q_{AB}$. Therefore, the open circuit output voltage is

$$V_o = \frac{2(V_P - V_F)\cdot(C_S - C_R)}{C_S + C_R + 2C_P} \tag{5.1.2}$$

Obviously, the output voltage cannot exceed $2V_F$.

The apparent advantages of the sensing circuit are: (1) The transducer is conveniently grounded; and (2) the circuit allows for a differential output as well as a single-ended output, an advantage for common mode rejection of the carrier voltage.

It may be mentioned that the voltage drop on diodes, V_F, reduces the effective excitation voltage to $(V_P - V_F)$ and the temperature dependence of V_F causes temperature drift. To eliminate these effects, analog switches are used to replace the diodes [1].

(2) Nonlinearity

For a displacement x of the movable plate, we have $C_S=dC_o/(d-x)$. If $C_R=C_o=A\varepsilon\varepsilon_o/d$, the output voltage is

$$V_o = \frac{2(V_P - V_F) \cdot C_o \tilde{x}}{2C_o + 2C_P - (C_o + 2C_P)\tilde{x}} \cong (V_P - V_F)\frac{C_o \tilde{x}}{C_o + C_P}\left(1 + \frac{C_o + 2C_P}{2C_o + 2C_P}\tilde{x}\right)$$

where $\tilde{x} = x/d$. If the ratio of C_P to C_o is designated as η, we have

$$V_o \cong (V_P - V_F)\frac{\tilde{x}}{1+\eta}\left(1 + \frac{0.5+\eta}{1+\eta}\tilde{x}\right) \qquad (5.1.3)$$

From Eq. (5.1.3), we conclude that the larger the parasitic capacitor C_P (i.e., the larger η), the smaller the sensitivity, and the larger the nonlinearity. For the definition of nonliniatlity commonly used, readers are referred to §6.6.3.

A common problem for the capacitive measurement is the side effect of the electrostatic force caused by the excitation signal. For the diode-quad measurement scheme shown in Fig. 5.1.2, the voltage on the capacitor is V_p–V_F. The electrostatic force causing the capacitance to change is $\Delta F \approx A\varepsilon_o (V_P - V_F)^2 / 2d^2$. This will introduce error, which could be large. This effect will be discussed in §5.2 and §5.3.

To reduce electrostatic force, a symmetric structure for differential capacitance measurement is shown in Fig. 5.1.3. The electrostatic forces on both sides of the movable electrodes cancel with each other mostly as shown in the equation below

$$\Delta F = \frac{A\varepsilon_o(V_P - V_F)^2}{2d^2}\left[\frac{1}{(1-\tilde{x})^2} - \frac{1}{(1+\tilde{x})^2}\right] \approx \frac{A\varepsilon_o(V_P - V_F)^2}{2d^2} \cdot 4\tilde{x} \cdot (1 + 2\tilde{x}^2) \qquad (5.1.4)$$

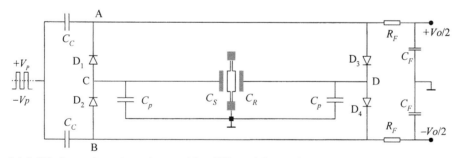

Fig. 5.1.3. Diode-quad sensing scheme with a differential capacitance

An additional advantage of the differential sensing scheme is the significantly improved linearity, as C_S and C_R change in opposite directions [7]. For a displacement, x, of the movable plate, we have

$$C_S = C_o\frac{1}{1-\tilde{x}} , \quad C_R = C_o\frac{1}{1+\tilde{x}}$$

By substituting them into Eq. (5.1.2), the open circuit output voltage is

$$V_o = \frac{2(V_P - V_F) \cdot C_o}{C_o + C_P(1 - \tilde{x}^2)} \tilde{x}$$

For small displacement, the output is approximated as

$$V_o \cong 2(V_P - V_F) \frac{\tilde{x}}{1+\eta}\left(1 + \frac{\eta}{1+\eta}\tilde{x}^2\right) \tag{5.1.5}$$

From Eq. (5.1.5), we conclude that: (i) The sensitivity is doubled for the differential sensing when compared with the single-sided sensing; (ii) The nonlinearity is reduced significantly due to the absence of quadric term in Eq. (5.1.5); and (iii) The nonlinearity caused by parasitic capacitance C_P is reduced, especially for $C_P \le C_o$.

(3) Loading Effect

Now let us consider a diode-quad circuit with outside loading as shown in Fig. 5.1.4 and discuss the dependence of loading effect on the excitation frequency [6]. According to the working principle described above, the net current flowing from B to A is

$$\Delta I_{BA} = I_{BA} - I_{AB} = [2 \cdot (V_p - V_F) \cdot (C_S - C_R) - V_o \cdot (C_S + C_R + 2C_p)] \cdot f$$

where f is the frequency of the excitation signal. Meanwhile, the current on R_L is

$$I_L = \frac{V_o}{2(R_F + R_L)}$$

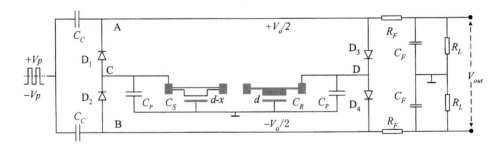

Fig. 5.1.4. The diode-quad circuit with outside loading

For balance, these two currents should be the same. Therefore, we have

$$V_o = \frac{2(V_p - V_F) \cdot (C_S - C_R)}{C_S + C_R + 2C_p} \cdot \frac{2(R_F + R_L) \cdot (C_S + C_R + 2C_p) \cdot f}{1 + 2(R_F + R_L) \cdot (C_S + C_R + 2C_p) \cdot f} \tag{5.1.6}$$

And, the output voltage on the outside loading resistors is

$$V_{out} = \frac{R_L}{R_L + R_F} V_o$$

Eq. (5.1.6) implies that the output impedance of the diode-quad circuit (excluding R_F, C_F and R_L) is

$$R_{out} = \frac{1}{2(C_S + C_R + 2C_p) \cdot f}$$

The fact that the output impedance of the diode-quad is inversely proportional to the excitation frequency is found true for most capacitive sensing.

In fact, the output voltage increases with the excitation frequency at low frequency and levels off at higher frequencies in according with Eq. (5.1.6). The output drops again for very high frequencies as the diodes do not work properly then. For these reasons, the frequency of the excitation signal is usually in the range of 100kHz to 10 MHz.

§5.1.3. Opposite Excitation Sensing

With a differential sensing structure, the displacement of the movable plate can be measured using an opposite excitation scheme as shown in Fig. 5.1.5, where C_P is the parasitic capacitance and C_i is the input capacitance of the buffer amplifier. The measurement of capacitance is made by the two equal amplitude, out-of-phase sinusoidal excitation signals, $+V_R$ and $-V_R$, with the same frequency ω [8]. When C_R and C_S are exactly the same, the input voltage of the amplifier is zero. If C_S and C_R are different, the input voltage, V_i, of the buffer amplifier can be found by the condition of zero input current of the amplifier

$$(V_R - V_i) \cdot C_S \omega = (V_R + V_i) \cdot C_R \omega + V_i (C_P + C_i) \cdot \omega$$

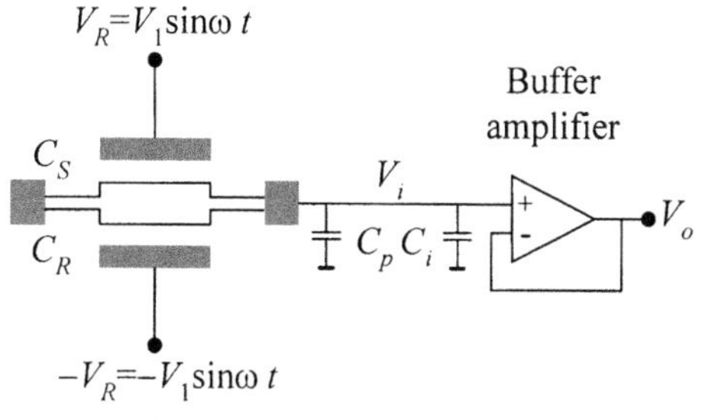

Fig. 5.1.5. Opposite excitation sensing scheme

Thus, the signal voltage on the movable plate is

$$V_i = \frac{C_S - C_R}{C_S + C_R + C_P + C_i} V_R$$

For a small displacement x, we have

$$V_i = \frac{2C_o V_R}{2C_o + (C_p + C_1)(1 - \tilde{x}^2)} \tilde{x} \tag{5.1.7}$$

As V_i is dependent on C_P and C_i, the accuracy and stability of the measurement deteriorates. To alleviate this problem, a feedback scheme can be used as shown in Fig. 5.1.6, where C_{FB} is a feedback capacitor. By using the condition of current balance as before, we have

$$(V_R - V_i) \cdot C_S \omega = (V_R + V_i) \cdot C_R \omega + V_i \omega \cdot (C_P + C_i) + (V_i + A_1 V_i) \cdot C_{FB} \omega$$

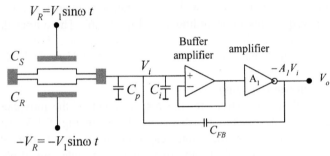

Fig. 5.1.6. Opposite excitation with capacitive feedback

The signal voltage on the movable plate is

$$V_i = \frac{(C_S - C_R) \cdot V_R}{C_S + C_R + C_P + C_i + (1 + A_1) \cdot C_{FB}}$$

where A_1 is the open-loop gain of the amplifiers A_1. The output of the circuit is

$$V_o = -A_1 V_i = \frac{-A_1(C_S - C_R) \cdot V_R}{C_S + C_R + C_P + C_i + (1 + A_1) \cdot C_{FB}} \tag{5.1.8}$$

If A_1 is large enough, we have the approximate result

$$V_o \doteq -\frac{C_S - C_R}{C_{FB}} V_R = -\frac{2C_o \tilde{x}}{C_{FB}(1 - \tilde{x}^2)} V_R \tag{5.1.9}$$

The output V_o is linearly dependent on the differential capacitance, $C_S - C_R$, but not dependent on the parasitic capacitance C_p.

§5.1.4. Force-balanced Sensing

Suppose that the feedback capacitor, C_{FB}, in Fig. 5.1.6 is replaced by a large-valued resistor, R_{FB}, so that a dc voltage can be fed back through the resistor. In the meantime, two equal-valued but opposite-signed dc voltages $+V_o$ and $-V_o$ are added to the excitation signals in the top and bottom, respectively [9], and a synchronous demodulator is inserted between the buffer and the amplifier A_1, as shown in Fig. 5.1.7. As the dc feedback voltage changes the electrostatic forces between the electrodes, the feedback is an electromechanical one. Therefore, the measurement is not a pure electrical measurement for capacitance, but a close-looped electro-mechanical operation based on the capacitive sensing that gives result corresponding to the measurand. The operation is described as follows.

Originally, the movable plate is at its balanced position between two fixed plates. The gap distances on both sides are d. In this case, the excitation signal creates no signal at the input of amplifier A_o ($V_i=0$) and no net force on the central plate. If the movable plate is then forced to move up (i.e., in a positive direction) for a distance x by the action of a measurand (for example, an upward inertial force, ma, caused by a downward acceleration, a), the alternative signal V_i appearing at the input of amplifier A_o (for small C_p and C_i) is

$$V_i = \tilde{x} V_1 \sin \omega t \tag{5.1.10}$$

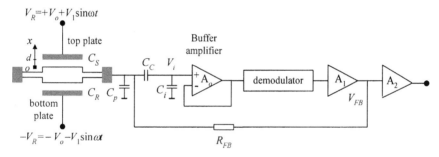

Fig. 5.1.7. Force-balanced sensing technique

This signal is amplified and processed to feedback a dc voltage, V_{FB}, to the input via R_{FB}. V_{FB} is

$$V_{FB} = +A_{op} V_1 \tilde{x} \tag{5.1.11}$$

where A_{op} is the open loop gain resulting from the buffer, the demodulator and the operational amplifier, A_1. Supposing that A_{op} is positive and large in value. The positive dc voltage at the central plate differentiates electrostatic forces from the top plate and the bottom plate. The net electrostatic force on the movable plate caused by the feedback is negative (i.e., in the direction opposite to the inertial force, ma).

Therefore, the displacement of the central plate is reduced by the electromechanical feedback and the reduction is significant if A_{op} is large enough. In fact, the central plate stays almost at its original balanced position. As the displacement of the central plate is very small, the measurand, a, is measured by the feedback voltage, which creates an electrostatic force to balance the external force *ma*. Therefore, this measurement technique is referred to as a force-balanced sensing scheme. In essence, however, the measurements is made through the capacitive sensing.

Now let us establish the relation between the feedback voltage and the measurand, *a*. For simplicity, let us assume that V_1 is much smaller than V_o so that the electrostatic force caused by the high frequency component is negligible, the electrostatic force applied on the movable plate is

$$F_e = \frac{A\varepsilon\varepsilon_o}{2d^2}\left[\frac{\left(V_o - A_{op}V_1\tilde{x}\right)^2}{\left(1-\tilde{x}\right)^2} - \frac{\left(V_o + A_{op}V_1\tilde{x}\right)^2}{\left(1+\tilde{x}\right)^2}\right]$$

where A is the area of the plates. By using the notations of $V_1 = \alpha V_o$, we have

$$F_e = \frac{A\varepsilon\varepsilon_o V_o^2}{2d^2}\left[\frac{\left(1-\alpha A_{op}\tilde{x}\right)^2}{\left(1-\tilde{x}\right)^2} - \frac{\left(1+\alpha A_{op}\tilde{x}\right)^2}{\left(1+\tilde{x}\right)^2}\right]$$

For $\alpha A_{op}\tilde{x} << 1$ and $\alpha A_{op} >> 1$, we have

$$F_e \doteq -\frac{2A\varepsilon\varepsilon_o V_o^2 \alpha A_{op}\tilde{x}}{d^2}$$

The displacement is determined by the equation

$$ma - \frac{2A\varepsilon\varepsilon_o V_o^2 \alpha A_{op}}{d^3}x - kx = 0 \tag{5.1.12}$$

Therefore, we have

$$\tilde{x} = \frac{ma}{kd + 2A\varepsilon\varepsilon_o V_o^2 \alpha A_{op}/d^2}$$

If A_{op} is large enough so that kd in the denominator can be neglected, $\tilde{x}$ is

$$\tilde{x} = \frac{d^2 ma}{2A\varepsilon\varepsilon_o V_o^2 \alpha A_{op}}$$

Obviously, $\tilde{x}$ is small. From Eq. (5.1.11), the feedback voltage V_{FB} is

$$V_{FB} = \frac{d^2 ma}{2A\varepsilon_o V_o} \qquad (5.1.13)$$

As V_{FB} is in direct proportion to ma, it can be used as a measure of the acceleration.

With the electrostatic force, the equation for the damped vibration of the mass is

$$m\ddot{x} + c\dot{x} + kx + \frac{2A\varepsilon\varepsilon_o V_o^2 \alpha A_{op}}{d^3} x = 0 \qquad (5.1.14)$$

This means that the system has an effective elastic constant k_{eff}

$$k_{eff} = k + \frac{2A\varepsilon\varepsilon_o V_o^2 \alpha A_{op}}{d^3}$$

Therefore, the resonant frequency of the system is increased by the electro-mechanical feedback. The frequency of the system is now $\omega_o' = \sqrt{k_{eff}/m}$ instead of $\omega_o = \sqrt{k/m}$. While the resonant frequency is increased by the electro-mechanical feedback, the damping ratio of the system is reduced to $\zeta' = \zeta\omega_o/\omega_o'$.

§5.1.5. Switched Capacitor Sensing

Switched capacitor sensing scheme is a useful method in measuring small capacitance [1, 10,11,12]. The basic principle of switched capacitor sensing is explained below with reference to Fig. 5.1.8, where switches S_1 and S_2 are usually CMOS analog switches, V_R is a reference voltage and A an ideal operation amplifier without offset voltage.

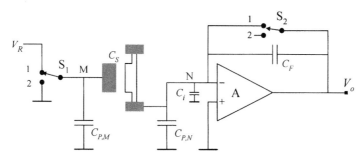

Fig. 5.1.8. Switched capacitor sensing

At time $t=t_1$, both S_1 and S_2 are in "1" positions. C_S and $C_{P,M}$ are charged by V_R. As point N is virtually grounded, the charge in C_S is $C_S V_R$ and the charge in C_F is 0. At time $t=t_2$, switches S_1 and S_2 are switched to their respective positions "2". As the left plate of C_S is grounded, the charge stored on the movable plate flows through point N and charges the capacitor C_F. The voltage at output V_o can be found by the relation for charge conservation

$$C_S V_R = \left(C_S + C_i + C_{P,N} \right) \cdot V_N + \left(V_N - V_o \right) \cdot C_F \qquad (5.1.15)$$

If the open loop gain of the amplifier is $-A_o$, we have

$$V_o = -A_o V_N \qquad (5.1.16)$$

From Eqs. (5.1.15) and (5.1.16), we have

$$C_S V_R = -\left(C_S + C_i + C_{P,N} \right) \cdot \frac{1}{A_o} V_o - C_F \frac{1+A_o}{A_o} V_o$$

and

$$V_o = \frac{A_o C_S V_R}{\left(A_o + 1 \right) \cdot C_F + C_S + C_i + C_{P,N}}$$

As A_o is large, the effects of C_i and $C_{P,N}$ are minimized. Thus, the output is directly proportional to C_S

$$V_o = \frac{C_S}{C_F} V_R \qquad (5.1.17)$$

For a changing C_S, S_1 and S_2 have to be switched between positions 1 and 2 at a high frequency and a "sample-and-hold" circuit has to be added at the output of the operational amplifier to facilitate further signal processing.

Alternatively, C_S and C_F in Fig. 5.1.8 may exchange their positions [10] if the electric connection does not contradict with the device structure. In this configuration, the output signal is inversely proportional to the sensing capacitance C_S, and directly proportional to the displacement of the movable plate. This is quite advantageous for a linear measurement. An additional advantage of this configuration is that the charge transferred to the sensing capacitor is constant in quantity. Thus, the operation range of the sensing configuration can be expanded due to the constant electrostatic force of the charge driving.

Fig. 5.1.9 shows a sensing scheme that compares the sensing capacitor with a reference capacitor [11]. For measurement, S_1 is first used to reset the circuit by discharging the feedback capacitor C_F. Then, S_2 and S_3 are switched from the central positions to the positions shown in the figure for charging C_S and discharging C_R. Then, S_2 and S_3 are switched to the opposite positions as shown by the arrows. Now the output voltage is

$$V_o = \frac{C_S - C_R}{C_F} V_{ref} \qquad (5.1.18)$$

Another switched capacitor circuit for sensing the differential capacitance is schematically shown in Fig. 5.1.10 [12]. The fixed electrodes of capacitors C_S and C_R are periodically switched between reference voltage $+V_{ref}/-V_{ref}$ and the output voltage V_o, resulting in the transfer of charges Q_1 and Q_2 to capacitors C_S and C_R, respectively. Their difference ($\Delta Q = Q_1 - Q_2$) is given by

$$\Delta Q = (C_S + C_p) \cdot (V_{ref} - V_o) - (C_R + C_p) \cdot (V_{ref} + V_o) \tag{5.1.19}$$

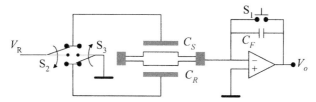

Fig. 5.1.9. Switched capacitor method for differential capacitive sensing

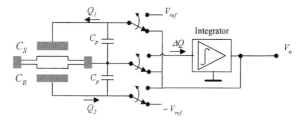

Fig. 5.1.10. The self-balancing bridge for differential capacitive sensing

The integrator circuit samples ΔQ from the movable plate and integrates it on a filter capacitor, thereby generating the output voltage V_o that is fed back to the excitation network in order to establish the equilibrium condition $\Delta Q = 0$. For this reason, the method is referred to as self-balancing bridge. From Eq. (5.1.19) the output voltage is

$$V_o = \frac{C_S - C_R}{C_S + C_R + 2C_p}$$

Note that the discussions above on the switched capacitive sensing scheme are merely conceptual. The switches are actually implemented by CMOS analog switches and the measurement involves a rather complicated circuit and the control of the switches.

§5.1.6. Frequency Sensing

A frequency output is easy to interface with a digital system. Therefore, it has long been considered for the capacitive sensing. The simplest frequency sensing method uses a Schmitt oscillator as shown in Fig. 5.1.11 [1, 13, 14].

If the hysteresis voltage of the oscillator is V_h, the oscillation frequency is

$$f = \frac{I_{o1} \cdot I_{o2}}{(I_{o1} + I_{o2})V_h} \cdot \frac{1}{C_S} = \frac{I_{o1} \cdot I_{o2}}{(I_{o1} + I_{o2})V_h C_o} \cdot (1 - \widetilde{x}) \tag{5.1.20}$$

As shown by Eq. (5.1.20), the variation of the output frequency is linearly related to the displacement of the plate.

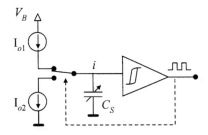

Fig. 5.1.11. A frequency sensing method using a Schmitt Oscillator

For the simple circuit described in the above, the stability of the frequency at original state may be a problem for many devices. To mitigate the difficulties, schemes measuring the differential capacitance are preferred. One of them is shown in Fig. 5.1.12 [14]. When the sensing and reference capacitors are connected to the oscillator alternatively, the difference of the frequencies is a more stable indication of the measurands.

$$\Delta f = f_S - f_R = \frac{I_o}{2V_h}\left(\frac{1}{C_S} - \frac{1}{C_R}\right) \tag{5.1.21}$$

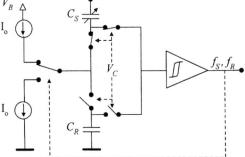

Fig. 5.1.12. A frequency sensing method comparing the sensing and the reference capacitors

§5.2. Effects of Electrical Excitation — Static Signal

As shown in §5.1, electrical excitation is necessary for the measurement of capacitance. The excitation voltage causes an electrostatic force between electrodes and causes the movement of the movable electrode. Therefore, the accuracy of the measurement or even the normal operation of the capacitive sensor is affected by the excitation voltage [15].

Generally, the excitation voltage consists of an ac component and a dc component [8,9]. A commonly used form is $(\pm V_o \pm V_1 \sin \omega t)$, where frequency ω (in the order of 10^6) is usually much larger than the frequency of the measurand signal and the natural vibration frequency of the mechanical structure (both are in the order of $10^3 \sim 10^4$). Therefore, the force on the movable electrode is the average of the electrostatic force of the voltage

$$F_e = \frac{A\varepsilon\varepsilon_o}{2(d-x)^2}(V_o^2 + \frac{1}{2}V_1^2)$$

If an effective voltage is defined as $V_{eff} = \sqrt{(V_o^2 + V_1^2/2)}$, the general form of electrostatic force on the electrode is

$$F_e = \frac{A\varepsilon\varepsilon_o}{2(d-x)^2}V_{eff}^2 \tag{5.2.1}$$

Now let us consider the effect of the excitation voltage on a single-sided capacitor, a double-sided capacitor and a double-sided capacitor with an electromechanical feedback. For specific, capacitive accelerometers are used as examples and, for simplicity, V is used in the following discussions in place of V_{eff}.

§5.2.1. Single-sided Excitation

For a single-sided capacitive accelerometer as shown in Fig. 5.2.1, the excitation is also single-sided. The equation to determine the displacement of the movable plate is

$$\frac{A\varepsilon\varepsilon_o V^2}{2(d-x)^2} + ma - kx = 0 \tag{5.2.2}$$

where ma is the inertial force in the x-direction and d the original gap distance between the electrodes.

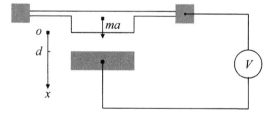

Fig. 5.2.1. Schematic structure for a single-sided capacitive accelerometer

By using the dimensionless notations of $\tilde{x} = x/d$, $p = A\varepsilon\varepsilon_o V^2/2kd^2$ and $q = ma/kd$, Eq. (5.2.2) can be written as

$$\tilde{x} - \frac{p}{(1-\tilde{x})^2} = q \tag{5.2.3}$$

Based on Eq. (5.2.3), the following conditions are discussed:

(1) Zero V, i.e., $p=0$

We have $\tilde{x} = q$. The relation between displacement and acceleration is linear.

(2) Nonzero V but Small p, q and x

For small $\tilde{x}$, by using the approximation of $(1+y)^{-1} = 1 - y + y^2 - y^3 + \cdots$, Eq. (5.2.3) is developed to the second power of $\tilde{x}$ as

$$3p\tilde{x}^2 - (1-2p)\tilde{x} + (p+q) = 0 \qquad (5.2.4)$$

The solution to the equation is

$$\tilde{x} = \frac{(1-2p) - \sqrt{(1-2p)^2 - 12p(p+q)}}{6p}$$

By using of the approximatin of $\sqrt{1-y} = 1 - y/2 - y^2/8 + \cdots$, for small p and q, the solution to Eq. (5.2.4) can be approximated as

$$\tilde{x} = \frac{p(1-4p+7p^2)}{(1-2p)^3} + \frac{1-4p+10p^2}{(1-2p)^3} q\left(1 + \frac{3p}{1-4p+10p^2} q\right) \qquad (5.2.5)$$

From this equation, we conclude that

(i) The excitation voltage causes an offset displacement at zero acceleration as indicated by the first term on the right side. Obviously, the larger the value of p, the larger the offset displacement.

(ii) The sensitivity of the accelerometer is proportional to $(1-4p+10p^2)/(1-2p)^3$, which is related to the excitation voltage. Therefore, the larger the voltage, the larger the sensitivity.

(iii) The excitation voltage causes nonlinearity between displacement and acceleration. The nonlinearity is $NL = -3pq_{max}/4(1-4p+10p^2)$, where q_{max} is the q value corresponding to the maximum operation range of acceleration. Readers are referred to §6.6.3 for the definition of nonlinearity.

For example, for $p=0.05$, we have $\tilde{x} = 0.056 + 1.13q(1+0.18q)$. If q_m is 0.1, the nonlinearity caused by the excitation is -0.45%.

(3) General Situation

In general, Eq. (5.2.3) can be solved using a graphic method. By defining a function

$$f(\tilde{x}, p) = \tilde{x} - \frac{p}{(1-\tilde{x})^2} \qquad (5.2.6)$$

the curves of *f* as a function of *x* and *p* are shown in Fig. 5.2.2. The solutions to Eq. (5.2.3) for a set of *p* and *q* values can be found by the intersection between a horizontal line *f=q* and the curve for the *p* value.

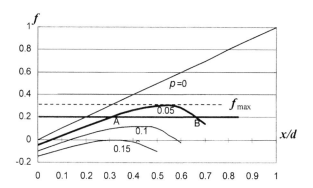

Fig. 5.2.2. Graphic solution to single-sided excitation

For example, for *p*=0.05 and $q = 0.2$, there are two intersections, A and B, between the $f \sim \tilde{x}$ curve for *p*=0.05 and the line of $q = 0.2$. The displacement corresponding to point A (at $x \cong 0.3d$) represents a stable solution but the displacement corresponding to point B (at $x \cong 0.67d$) is an unstable solution.

We can find that, for a specific *p*, there is a maximum *f* value, f_{max}, corresponding to a critical acceleration, $a_c = f_{max}kd / m$. For a *q* value larger than f_{max} (i.e., for an acceleration, *a*, larger than a_c), there would be no stable solution for the equation. This means that the pull-in effect occurs due to the combined effect of the electrostatic force and the inertial force. Thus, the larger the *p* value (i.e., the larger the excitation voltage), the smaller the f_{max}. For larger reliable operation range, *p* should be as small as possible. However, the smaller the *p*, the smaller the sensitivity. Therefore, there should be a trade-off for a practical design.

From the condition of $\partial f / \partial \tilde{x} = 0$, the expression of f_{max} can be found

$$f_{max} = 1 - \frac{3}{2}\sqrt[3]{2p}$$ (5.2.7)

For example, for *p*=0.05, f_{max} is 0.304, i.e., the maximum acceleration that does not cause pull-in is 0.304*kd/m* (instead of *kd/m*). For *p*=0.1, the maximum acceleration is reduced to 0.122*kd/m*. The accelerometer fails to work if *p* is larger than 4/27.

§5.2.2. Double-sided Excitation

For a double-sided capacitive accelerometer with a double-sided excitation as shown in Fig. 5.2.3, the equation for the displacement of the movable plate is

$$\frac{A\varepsilon\varepsilon_o V^2}{2d^2}\left[\frac{1}{(1-\tilde{x})^2}-\frac{1}{(1+\tilde{x})^2}\right]+ma-kx=0 \tag{5.2.8}$$

By using the dimensionless notations $\tilde{x}$, p and q, Eq. (5.2.8) is written as

$$\tilde{x}\left[1-\frac{4p}{(1-\tilde{x}^2)^2}\right]=q \tag{5.2.9}$$

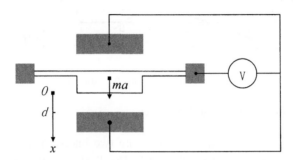

Fig. 5.2.3. Schematic structure for a double-sided accelerometer

Based on Eq. (5.2.9), we can have the discussion as follows:

(1) Zero V, i.e., $p=0$,
　　We have $\tilde{x}=q$. The relationship between displacement and acceleration is linear.

(2) Non-zero V but Small p, q and x
　　From Eq. (5.2.9), we conclude:
　　(i) There is no offset displacement caused by the excitation voltage due to the symmetric electrostatic forces on the movable plate.
　　(ii) For very small $\tilde{x}$, $\tilde{x}\cong q/(1-4p)$. This means that the larger the excitation voltage, the larger the sensitivity of the accelerometer. And the maximum p value for stable operation is 0.25.

(3) General Situation
　　For the general situation, the equation can be solved using a graphic method. By defining a function of $f(\tilde{x},p)$ below and drawing the $f\sim\tilde{x}$ curves with p as a parameter, we obtain the curves in Fig. 5.2.4.

$$f(\tilde{x},p)=\tilde{x}\left[1-\frac{4p}{(1-\tilde{x}^2)^2}\right] \tag{5.2.10}$$

Solutions to Eq. (5.2.10) for a set of p and q values can be found by the intersections between the horizontal line $f=q$ for the q value and the $f \sim \tilde{x}$ curve for the p. For example, for $p=0.05$ and $q=0.3$, there are two intersections between the line $f=0.3$ and the $f \sim \tilde{x}$ curve for $p=0.05$ as shown by the bold line and curve in Fig. 5.2.4, respectively. The left point intersection (at $x \approx 0.42d$) is a stable solution but the right intersection (at $x \approx 0.61d$) is an unstable solution.

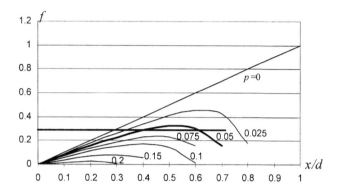

Fig. 5.2.4. Graphic solution to double-sided excitation

Also we can see that, for a specific p, there is a maximum value f_{max} for the curve of $f \sim \tilde{x}$. For a q value larger than f_{max}, there is no real-valued solution. This means that the mass is pulled-in by the combined effects of electrostatic and inertial forces. The larger the p values (i.e., the larger the V) the smaller the f_{max} values. For example, for the curve of $p=0.05$, f_{max} is about 0.32. Therefore, the critical acceleration a_c for causing pull-in effect is $0.32kd/m$. For $p=0.1$, the critical acceleration is reduced to about $0.17kd/m$. There would be no stable solution if p is larger than 0.25. The mass will not be stable even in its original position and will be pulled into contact with one of the fixed electrodes.

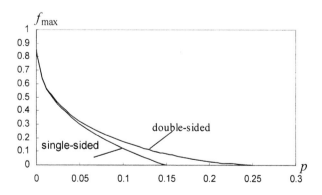

Fig. 5.2.5. The dependence of f_{max} on p for single-sided and double-sided capacitive accelerometers

In summary, the double-sided capacitive accelerometer is better than the single-sided capacitive accelerometer in that no offset is caused by the excitation voltage, the linearity is better and the excitation voltages can be larger without causing the pull-in effect.

The curves in Fig. 5.2.5 show the dependence of f_{max} on p for single-sided and double-sided capacitive accelerometers. We can conclude that, for reliable operation, p should be small enough. However, a smaller p means a smaller sensitivity of the accelerometer.

§5.2.3. Force-balanced Configuration

For a force-balanced accelerometer with a feedback voltage V_r, the simplified model is shown in Fig. 5.2.6 [9]. For a small displacement, the feedback voltage is proportional to the displacement of the mass: $V_r = \beta V_1 \tilde{x}$, where β is the feedback coefficient of the circuit. Obviously, both V_o and V_r are restricted by the supply voltage, V_S, of the electronic system (say, $V_o, V_r \le V_S/2$).

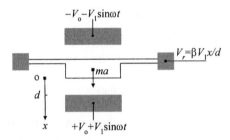

Fig. 5.2.6. A simplified model for force-balanced configuration

With the feedback voltage, the electrostatic force on the mass is

$$F_e = \frac{A\varepsilon\varepsilon_o}{2d^2}\left[\frac{(V_o + V_1\sin\omega t - V_r)^2}{(1-\tilde{x})^2} - \frac{(V_o + V_1\sin\omega t + V_r)^2}{(1+\tilde{x})^2}\right] \tag{5.2.11}$$

As the excitation frequency, ω, is much larger than the signal frequency and the natural vibration frequency of the mechanical structure, the average force on the mass is

$$\overline{F}_e = \frac{A\varepsilon\varepsilon_o}{2d^2(1-\tilde{x}^2)^2}\left[4(V_o^2 + \frac{1}{2}V_1^2)\tilde{x} + 4V_r^2\tilde{x} - 4V_rV_o - 4V_rV_o\tilde{x}^2\right] \tag{5.2.12}$$

By using $V_r = \beta V_1\tilde{x}$ and designating $V_1 = \alpha V_o$, we have

$$\overline{F}_e = \frac{2A\varepsilon\varepsilon_oV_o^2}{d^2(1-\tilde{x}^2)^2}\left[(1+\frac{1}{2}\alpha^2)\tilde{x} - (\alpha\beta\tilde{x}) + (\alpha\beta\tilde{x})^2\tilde{x} - (\alpha\beta\tilde{x})\tilde{x}^2\right] \tag{5.2.13}$$

Therefore, the force balance equation for the mass subjected to an acceleration, a, is

$$ma - kd\widetilde{x} + \frac{2A\varepsilon\varepsilon_o V_o^2}{d^2(1-\widetilde{x}^2)^2}\left[(1+\frac{1}{2}\alpha^2)\widetilde{x} - (\alpha\beta\widetilde{x}) + (\alpha\beta\widetilde{x})^2\widetilde{x} - (\alpha\beta\widetilde{x})\widetilde{x}^2\right] = 0 \qquad (5.2.14)$$

Using notations $p_o = A\varepsilon\varepsilon_o V_o^2/2kd^3$ and $q = ma/kd$, we find

$$q = \widetilde{x} - \frac{4p_o}{(1-\widetilde{x}^2)^2}\left[(1+\frac{1}{2}\alpha^2)\widetilde{x} - (\alpha\beta\widetilde{x}) + (\alpha\beta\widetilde{x})^2\widetilde{x} - (\alpha\beta\widetilde{x})\widetilde{x}^2\right] \qquad (5.2.15)$$

According to Eq. (5.2.15), the $q \sim \widetilde{x}$ relation can be discussed as follows:

(1) Zero V, i.e., $p_o=0$
   As $q = \widetilde{x}$, the force-displacement relation is linear.

(2) Nonzero V but Small p_o, q and x
   In this case, we have $\widetilde{x} \cong \dfrac{q}{1 - 4p(1+\alpha^2/2) + 4p_o\alpha\beta}$. The $\widetilde{x} \sim q$ relation is related to the value of β. For $\beta \neq 0$, the larger the value of β, the smaller the displacement due to the electro-mechanical feedback, which is negative in nature.

(3) General Conditions
   Generally, the $q \sim \widetilde{x}$ relation can be found using a graphical method. By defining the function of $f(\widetilde{x},\beta,p_o,\alpha)$

$$f(\widetilde{x},\beta,p_o,\alpha) = \widetilde{x} - \frac{4p_o}{(1-\widetilde{x}^2)^2}\left[(1+\frac{1}{2}\alpha^2)\widetilde{x} - (\alpha\beta\widetilde{x}) + (\alpha\beta\widetilde{x})^2\widetilde{x} - (\alpha\beta\widetilde{x})\widetilde{x}^2\right]$$

and drawing curves for $f(\widetilde{x},\beta,p_o,\alpha)$ with different β values for specific p_o and α values, we obtain the curves of $f(\widetilde{x})$ as shown in Fig. 5.2.7. In the calculation, a ceiling of V_o is set for V_r.
   It can be seen from Fig. 5.2.7 that, for a small β value, the maximum of the curve, f_{max}, is small due to the electrostatic force of the excitation signal. This means that the critical acceleration which causes the pull-in effect, i.e., $a_c = f_{max}kd/m$, is small. f_{max} increases with β until β is about 5 when the curve has the highest linearity. All the curves with β larger than 5 merge together at large x because the feedback voltages are restricted to a maximum value of V_o.
   By comparing Fig. 5.2.7(a) and (b), it can be seen that, for a small β value, the larger the p_o value, the smaller the maximums of the curves (due to the larger electrostatic force of the excitation signal). This means that the larger the p value, the smaller the critical acceleration, a_c, which causes the pull-in effect. However, if the feedback is large enough

(say, $\beta>5$ for the conditions considered), f_{max} can be quite close to 0.9. This means that the pull-in effect caused by the excitation signal is mostly eliminated by the feedback.

It can be concluded that, of the three configurations considered, the configuration of double-sided excitation with an electro-mechanical feedback is the one least affected by the side effects of the excitation signal and has the best performance.

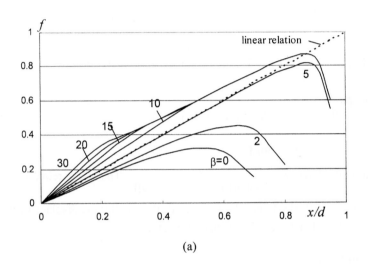

(a)

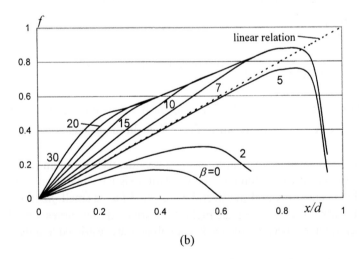

(b)

Fig. 5.2.7. Graphic solutions to double-sided excitation with electromechanical feedback (a) p_o=0.05, α=0.2 and $V_r{\leq}V_o$; (b) p_o=0.1, α=0.2 and $V_r{\leq}V_o$

§5.3. Effects of Electrical Excitation — Step Signal

In last section, the reliable operation range for a static acceleration signal was analyzed. As a matter of fact, the inertial signals are often dynamic ones and the damping ratio of the mechanical structure might be much smaller than unity. Under these conditions, the inertial capacitive accelerometer is more susceptible to "pull-in" failure due to the overshooting movement of seismic mass. In the worst condition, the damping effect is close to zero, resulting in the smallest reliable operation range. Though damping is always exists for a practical working condition, the worst condition of zero damping is preferred for design considerations when the effect of damping is difficult to estimate. Therefore, reliable operation conditions of capacitive accelerometers with light damping for step and pulse signals [16] are investigated in this and next sections, respectively. The results in §5.3 and §5.4 will be of more practical use for design application than that in §5.2.

§5.3.1. Single-sided Excitation

For a single-sided capacitive sensor with an excitation signal $V=V_o \pm V_1 \sin \omega t$ as schematically shown in Fig. 5.3.1, the equilibrium position of the seismic mass x_o before the inertial signal is applied is determined by Eq. (4.2.4).

With a step inertial force ma (caused by acceleration a starting at time $t=0$) on the mass at its balanced position x_0, the kinetic energy of the mass is

$$\frac{1}{2}m\dot{x}^2 = \int_{x_o}^{x} \left[\frac{A\varepsilon\varepsilon_o V^2}{2(d-x)^2} + ma - kx \right] dx \qquad (5.3.1)$$

As the maximum displacement the seismic mass can reach is determined by the condition of $\dot{x}=0$, the equation for the maximum displacement x_1 is

$$p\frac{1}{1-\tilde{x}_1} + q\tilde{x}_1 - \frac{1}{2}\tilde{x}_1^2 = p\frac{1}{1-\tilde{x}_o} + q\tilde{x}_o - \frac{1}{2}\tilde{x}_o^2 \qquad (5.3.2)$$

where $q = ma/kd$ and $\tilde{x} = x/d$. A solution to the equation for $\tilde{x}_1$ can be found if q is small. However, no real-valued solution can be found in the region of $(0 \sim 1)$ when q reaches beyond a critical value q_c. This means that the displacement of the seismic mass will increase continuously until it is pulled-in by the electrode.

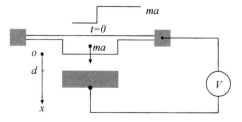

Fig. 5.3.1. The schematic structure of a single-sided capacitive accelerometer

In order to find the critical value q_c, q is expressed as a function of $\tilde{x}_1$

$$q(\tilde{x}_1) = \frac{1}{2}(\tilde{x}_1 + \tilde{x}_o) - \frac{p}{(1-\tilde{x}_o)(1-\tilde{x}_1)}$$

There is a critical value of q_c, which corresponds to the maximum value of $\tilde{x}_1$, which is designated as $\tilde{x}_m$. By the condition of $\partial q / \partial \tilde{x}_1 = 0$, $\tilde{x}_m$ is found to be

$$\tilde{x}_m = 1 - \sqrt{\frac{2p}{1-\tilde{x}_o}} \qquad (5.3.3)$$

Therefore, q_c is

$$q_c = \frac{1}{2}(\tilde{x}_m + \tilde{x}_o) - \frac{p}{(1-\tilde{x}_o)(1-\tilde{x}_m)} \qquad (5.3.4)$$

The dependence of q_c on p for a step signal is shown in Fig. 5.3.2. For comparison, the dependence of q_c on p for a static signal is also shown in the figure. Typically, q_c for a step signal is smaller than that for a static signal by a factor of 0.5 to 0.7 (for the same p).

The critical signal level of accelerometer for reliable operation is $a_c = \omega_o^2 d q_c$. As q_c is strongly dependent on p, a_c is also strongly dependent on the excitation voltage.

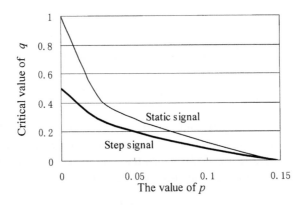

Fig. 5.3.2. The dependence of q_c on p for a single-sided capacitive sensor

§5.3.2. Double-sided Excitation

For a double-sided capacitive accelerometer as shown in Fig. 5.3.3, the original equilibrium position of the seismic mass is at the central position between the two electrodes (x=0). If a step acceleration of a is applied starting at t=0, the equation governing the dynamic behavior of the seismic mass is

$$\frac{1}{2}m\dot{x}^2 = \int_0^y \left\{ \frac{A\varepsilon_o V^2}{2d_o{}^2} \left[\frac{1}{(1-\tilde{x})^2} - \frac{1}{(1+\tilde{x})^2} \right] + ma - kx \right\} dx$$

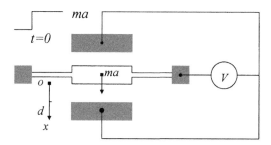

Fig. 5.3.3. A double-sided capacitive inertial sensor with a step inertial signal

Or, we have

$$\frac{1}{2}m\dot{x}^2 = 2\frac{A\varepsilon_o V^2}{2d_o}\frac{\tilde{x}^2}{1-\tilde{x}^2} + mad_o\tilde{x} - \frac{1}{2}kd_o{}^2\tilde{x}^2$$

By letting $\dot{x} = 0$, the expression for q is found

$$q = \frac{1}{2}\tilde{x}_1 - 2p\frac{\tilde{x}_1}{1-\tilde{x}_1{}^2}$$

By the same argument given in §5.3.1, the critical value q_c corresponding to $\tilde{x}_m$ is determined by the condition of $\partial q / \partial y_1 = 0$. Thus, we have

$$\tilde{x}_m = \sqrt{(1+2p) - \sqrt{4p(2+p)}} \tag{5.3.5}$$

For a specific p, $\tilde{x}_m$ can be found from Eq. (5.3.5) and then q_c is obtained

$$q_c = \frac{1}{2}\tilde{x}_m - 2p\frac{\tilde{x}_m}{1-\tilde{x}_m{}^2} \tag{5.3.6}$$

The critical acceleration for reliable operation of the step signal is $a_c = \omega_o{}^2 d_o q_c$.

The dependence of q_c on p is shown by the curve with diamond marks for double-sided capacitor in Fig. 5.3.4. For comparison, the dependence of q_c on p for a static signal is also shown by a dotted line. The value of q_c for a step signal is smaller than that for a static signal by a factor of 0.5~0.7 for the same excitation level.

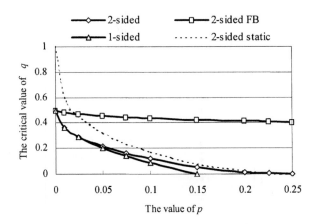

Fig. 5.3.4. The dependence of q_c on p for three configurations for step signals
(For comparison, q_c for static signal for 2-sided configuration is also shown)

§5.3.3. Force-balanced Configuration

For a force-balanced accelerometer with a feedback voltage V_r, the simplified model is shown in Fig. 5.2.6. For a mass displacement, the feedback voltage is $V_r = \beta V_1 \tilde{x}$, where β is a feedback coefficient. Note that V_r has a ceiling value determined by the supply voltage V_S of the electronic circuit (usually, $V_r \leq V_S/2$).

According to the model, the average electrostatic force applied on the seismic mass is

$$\overline{F}_e = \frac{2A\varepsilon\varepsilon_o V_o^2}{d^2(1-\tilde{x}^2)^2}\left[(1+\frac{1}{2}\alpha^2)\tilde{x} - (\alpha\beta\tilde{x}) + (\alpha\beta\tilde{x})^2\tilde{x} - (\alpha\beta\tilde{x})\tilde{x}^2\right]$$

With a step acceleration, the kinetic energy of the seismic mass is

$$\frac{1}{2}mx^2 = \int_0^x \left(\overline{F}_e + ma - kx\right)dx$$

$$= \int_0^x \left(\frac{2A\varepsilon\varepsilon_o}{(1-\tilde{x}^2)^2}\left[(1+\frac{1}{2}\alpha^2)V_o^2\tilde{x} - (\alpha\beta\tilde{x})V_o + (\alpha\beta\tilde{x})^2V_o^2\tilde{x} - (\alpha\beta\tilde{x})V_o\tilde{x}^2\right] + ma - kx\right)dx$$

Further analysis is a little complicated, as the feedback voltage V_r equals $\alpha\beta\tilde{x}V_o$ for small mass displacement and $V_r = V_o$ (i.e., $\alpha\beta\tilde{x} = 1$) for $\tilde{x} \geq 1/\alpha\beta$ due to the restriction by the supply voltage of the electronic circuit. The equation has to be considered in two regions and the critical value q_c is the larger one of the two maximums in the two regions. As an example, the dependence of q_c on p for the condition of $\alpha=0.2$, $\beta=5$ is given by the curve with square marks in Fig. 5.3.4. When this curve is compared with that withouot a feedback (diamond marks), we can find that the reliable operation range is improved by the electro-mechanical feedback configuration.

§5.4. Effects of Electrical Excitation — Pulse Signal

Due to the combined effect of the inertial force, the elastic force and the electrostatic force of the excitation signal, there is also restriction to a pulse signal for the reliable operation of a capacitive sensor. The seismic mass of the sensor will be pulled-in if the level and/or the duration of the acceleration pulse is too large. The discussion on the conditions of reliable operation for a pulse acceleration signal is given in this section [16].

§5.4.1. Single-sided Excitation

According to Eq. (4.1.4), with the excitation voltage V, the potential energy (referring to the state of $x=0$) of a single-sided capacitive sensor (as shown in Fig. 5.2.1) is

$$E(x) = \frac{1}{2}kx^2 - \frac{1}{2}\left(\frac{A\varepsilon_o}{d-x} - \frac{A\varepsilon_o}{d}\right)V^2 = \frac{1}{2}kd^2\left(\tilde{x}^2 - 2p\frac{\tilde{x}}{1-\tilde{x}}\right)$$

The condition for the extrema of $E(x)$ is the well known equation for balanced positions: $\tilde{x}(1-\tilde{x})^2 = p$. Under the condition of $p<0.148$, there are two solutions for x in the range of 0 to d; a stable balanced position x_o corresponding to an energy minimum and an unstable balanced position x_1 corresponding to an energy maximum. Generally, x_o and x_1 are found through numerical calculation. For example, for $p = 0.05$, we have $x_o = 0.05612d$ and $x_1 = 0.74d$.

The potential energy of the system at the stable balanced position is $E(x_o)$. If the energy supplied by the pulse inertial force signal is larger than the potential barrier $E(x_1) - E(x_o)$, the seismic mass can pass over position x_1 and be pulled-in by the static electrode, i.e., the system is not stable.

If the amplitude of the acceleration pulse, a, is large enough and the time duration of the pulse, Δt, is small so that the velocity and the mass displacement at the end of the pulse is small, the initial conditions can be approximated as

$$\dot{x}_o = a(\Delta t), \, x_o = \frac{1}{2}a(\Delta t)^2 \tag{5.4.1}$$

Therefore, the condition for a reliable operation is $\frac{1}{2}ma^2(\Delta t)^2 \leq E(\tilde{x}_1) - E(\tilde{x}_o)$, or,

$$a(\Delta t) \leq \omega_o d \sqrt{(\tilde{x}_1^2 - \tilde{x}_o^2) - 2p\left(\frac{\tilde{x}_1}{1-\tilde{x}_1} - \frac{\tilde{x}_o}{1-\tilde{x}_o}\right)} \equiv \omega_o df_1(p) \tag{5.4.2}$$

where

$$f_1(p) = \sqrt{(\tilde{x}_1^2 - \tilde{x}_o^2) - 2p\left(\frac{\tilde{x}_1}{1-\tilde{x}_1} - \frac{\tilde{x}_o}{1-\tilde{x}_o}\right)} \tag{5.4.3}$$

The results show that the critical value of $(a\Delta t)$ is proportional to the factor $f_1(p)$ as well as $d\omega_o$. The dependence of $f_1(p)$ on p is given by the curve with square marks in Fig. 5.4.1.

The above analysis is based on the condition that the amplitude of the pulse acceleration is large and the time duration of the pulse is short so that the initial conditions are given in Eq. (5.4.1). The requirement can be discussed as follows.

(i) The inertial force should be much larger than the elastic force in the time duration of pulse. This gives $ma >> ka(\Delta t)^2/2$. Therefore, this condition requires $\Delta t << \sqrt{2}/\omega_o$.

(ii) The inertial force should be much larger than damping force at the end of the pulse. This means that $ma >> ca(\Delta t)$ (where c is the coefficient of damping force). This condition requires $\Delta t << m/c = 1/2n$ or $\Delta t << Q/\omega_o$.

Therefore, the requirement is $\Delta t << \sqrt{2}/\omega_o$, if Q is large.

§5.4.2. Double-sided Excitation

According to Eq. (4.2.40), with the excitation voltage V, the potential energy (referring to the original state of $x=0$) of the double-sided capacitor shown in Fig. 5.2.3 is

$$E(\tilde{x}) = \frac{1}{2}kd^2\tilde{x}^2 - \frac{A\varepsilon\varepsilon_o V^2 \tilde{x}^2}{d(1-\tilde{x}^2)} \tag{5.4.4}$$

From the condition of $\partial E(x)/\partial x = 0$, the positions for the extrema of $E(x)$ are

$$\tilde{x}_1 = 0, \quad \tilde{x}_{2,3} = \pm\sqrt{1 - \sqrt{\frac{C_o V^2}{kd^2}}} = \pm\sqrt{1 - \sqrt{4p}}$$

Further examination shows that, for an originally stable condition, (i.e., $V < \sqrt{kd^3/2A\varepsilon\varepsilon_o}$), $x_1=0$ is the position for a minimum energy (i.e., a stable balanced position) while x_2 and x_3 are positions for maximum energy (i.e., unstable balanced positions).

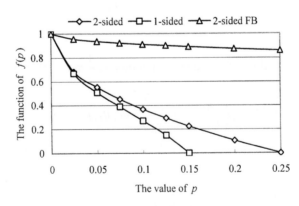

Fig.5.4.1. Dependence of $f_i(p)$ on p for 1-sided, 2-sided and 2-sided with feedback configurations

Therefore, if the energy supplied by the pulse inertial force signal is larger than the potential barrier between x_2 (or x_3) and x_1, the seismic mass will go beyond position x_2 (or x_3) and be pulled-in by a static electrode, causing failure of the system. From Eq. (5.4.4), the potential barrier is

$$E(x_2) = \frac{1}{2} kd^2 (1 - \sqrt{4p})^2$$ (5.4.5)

Therefore, the condition for reliable operation is $ma^2(\Delta t)^2 / 2 = E(x_2)$, or,

$$a\Delta t = d\omega_o(1 - 2\sqrt{p}) = d\omega_o f_2(p)$$ (5.4.6)

where

$$f_2(p) = 1 - 2\sqrt{p}.$$

The dependence of $f_2(p)$ on p is shown by the curve in Fig. 5.4.1 with diamond marks. As $f_2(p)$ is consistently larger than $f_1(p)$, the double-sided configuration has larger reliable operation range than the single-sided configuration..

§5.4.3. Force-balanced Configuration

For sensor with force-balanced configuration as shown in Fig. 5.2.6, the dependence of the system energy on the displacement is

$$E(\tilde{x}) = \frac{1}{2} kd^2 \tilde{x}^2 - \frac{A\varepsilon_o}{2d} \tilde{x} \left(\frac{V^2 - 2VV_r + V_r^2}{1 - \tilde{x}} - \frac{V^2 + 2VV_r + V_r^2}{1 + \tilde{x}} \right)$$

where $V = V_o + V_1 \sin \omega t$, $V_1 = \alpha V_o$, $V_r = \alpha\beta V_o \tilde{x}$ for $\tilde{x} < 1/\alpha\beta$ but $V_r = V_o$ for $\tilde{x} \geq 1/\alpha\beta$. For simplicity, we consider the condition of $\alpha\beta < 1$ so that $\tilde{x} < 1/\alpha\beta$. As ω is much larger than the natural vibration frequency of the mechanical structure, we have $\overline{V^2} = V_o^2 + V_1^2/2$ and $\overline{V} = V_o$. Therefore, the average energy over time for short term is

$$E(\tilde{x}) = \frac{1}{2} kd^2 \tilde{x}^2 - \frac{A\varepsilon_o V_o^2}{d(1 - \tilde{x}^2)} \tilde{x}^2 \left[\left(1 + \frac{1}{2}\alpha^2 - 2\alpha\beta \right) + \alpha^2\beta^2\tilde{x}^2 \right]$$ (5.4.7)

Further, for small $\tilde{x}$, the above equation can be approximated as

$$E(\tilde{x}) = \frac{1}{2} kd^2 \tilde{x}^2 - \frac{A\varepsilon_o V_o^2}{d} \left(1 + \frac{1}{2}\alpha^2 - 2\alpha\beta \right) \left(\frac{\tilde{x}^2}{1 - \tilde{x}^2} \right)$$

If the energy is normalized to the maximum mechanical potential energy, $kd^2/2$, we have

$$\widetilde{E}(\widetilde{x}) = \frac{2E(\widetilde{x})}{kd^2} = \widetilde{x}^2 - 4p_o\left(1 + \frac{1}{2}\alpha^2 - 2\alpha\beta\right)\left(\frac{\widetilde{x}^2}{1-\widetilde{x}^2}\right)$$

where $p_o = \dfrac{A\varepsilon_o V_o^2}{2kd^3} = p/(1+\dfrac{1}{2}\alpha^2)$. Obviously, p_o is slightly smaller than p.

The extrema of the energy can be determined by the condition of $\partial\widetilde{E}(\widetilde{x})/\partial\widetilde{x} = 0$. Three solutions can be found

$$\widetilde{x}_1 = \widetilde{0} \ , \ \widetilde{x}_{2,3} = \pm\sqrt{1-\sqrt{4p_o(1+\alpha^2/2-2\alpha\beta)}}$$

In condition of $1 > 4p_o(1+\alpha^2/2-2\alpha\beta) > 0$, $\widetilde{x}_1$ corresponds to an energy minimum and $\widetilde{x}_2$ and $\widetilde{x}_3$ correspond to two energy maximums with the same value.

The energy difference between $\widetilde{x}_1$ and $\widetilde{x}_2$ (or $\widetilde{x}_3$) is

$$\Delta E = E(\widetilde{x}_2) - E(\widetilde{x}_1) = \frac{1}{2}kd_o^2\left(1 - \sqrt{4p_o(1+\frac{1}{2}\alpha^2 - 2\alpha\beta)}\right)^2 \tag{5.4.8}$$

Therefore, for a pulse acceleration signal of amplitude a and time duration Δt, the critical condition for reliable operation is $ma^2(\Delta t)^2/2 = \Delta E$, or,

$$a(\Delta t_c) = \omega_o d_o f_3(p,\alpha,\beta) \tag{5.4.9}$$

where $f_3(p,\alpha,\beta) = 1 - \sqrt{4p_o(1+\alpha^2/2-2\alpha\beta)}$.

For a specific set of p_o and α, the maximum of the function $f_3(p,\alpha,\beta)$ appears at $\beta = 1/\alpha$ and the maximum value is $f_{3\max} p_o,\alpha,\beta) = 1-\alpha\sqrt{2p_o}$. The value is quite close to unity for small p_o and α. The dependence of $f_3(p,\alpha,\beta)$ on p for $\alpha = 0.2$ and $\beta=5$ is shown in Fig. 5.4.1 by the curve with triangle marks. $f_3(p,\alpha,\beta)$ is generally much larger than $f_1(p)$ and $f_2(p)$. This means that the reliable operation range of capacitive accelerometer is increased significantly by using a feedback configuration.

§5.5. Problems

Problems on Capacitive Sensing Schemes

1. A micro mechanical silicon microphone consists of a parallel-plate condenser with a signal pre-conditioning circuit as shown in Fig. 5.1.1. The effective area of the movable plate is 1mm^2, the thickness of the plate is 1 μm, the effective elastic constant is k=30N/m

and the nominal gap distance is 2 μm. (1) If the bias voltage is half of the pull-in voltage of the parallel-plate capacitor, find the mechanical sensitivity S_m, the open-circuit electric sensitivity S_e and the overall open circuit sensitivity S_o of the microphone. (2) If the parasitic capacitance in parallel with the microphone is C_p=2pF and the input capacitance of the buffer (follower) is C_i=1pF, find the sensitivity S_F at the output of the buffer amplifier.

2. For a capacitive sensor with diode-quad sensing circuit as shown in Fig. 5.1.2, the original capacitance of the sensing capacitor is C_S=C_R=C_o=2pF. Find (1) the open circuit output voltage if C_S =2.1pF, C_p =1pF, V_p=1.5V and V_F=0.5V, and (2) the output voltage V_{out} on load resistors (as shown in Fig. 5.1.4) if R_F=100kΩ, R_L=500kΩ and the frequency of the excitation signal is f=500kHz.

3. For a capacitive sensor with a opposite driving circuit as shown in Fig. 5.1.5, the parameters are C_o=2pF, C_p =1pF, C_i =1pF and the amplitude of the excitation signal V_1 =0.2V. If the nominal gap distance for C_R and C_S is 2 μm, for a mass displacement of 0.2μm, find V_i at the input of the buffer amplifier.

4. The force-balanced sensing circuit as shown in Fig. 5.1.7 can be modeled by Fig. 5.2.6. If V_1=αV_o, find the vibration frequency of the seismic mass at original position x=0.

5. For a capacitive sensor with a frequency output circuit as shown in Fig. 5.1.11, the parameters are I_{o1}=0.1μA, I_{o2}=2.5μA, V_h=0.6V, C_S = C_o /(1 – $\tilde{x}$) and C_o=10pF. Find (1) the dependence of output frequency on the displacement of the movable plate, and (2) the dependence of output frequency on the displacement of the movable plate, if there is a parasitic capacitor C_p=2pF in parallel with sensing capacitor C_S.

Problems on the Effects of Excitation Signals

6. If the excitation signal for capacitive sensing is a square wave $V_o \pm V_1$ with 50% duty cycle as shown in the figure below, find the effective voltage V_{eff} of the excitation signal.

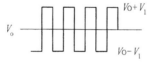

7. A silicon capacitive structure as shown in the figure below is used as a capacitive accelerometer. The parameters of the structure are A=B=2mm, b=40μm, h=10μm, H=100μm, and d=3μm. If the free vibration frequency of the mass in its normal direction is f_o=425Hz and the excitation voltage for the sensing is V=(1V)·sinωt (with ω>>2πf_o), (1) find the maximum reliable operation acceleration for a quasi-static signal, and (2) find the maximum reliable operation acceleration for a step signal.

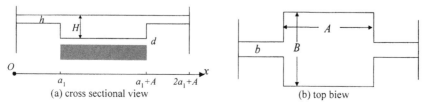

(a) cross sectional view   (b) top biew

8. For a silicon capacitive structure with the same structure and parameters as last problem, find the maximum time duration for the reliable operation of a pulse acceleration signal with amplitude of 1000g.

9、 For a double-sided silicon capacitive structure shown in the figure below, the parameters are $A=B=3$mm, $b=40\mu$m, $h=10\mu$m, $l=200\mu$m, $H=200\mu$m, $d=5\mu$m and the excitation voltage for the sensing is $V=(1V)\cdot\sin\omega t$ (with $\omega>>2\pi f_0$, where f_0 is the natural vibration frequency of the mechanical structure). Find (1) the maximum reliable operation acceleration for a quasi-static signal, (2) the maximum reliable operation acceleration for a step signal, and (3) the maximum time duration for the reliable operation of a pulse acceleration signal with an amplitude of 2000g.

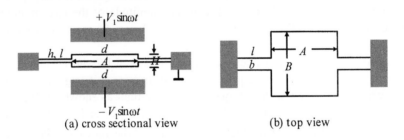

(a) cross sectional view   (b) top view

References

[1] W. Ko, Solid-state capacitive pressure transducers, Sensors and Actuators 10 (1986) 303-320

[2] R. Puers, Capacitive sensors: when and how to use them, Sensors and Actuators A37-38 (1993) 93-105

[3] P. Scheeper, A. Van der Donk, W. Olthuis, P. Bergveld, A review of silicon microphones, Sensors and Actuators A44 (1994), 1-11

[4] X. Li, R. Lin, H. Kek, J. Miao, Q. Zou, Sensitivity-improved silicon condenser microphone with a novel single deeply corrugated diaphragm, Sensors and Actuators A92 (2001) 257-262

[5] D.R. Harrison, J. Dimeff, A diode-quad bridge for use with capacitive transducers, Rev. Sci. Instrum., Vol.44 (1973) 1468- 1477

[6] W. Ko, M. Bao, Y. Hong, A high sensitivity integrated-circuit capacitive pressure transducer, IEEE Trans. On Electron Devices, Vol. ED-29 (1982) 48-56

[7] X. Li, M. Bao, S. Shen, Study on linearization of silicon capacitive pressure sensors, Sensors and Actuators A 63 (1997) 1-6

[8] B. Boser, Electronics for micromachined inertial sensors, Digest of Technical Papers, the 9th Intl. Conf. on Solid-State Sensors and Actuators, Chicago, IL, USA, June 16-19, 1997 (Transducers'97) 1169-1172

[9] W. Kuehnel, S. Sherman, A surface micromachined silicon accelerometer with on chip detection circuitry, Sensors and Actuators A45 (1994) 7-16

[10] B. Puers, A capacitive pressure sensor with low impedance output and active suppression of parasitic effects, Sensors and Actuators A21-23 (1990) 108-114

[11] Y. Park, K. Wise, An MOS switched-capacitor readout amplifier for capacitive pressure sensors, IEEE Proc. Custom IC Conf., Rochester, May 23, 1983, 380-384

[12] H. Leuthold, F. Ruldolf, An ASIC for high-resolution capacitive microaccelerometers, Sensors and Actuators A21-23 (1990) 278-281

[13] C. Sander, J. Knutt, J. Meindl, A monolithic capacitive pressure sensor with pulse-periond output, IEEE Trans. On Electron Devices, Vol. ED-27 (1980) 927-930

[14] Y. Matsumoto, M. Esashi, Low drift integrated capacitive accelerometer with PLL servo technique, Digest of Technical Papers, the 7th Intl. Conf. on Solid-State Sensors and Actuators, Yokohama, Japan, June 7-10, 1993 (Transducers'93) 826-829

[15] M. Bao, H. Yin, H. Yang, S. Shen, Effects of electrostatic forces generated by the driving signal on capacitive sensing devices, Sensors and Actuators A84 (2000) 213-219

[16] M. Bao, Y. Huang, H. Yang, Y. Wang, Reliable operation conditions of capacitive inertial sensor for step and shock signals, Sensors and Actuators A114 (2004) 41-48

Chapter 6

Piezoresistive sensing

Silicon pressure transducers and accelerometers based on the piezoresistance effect in silicon have been the most widely used micro mechanical transducers over the past decades. Piezoresistive silicon transducers and accelerometers have grown up into multi-million industries and played an important role in driving micro mechanical technologies forward. Therefore, it is widely accepted that the micro mechanical transducer technology begun with the discovery of the piezoresistance effect in silicon by C. S. Smith in 1954 [1].

Though more and more sensing schemes have been developed in recent years, piezoresistive sensing remains one of the most important sensing schemes for MEMS technologies.

The mathematical analysis, the design methods and the related properties of piezoresistance effect and piezoresistive sensors will be studied in this chapter.

§6.1. Piezoresistance Effect in Silicon

§6.1.1. Resistivity Tensor

In a metallic material the relations between the electric field $\vec{E}$ and the electric current density $\vec{J}$ are isotropic. The relations are $\vec{J} = \sigma\vec{E}$ and $\vec{E} = \rho\vec{J}$, where $\vec{E}$ and $\vec{J}$ are vectors and the conductivity σ and the resistivity ρ are scalars. If the relations are expressed in the Cartesian coordinate system, we have

$$J_X = \sigma E_X, \ J_Y = \sigma E_Y, \ J_Z = \sigma E_Z, \text{ or, } \ E_X = \rho J_X, \ E_Y = \rho J_Y, \ E_Z = \rho J_Z$$

However, in a single crystal material, the conductivity and resistivity are generally anisotropic. The general relation between the electric field and the current density is

$$\begin{pmatrix} J_X \\ J_Y \\ J_Z \end{pmatrix} = \begin{pmatrix} \sigma_{xx} & \sigma_{xy} & \sigma_{xz} \\ \sigma_{yx} & \sigma_{yy} & \sigma_{yz} \\ \sigma_{zx} & \sigma_{zy} & \sigma_{zz} \end{pmatrix} \begin{pmatrix} E_X \\ E_Y \\ E_Z \end{pmatrix} \text{ or } \begin{pmatrix} E_X \\ E_Y \\ E_Z \end{pmatrix} = \begin{pmatrix} \rho_{xx} & \rho_{xy} & \rho_{xz} \\ \rho_{yx} & \rho_{yy} & \rho_{yz} \\ \rho_{zx} & \rho_{zy} & \rho_{zz} \end{pmatrix} \begin{pmatrix} J_X \\ J_Y \\ J_Z \end{pmatrix} \quad (6.1.1)$$

Eq. (6.1.1) can be written as $(J)=(\sigma)(E)$ or $(E)=(\rho)(J)$, where J and E are first rank tensors (vectors) and ρ and σ are second rank tensors. ρ and σ are called the resistivity tensor and the conductivity tensor, respectively.

According to Onsager's theorem [2], the conductivity tensor is a symmetrical tensor (i.e., $\sigma_{ij}=\sigma_{ji}$). The same is true for the resistivity tensor. Therefore, there are only six independent components for ρ or σ. By using the notations of $\rho_1=\rho_{xx}$, $\rho_2=\rho_{yy}$, $\rho_3=\rho_{zz}$, $\rho_4=\rho_{yz}$ $\rho_5=\rho_{zx}$, and $\rho_6=\rho_{xy}$, the resistivity tensor takes the form

$$\rho = \begin{pmatrix} \rho_1 & \rho_6 & \rho_5 \\ \rho_6 & \rho_2 & \rho_4 \\ \rho_5 & \rho_4 & \rho_3 \end{pmatrix}$$

For a single crystalline material with cubic symmetry like a silicon crystal, it can be proved that $\rho_1=\rho_2=\rho_3=\rho_o$ and $\rho_4=\rho_5=\rho_6=0$ when the material is free from any deformation. This means that the relation between $\vec{E}$ and $\vec{J}$ in the material is isotropic, i.e.,

$$\begin{pmatrix} E_X \\ E_Y \\ E_Z \end{pmatrix} = \begin{pmatrix} \rho_o & 0 & 0 \\ 0 & \rho_o & 0 \\ 0 & 0 & \rho_o \end{pmatrix} \begin{pmatrix} J_X \\ J_Y \\ J_Z \end{pmatrix} \tag{6.1.2}$$

Eq. (6.1.2) can also be expressed in a simple form as: $\vec{E} = \rho_o \vec{J}$.

This relation means that ρ is isotropic in a single crystalline silicon material when the material is free from stress (or free of deformation) so that the crystal remains cubic-symmetric. Once the crystal is deformed by a stress, Eq. (6.1.2) is no longer valid and Eq. (6.1.1) has to be used to define the relations between $\vec{E}$ and $\vec{J}$.

§6.1.2. Piezoresistive Coefficient Tensor
As discussed in Chapter 2, a stress tensor is a second rank tensor with six independent components. It can be expressed as

$$T = \begin{pmatrix} T_1 & T_6 & T_5 \\ T_6 & T_2 & T_4 \\ T_5 & T_4 & T_3 \end{pmatrix}$$

where $T_1 = T_{XX}$, $T_2 = T_{YY}$, $T_3 = T_{ZZ}$, $T_4 = T_{YZ}$, $T_5 = T_{ZX}$ and $T_6 = T_{XY}$.

The piezoresistance effect indicates that the stress tensor in a crystalline material causes the change of resistivity tensor. The general relation is

$$\begin{pmatrix} \rho_1 \\ \rho_2 \\ \rho_3 \\ \rho_4 \\ \rho_5 \\ \rho_6 \end{pmatrix} = \begin{pmatrix} \rho_o \\ \rho_o \\ \rho_o \\ 0 \\ 0 \\ 0 \end{pmatrix} + \rho_o \begin{pmatrix} \pi_{11} & \pi_{12} & \pi_{13} & \pi_{14} & \pi_{15} & \pi_{16} \\ \pi_{21} & \pi_{22} & \pi_{23} & \pi_{24} & \pi_{25} & \pi_{26} \\ \pi_{31} & \pi_{32} & \pi_{33} & \pi_{34} & \pi_{35} & \pi_{36} \\ \pi_{41} & \pi_{42} & \pi_{43} & \pi_{44} & \pi_{45} & \pi_{46} \\ \pi_{51} & \pi_{52} & \pi_{53} & \pi_{54} & \pi_{55} & \pi_{56} \\ \pi_{61} & \pi_{62} & \pi_{63} & \pi_{64} & \pi_{65} & \pi_{66} \end{pmatrix} \begin{pmatrix} T_1 \\ T_2 \\ T_3 \\ T_4 \\ T_5 \\ T_6 \end{pmatrix} \tag{6.1.3}$$

If the relative change of the resistivity component is defined as

$$\Delta_i = (\rho_i - \rho_o)/\rho_o \ (i=1,2,3); \ \Delta_j = \rho_j/\rho_o \ (j=4,5,6) \tag{6.1.4}$$

we have

$$\begin{pmatrix} \rho_1 & \rho_6 & \rho_5 \\ \rho_6 & \rho_2 & \rho_4 \\ \rho_5 & \rho_4 & \rho_3 \end{pmatrix} = \rho_o \begin{pmatrix} 1 & 0 & 0 \\ 0 & 1 & 0 \\ 0 & 0 & 1 \end{pmatrix} + \rho_o \begin{pmatrix} \Delta_1 & \Delta_6 & \Delta_5 \\ \Delta_6 & \Delta_2 & \Delta_4 \\ \Delta_5 & \Delta_4 & \Delta_3 \end{pmatrix}$$

Or, in a simplified form, we have $(\rho)=\rho_o(I)+\rho_o(\Delta)$, where (Δ) is a second rank tensor determined by the piezoresistive coefficient tensor (π) and the stress (T)

$$\begin{pmatrix} \Delta_1 \\ \Delta_2 \\ \Delta_3 \\ \Delta_4 \\ \Delta_5 \\ \Delta_6 \end{pmatrix} = \begin{pmatrix} \pi_{11} & \pi_{12} & \pi_{13} & \pi_{14} & \pi_{15} & \pi_{16} \\ \pi_{21} & \pi_{22} & \pi_{23} & \pi_{24} & \pi_{25} & \pi_{26} \\ \pi_{31} & \pi_{32} & \pi_{33} & \pi_{34} & \pi_{35} & \pi_{36} \\ \pi_{41} & \pi_{42} & \pi_{43} & \pi_{44} & \pi_{45} & \pi_{46} \\ \pi_{51} & \pi_{52} & \pi_{53} & \pi_{54} & \pi_{55} & \pi_{56} \\ \pi_{61} & \pi_{62} & \pi_{63} & \pi_{64} & \pi_{65} & \pi_{66} \end{pmatrix} \begin{pmatrix} T_1 \\ T_2 \\ T_3 \\ T_4 \\ T_5 \\ T_6 \end{pmatrix} \tag{6.1.5}$$

Eq. (6.1.5) can be expressed in a simplified form as $(\Delta)=(\pi)(T)$. As (π) defines the relation between two second rank tensors, it is a forth rank tensor.

§6.1.3. Piezoresistive Coefficient of Silicon

The silicon crystal has a diamond structure. The basic unit of the structure is a face-centered cubic cell that has the symmetry of O_h group. By the symmetric operations of O_h group, it can be verified (see §6.3.2) that, in the crystallographic coordinate system of the crystal, there are only three non-zero independent components for the piezoresistive coefficient tensor. They are $\pi_{11}=\pi_{22}=\pi_{33}$, $\pi_{12}=\pi_{21}=\pi_{13}=\pi_{31}=\pi_{23}=\pi_{32}$ and $\pi_{44}=\pi_{55}=\pi_{66}$. Thus, the piezoresistive coefficient tensor of silicon has a simple form in a crystallographic coordinate system

$$\pi = \begin{pmatrix} \pi_{11} & \pi_{12} & \pi_{12} & 0 & 0 & 0 \\ \pi_{12} & \pi_{11} & \pi_{12} & 0 & 0 & 0 \\ \pi_{12} & \pi_{12} & \pi_{11} & 0 & 0 & 0 \\ 0 & 0 & 0 & \pi_{44} & 0 & 0 \\ 0 & 0 & 0 & 0 & \pi_{44} & 0 \\ 0 & 0 & 0 & 0 & 0 & \pi_{44} \end{pmatrix} \tag{6.1.6}$$

The three non-zero independent components were found experimentally by C. S. Smith for high resistivity silicon material [1]. The data is given in Table 6.1.1. According to the data in the table, the approximation often used is: $\pi_{11}=\pi_{12}=0$ for p-Si, and $\pi_{44}=0$ and $\pi_{11}=-2\pi_{12}$ for n-Si.

Table 6.1.1. Components of the piezoresistive coefficient tensor in silicon (in 10^{-11}/Pa)

	π_{11}	π_{12}	π_{44}
p-Si (7.8Ω·cm)	6.6	−1.1	138.1
n-Si (11.7Ω·cm)	−102.2	53.4	−13.6

For example, we have an n-Si filament with a cross sectional area of $A=1\text{mm}^2$. If the filament is in the <100> direction and a tensile force of 10N is applied on the filament along the <100> direction, let us find the relative change of resistance and the gauge factor of the filament.

The stress components caused by the force in the filament are $T_1=10^7$Pa and $T_2=T_3=T_4=T_5=T_6=0$. According to Eq. (6.1.5), the relative variation in resistance is $\Delta R/R_o=\Delta\rho_1/\rho_o=\pi_{11}T_1\cong-0.01$. The gauge factor of the filament is $G=\pi_{11}E=(102\times10^{-11}\text{/Pa})\times(1.7\times10^{11}\text{Pa})=173$.

The result shows that the resistance of the filament is reduced by 1% due to the tensile force and the gauge factor is about two orders of magnitude larger than that of a metal strain gauge.

§6.1.4. Dependence of Piezoresistance on Doping Level and Temperature
The data listed in Table 6.1.1 are measured in high resistivity silicon material at room temperature (300K). Generally, The piezoresistance is dependent on the doping level of the material and the temperature of measurement. It has been found that the magnitude of the piezoresistance coefficient decreases appreciably with the doping concentration and the temperature.

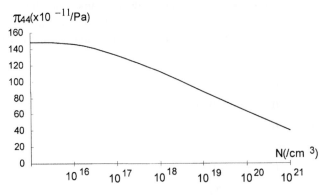

Fig. 6.1.1. Dependence of π_{44} (p-Si) on doping level

Systematic experimental studies on piezoresistance over a broad temperature range and doping concentration levels were carried out by Tufte and Stelzer [3,4] in the early 1960's, and then Pietrenko [5, 6] in 1970's. Other studies by many investigators have generally confirmed the trend of their results. Fig. 6.1.1 shows the dependence of π_{44} of p-Si on the doping level, Fig. 6.1.2 shows the temperature coefficient of π_{44} and Fig. 6.1.3 is the dependence of π_{11} of n-Si on the doping level.

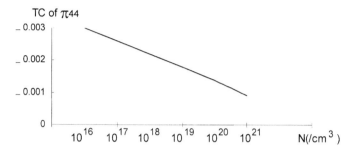

Fig. 6.1.2. Temperature coefficient of π_{44} (p-Si)

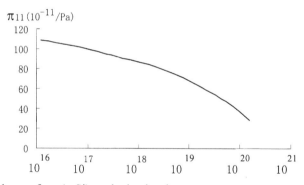

Fig. 6.1.3. Dependence of π_{11} (n-Si) on doping level

§6.1.5. Energy Band Theory of Piezoresistance Effect

The piezoresistnace effect of silicon can be explained using the energy band theory of solid-state physics. To understand the basic physics principle, we consider only the piezoresistance effect in n-type silicon; the band structure of p-type silicon is too complicated to be used as an example.

According to the energy band theory of solid-state physics, the conduction band of n-type silicon consists of six energy valleys. The minimums of the valleys have the same energy value, E_C. The six energy minimums are located on the k_x-, k_y- and k_z-axes with an equal distance to the origin of the k-space. In Fig. 6.1.4, the prolate ellipsoids of revolution that are aligned with the axes they are located show the constant-energy surface of the valleys schematically. Since the constant-energy surfaces possess principle axes of unequal length, the effective mass and, hence, the mobility of an electron in such a valley are different in the three principle directions; the mobility in the direction in parallel with the long axis μ_l is much smaller than that in the direction in perpendicular to the short axis, μ_t. The ratio of μ_l to μ_t for n-Si is

$$\gamma = \frac{\mu_t}{\mu_l} \cong 5 \tag{6.1.7}$$

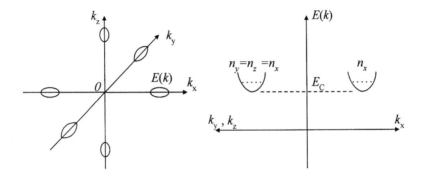

Fig. 6.1.4. Energy band structure of n-silicon

As the energy band structure has the same symmetric properties as the crystal lattice, all six valleys are identical. Therefore, the electron concentration for each valley is one sixth of the total electron concentration n. (Generally, at a room temperature, n is equal to the doping concentration of donor atoms, N_d.) Thus, the current density caused by an applied electric filed $\vec{E} = E_x \vec{i} + E_y \vec{j} + E_z \vec{k}$ is

$$\vec{j}_e = \frac{1}{6} ne \left[(2\mu_l + 4\mu_t) E_x \vec{i} + (2\mu_l + 4\mu_t) E_y \vec{j} + (2\mu_l + 4\mu_t) E_z \vec{k} \right] = \frac{1}{3} n\mu_l (1 + 2\gamma) \vec{E} \tag{6.1.8}$$

where $\vec{i}, \vec{j}, \vec{k}$ are unit vectors in the x-, y- and z-directions, respectively.

As the conductivity is independent of the direction of the electric field, the conductivity (and, hence, the resistivity) of the material is isotropic. The conductivity is

$$\sigma_o = \frac{1}{3} ne\mu_l (1 + 2\gamma) \tag{6.1.9}$$

Now, if the silicon material is applied with a normal stress, $T_1 = T_{XX}$, the lattice is stretched in the x-direction and contracts in the y- and z-directions. The change in the lattice reduces the symmetry of the lattice and causes the energy minimums on the x-axis to move up and those on the y- and z-axes to move down. Thus, part of the electron population in the k_x valleys move out of the valleys and relocated evenly in the other four valleys, as schematically shown by the dotted ellipsoids in Fig. 6.1.5. If the difference in energy between the minimums of the k_x-valleys and those of the k_y- and k_z-valleys is ΔE, we have

$$n_y = n_z = n_x e^{\frac{\Delta E}{kT}} \tag{6.1.10}$$

where n_x, n_y and n_z are electron concentrations in the k_x-, k_y- and k_z-valleys, respectively, k is the Boltzmann constant ($k=1.38\times10^{-23}$J/K, or, $k=8.62\times10^{-5}$eV/K) and T the absolute temperature. As, for small stress level, ΔE is proportional to the stress T_1, $\Delta E/kT$ is also very small and designated as αT_1. As the total electron concentration is unchanged, we have

$$n = 2(n_x + n_y + n_z) = 2n_x(1 + 2e^{\alpha T_1}).$$

Thus, we have

$$n_x = \frac{n}{2(1 + 2e^{\alpha T_1})}, n_y = n_z = \frac{n}{2(1 + 2e^{\alpha T_1})}e^{\alpha T_1} \tag{6.1.11}$$

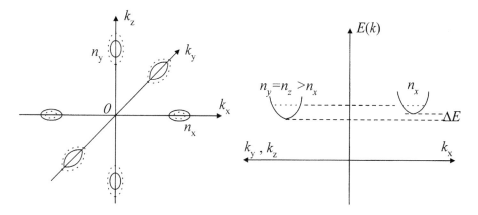

Fig. 6.1.5. Energy band structure of an n-silicon strained by T_{XX}

With the modification caused by the stress, the conductivity in the crystal changes. If the electric field is again $\vec{E} = E(l\vec{i} + m\vec{j} + n\vec{k})$, the current density is

$$\vec{j}_e = 2e[n_x(\mu_l E_x\vec{i} + \mu_t E_y\vec{j} + \mu_t E_z\vec{k}) + n_y(\mu_t E_x\vec{i} + \mu_l E_y\vec{j} + \mu_t E z\vec{k}) + n_z(\mu_t E_x\vec{i} + \mu_t E_y\vec{j} + \mu_l E_z\vec{k})]$$

By Eqs. (6.1.7) and (6.1.11), we have

$$\vec{j}_e = \frac{ne\mu_l}{1 + 2e^{\alpha T_1}}\left\{(1 + 2\gamma e^{\alpha T_1})\cdot E_x\vec{i} + (\gamma + e^{\alpha T_1} + \gamma e^{\alpha T_1})\cdot(E_y\vec{j} + E_z\vec{k})\right\} \tag{6.1.12}$$

With the approximation of $e^{\alpha T_1} \cong 1 + \alpha T_1$, we have

$$\vec{j}_e = \frac{ne\mu_l}{3+2\alpha T_1}\left\{[(1+2\gamma)+2\gamma\alpha T_1]E_x\vec{i} + [(1+2\gamma)+(1+\gamma)\alpha T_1]\cdot(E_y\vec{j}+E_z\vec{k})\right\} \qquad (6.1.13)$$

From Eq. (6.1.13), the components of conductivity are

$$\sigma_1 = ne\mu_l\frac{1+2\gamma+2\gamma\alpha T_1}{3+2\alpha T_1}, \sigma_2 = \sigma_3 = ne\mu_l\frac{(1+2\gamma)+(1+\gamma)\alpha T_1}{3+2\alpha T_1}, \sigma_4 = \sigma_5 = \sigma_6 = 0 \quad (6.1.14)$$

According to the definition given by Eq. (6.1.4), we have $\Delta_1 = \pi_{11}T_1 = (\rho_{11}-\rho_o)/\rho_o = (\sigma_o - \sigma_{11})/\sigma_{11}$. By using Eqs. (6.1.9) and (6.1.14), we have

$$\pi_{11} = -\frac{2(\gamma-1)}{3(1+2\gamma)}\alpha \qquad (6.1.15)$$

Similarly, we have

$$\pi_{12} = \frac{(\gamma-1)}{3(1+2\gamma)}\alpha \qquad (6.1.16)$$

It is interesting to find that the theoretical result, $\pi_{11}=-2\pi_{12}$, agrees very well with the Smith's experimental data given in Table 6.1.1.

§6.2. Coordinate Transformation of Second Rank Tensors

§6.2.1. Coordinate Transformation of a Vector

(1) Expression of a Vector in Different Coordinate Systems
In the coordinate system O-XYZ, a vector $\vec{r}$ as shown in Fig. 6.2.1 can be expressed as

$$\vec{r} = x\vec{i} + y\vec{j} + z\vec{k} \qquad (6.2.1)$$

where $\vec{i}$, $\vec{j}$ and $\vec{k}$ are unit vectors in the x-, y- and z-directions, respectively.

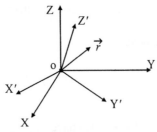

Fig.6.2.1. A vector in two coordinate systems

The same vector $\vec{r}$ can also be described in the Cartesian coordinate system O–X'Y'Z'

$$\vec{r}' = x'\vec{i} + y'\vec{j}' + z'\vec{k}' \tag{6.2.2}$$

where $\vec{i}'$, $\vec{j}'$ and $\vec{k}'$ are unit vectors in the x'-, y'- and z'-directions, respectively. As $\vec{r}$ and $\vec{r}'$ are the same vector, we have

$$x'\vec{i}' + y'\vec{j}' + z'\vec{k}' = x\vec{i} + y\vec{j} + z\vec{k} \tag{6.2.3}$$

The relationship between x', y', z' and x, y, z is referred to as the coordinate transformation relation of vector (a first rank tensor).

(2) The Relation Between Two Coordinate Systems

The relation between x', y', z' and x, y, z is related to the relation between $\vec{i}'$, $\vec{j}'$, $\vec{k}'$ and $\vec{i}$, $\vec{j}$, $\vec{k}$. Generally, the relation between $\vec{i}'$, $\vec{j}'$, $\vec{k}'$ and $\vec{i}$, $\vec{j}$, $\vec{k}$ can be expressed as

$$\begin{pmatrix} \vec{i}' \\ \vec{j}' \\ \vec{k}' \end{pmatrix} = \begin{pmatrix} l_1 & m_1 & n_1 \\ l_2 & m_2 & n_2 \\ l_3 & m_3 & n_3 \end{pmatrix} \begin{pmatrix} \vec{i} \\ \vec{j} \\ \vec{k} \end{pmatrix} \equiv (R) \begin{pmatrix} \vec{i} \\ \vec{j} \\ \vec{k} \end{pmatrix} \tag{6.2.4}$$

where l_1, m_1 and n_1 are direction cosines of $\vec{i}'$ on the X-, Y- and Z-axes, respectively. Similarly, l_2, m_2, n_2 and l_3, m_3, n_3 are direction cosines of $\vec{j}'$ and $\vec{k}'$. The matrix formed by $l's$, $m's$ and $n's$ is denoted as (R)

$$(R) = \begin{pmatrix} l_1 & m_1 & n_1 \\ l_2 & m_2 & n_2 \\ l_3 & m_3 & n_3 \end{pmatrix} \tag{6.2.5}$$

The inverse relation to Eq. (6.2.4) is

$$\begin{pmatrix} \vec{i} \\ \vec{j} \\ \vec{k} \end{pmatrix} = \begin{pmatrix} l_1 & l_2 & l_3 \\ m_1 & m_2 & m_3 \\ n_1 & n_2 & n_3 \end{pmatrix} \begin{pmatrix} \vec{i}' \\ \vec{j}' \\ \vec{k}' \end{pmatrix} \equiv (R^{-1}) \begin{pmatrix} \vec{i}' \\ \vec{j}' \\ \vec{k}' \end{pmatrix} \tag{6.2.6}$$

where

$$(R^{-1}) = \begin{pmatrix} l_1 & l_2 & l_3 \\ m_1 & m_2 & m_3 \\ n_1 & n_2 & n_3 \end{pmatrix} \tag{6.2.7}$$

From Eqs. (6.2.4) and (6.2.6), we have

$$\begin{pmatrix} \vec{i}' \\ \vec{j}' \\ \vec{k}' \end{pmatrix} = (R) \begin{pmatrix} \vec{i} \\ \vec{j} \\ \vec{k} \end{pmatrix} = (R)(R^{-1}) \begin{pmatrix} \vec{i}' \\ \vec{j}' \\ \vec{k}' \end{pmatrix}$$

where $(R)(R^{-1}) = I$ is a unity matrix. Thus, we have the relations among the direction cosines

$$\begin{aligned} l_i^2 + m_i^2 + n_i^2 &= 1 \quad (i = 1,2,3) \\ l_i l_j + m_i m_j + n_i n_j &= 0 \quad (i \neq j) \end{aligned}$$

(6.2.8)

(3) Coordinate Transformation of a Vector

According to Eqs. (6.2.1) and (6.2.6), we have

$$\vec{r} = x\vec{i} + y\vec{j} + z\vec{k} = (xl_1 + ym_1 + zn_1)\vec{i}' + (xl_2 + ym_2 + zn_2)\vec{j}' + (xl_3 + ym_3 + zn_3)\vec{k}' \quad (6.2.9)$$

From Eqs. (6.2.9) and (6.2.2), the relation between (x', y', z') and (x, y, z) is

$$\begin{pmatrix} x' \\ y' \\ z' \end{pmatrix} = \begin{pmatrix} l_1 & m_1 & n_1 \\ l_2 & m_2 & n_2 \\ l_3 & m_3 & n_3 \end{pmatrix} \begin{pmatrix} x \\ y \\ z \end{pmatrix} \equiv (R) \begin{pmatrix} x \\ y \\ z \end{pmatrix} \quad (6.2.10)$$

Or, inversely, we have

$$\begin{pmatrix} x \\ y \\ z \end{pmatrix} = \begin{pmatrix} l_1 & l_2 & l_3 \\ m_1 & m_2 & m_3 \\ n_1 & n_2 & n_3 \end{pmatrix} \begin{pmatrix} x' \\ y' \\ z' \end{pmatrix} \equiv (R^{-1}) \begin{pmatrix} x' \\ y' \\ z' \end{pmatrix} \quad (6.2.11)$$

Many physical parameters, such as current density, electric field, etc., are vectors. Eqs. (6.2.10) and (6.2.11) can be used for their transformation between different Cartesian coordinate systems.

(4) Euler's Angles

As mentioned in the above, the matrix (R) represents the relation between O–XYZ and O–X'Y'Z'. There are nine components in (R), but only three are independent ones because of the six restriction conditions given in Eq. (6.2.8). As a matter of fact, the relationship between O–XYZ and O–X'Y'Z' can be described by three rotation angles known as Euler's angles.

Let us start from the coordinate system O–XYZ with a vector $\vec{r}$ as shown in Fig. 6.2.2 and make rotation operations for the coordination system as follows:

(i) Rotating the coordinate system O–XYZ along the Z-axis by an angle of ϕ, counter-clockwise (note that $\vec{r}$ is not rotated with the axes). The original coordinate system O–XYZ becomes O–X'Y'Z' after the rotation. The vector $\vec{r}$ in O–X'Y'Z' is

$$\begin{pmatrix} x' \\ y' \\ z' \end{pmatrix} = \begin{pmatrix} \cos\phi & \sin\phi & 0 \\ -\sin\phi & \cos\phi & 0 \\ 0 & 0 & 1 \end{pmatrix} \begin{pmatrix} x \\ y \\ z \end{pmatrix}$$

(6.2.12)

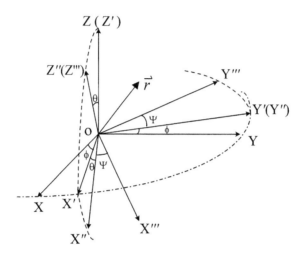

Fig. 6.2.2. Coordinate transformation by Euler's angles

(ii) Rotating the coordinate system O–X'Y'Z' along the Y'-axis by an angle θ to O–X''Y''Z''. The vector $\vec{r}$ in O–X''Y''Z'' is now

$$\begin{pmatrix} x'' \\ y'' \\ z'' \end{pmatrix} = \begin{pmatrix} \cos\theta & 0 & -\sin\theta \\ 0 & 1 & 0 \\ \sin\theta & 0 & \cos\theta \end{pmatrix} \begin{pmatrix} x' \\ y' \\ z' \end{pmatrix}$$

(6.2.13)

(iii) Rotating the coordinate system O–X''Y''Z'' along the Z''-axis by an angle ψ to the coordinate system O–X'''Y''Z'''. The vector $\vec{r}$ in the final coordinate system O–X'''Y''Z''' can be expressed as

$$\begin{pmatrix} x''' \\ y''' \\ z''' \end{pmatrix} = \begin{pmatrix} \cos\psi & \sin\psi & 0 \\ -\sin\psi & \cos\psi & 0 \\ 0 & 0 & 1 \end{pmatrix} \begin{pmatrix} x'' \\ y'' \\ z'' \end{pmatrix}$$

(6.2.14)

From Eqs. (6.2.12), (6.2.13) and (6.2.14), we have

$$
\begin{pmatrix} x''' \\ y''' \\ z''' \end{pmatrix} = \begin{pmatrix} \cos\psi\cos\theta\cos\phi - \sin\psi\sin\phi & \cos\psi\cos\theta\sin\phi + \sin\psi\cos\phi & -\cos\psi\sin\theta \\ -\sin\psi\cos\theta\cos\phi - \cos\psi\sin\phi & -\sin\psi\cos\theta\sin\phi + \cos\psi\cos\phi & \sin\psi\sin\theta \\ \sin\theta\cos\phi & \sin\theta\sin\phi & \cos\theta \end{pmatrix} \begin{pmatrix} x'' \\ y'' \\ z'' \end{pmatrix}
$$

$$
\equiv \begin{pmatrix} l_1 & m_1 & n_1 \\ l_2 & m_2 & n_2 \\ l_3 & m_3 & n_3 \end{pmatrix} \begin{pmatrix} x \\ y \\ z \end{pmatrix} \tag{6.2.15}
$$

If the Euler's Angles for the two coordinate systems are known, the direction cosines between the two coordinate systems can be found.

§6.2.2. Coordinate Transformation of Second Rank Tensors

Generally, the tensor relating two first rank tensors is a second rank tensor. A resistivity tensor relating an electric current density tensor (a first rank tensor, or a vector) and an electric field tensor (a first rank tensor, or a vector) is a second rank tensor. Other second rank tensors include conductivity, stress, strain, etc.

Take the resistivity tensor as an example. The relation between $\vec{E}$ and $\vec{J}$ in the O–XYZ coordinate system is

$$
(E) = (\rho) \cdot (J) \tag{6.2.16}
$$

while in the O–X'Y'Z' coordinate system, the same relationship can be expressed as

$$
(E') = (\rho') \cdot (J')
$$

As E and J are first rank tensors, we have $(E')=(R)\,(E)$ and $(J')=(R)\,(J)$. Therefore, we have

$$
(R) \cdot (E) = (\rho') \cdot (R) \cdot (J)
$$

By left-multiplying the equation by (R^{-1}) on both sides, we find

$$
(E) = (R^{-1}) \cdot (\rho') \cdot (R) \cdot (J) \tag{6.2.17}
$$

When Eq. (6.2.17) is compared with Eq. (6.2.16), we have

$$
(\rho) = (R^{-1}) \cdot (\rho') \cdot (R) \tag{6.2.18}
$$

Eq. (6.2.18) can also be expressed as

$$
(\rho') = (R) \cdot (\rho) \cdot (R^{-1}) \tag{6.2.19}
$$

The full expression of Eq. (6.2.19) is

$$
\begin{pmatrix} \rho'_{xx} & \rho'_{xy} & \rho'_{xz} \\ \rho'_{yx} & \rho'_{yy} & \rho'_{yz} \\ \rho'_{zx} & \rho'_{zy} & \rho'_{zz} \end{pmatrix} = \begin{pmatrix} l_1 & m_1 & n_1 \\ l_2 & m_2 & n_2 \\ l_3 & m_3 & n_3 \end{pmatrix} \cdot \begin{pmatrix} \rho_{xx} & \rho_{xy} & \rho_{xz} \\ \rho_{yx} & \rho_{yy} & \rho_{yz} \\ \rho_{zx} & \rho_{zy} & \rho_{zz} \end{pmatrix} \cdot \begin{pmatrix} l_1 & l_2 & l_3 \\ m_1 & m_2 & m_3 \\ n_1 & n_2 & n_3 \end{pmatrix} \tag{6.2.20}
$$

If Eq. (6.2.20) is developed and the simplified notations are used for the symmetrical resistivity tensor, we have

$$
\begin{pmatrix} \rho'_1 \\ \rho'_2 \\ \rho'_3 \\ \rho'_4 \\ \rho'_5 \\ \rho'_6 \end{pmatrix} = \begin{pmatrix} l_1^2 & m_1^2 & n_1^2 & 2m_1n_1 & 2n_1l_1 & 2l_1m_1 \\ l_2^2 & m_2^2 & n_2^2 & 2m_2n_2 & 2n_2l_2 & 2l_2m_2 \\ l_3^2 & m_3^2 & n_3^2 & 2m_3n_3 & 2n_3l_3 & 2l_3m_3 \\ l_2l_3 & m_2m_3 & n_2n_3 & m_2n_3+m_3n_2 & n_2l_3+n_3l_2 & m_2l_3+m_3l_2 \\ l_3l_1 & m_3m_1 & n_3n_1 & m_3n_1+m_1n_3 & n_3l_1+n_1l_3 & m_3l_1+m_1l_3 \\ l_1l_2 & m_1m_2 & n_1n_2 & m_1n_2+m_2n_1 & n_1l_2+n_2l_1 & m_1l_2+m_2l_1 \end{pmatrix} \begin{pmatrix} \rho_1 \\ \rho_2 \\ \rho_3 \\ \rho_4 \\ \rho_5 \\ \rho_6 \end{pmatrix} \tag{6.2.21}
$$

The simplified notation for Eq. (6.2.21) is $(\rho')=(\alpha)(\rho)$ or $\rho'=\alpha\,\rho$ [2], where

$$
\alpha = \begin{vmatrix} l_1^2 & m_1^2 & n_1^2 & 2m_1n_1 & 2n_1l_1 & 2l_1m_1 \\ l_2^2 & m_2^2 & n_2^2 & 2m_2n_2 & 2n_2l_2 & 2l_2m_2 \\ l_3^2 & m_3^2 & n_3^2 & 2m_3n_3 & 2n_3l_3 & 2l_3m_3 \\ l_2l_3 & m_2m_3 & n_2n_3 & m_2n_3+m_3n_2 & n_2l_3+n_3l_2 & m_2l_3+m_3l_2 \\ l_3l_1 & m_3m_1 & n_3n_1 & m_3n_1+m_1n_3 & n_3l_1+n_1l_3 & m_3l_1+m_1l_3 \\ l_1l_2 & m_1m_2 & n_1n_2 & m_1n_2+m_2n_1 & n_1l_2+n_2l_1 & m_1l_2+m_2l_1 \end{vmatrix} \tag{6.2.22}
$$

The inverse form of Eq. (6.2.21) is $(\rho)=(\alpha^{-1})(\rho')$, with

$$
\alpha^{-1} = \begin{vmatrix} l_1^2 & l_2^2 & l_3^2 & 2l_2l_3 & 2l_3l_1 & 2l_1l_2 \\ m_1^2 & m_2^2 & m_3^2 & 2m_2m_3 & 2m_3m_1 & 2m_1m_2 \\ n_1^2 & n_2^2 & n_3^2 & 2n_2n_3 & 2n_3n_1 & 2n_1n_2 \\ m_1n_1 & m_2n_2 & m_3n_3 & m_2n_3+m_3n_2 & m_3n_1+m_1n_3 & m_1n_2+m_2n_1 \\ n_1l_1 & n_2l_2 & n_3l_3 & n_2l_3+n_3l_2 & n_3l_1+n_1l_3 & n_1l_2+n_2l_1 \\ l_1m_1 & l_2m_2 & l_3m_3 & m_2l_3+m_3l_2 & m_3l_1+m_1l_3 & m_1l_2+m_2l_1 \end{vmatrix} \tag{6.2.23}
$$

As the stress T, the strain ε and the relative change of resistivity Δ are all second rank tensors, the coordinate transformation relationships for them are $T'=\alpha T$, $\varepsilon'=\alpha\varepsilon$ and $\Delta'=\alpha\Delta$, respectively.

For example, a stress is caused by a force in the x-direction so that the stress has a single non-zero component, $T_1 = T_{XX}$, in the coordinate system O–XYZ. Now we examine the stress in a new coordinate system O–X'Y'Z' that is formed by rotating the original coordinate system around the Z-axis by an angle of $\pi/4$. Therefore, we have

$$(R) = \begin{pmatrix} \sqrt{2}/2 & \sqrt{2}/2 & 0 \\ -\sqrt{2}/2 & \sqrt{2}/2 & 0 \\ 0 & 0 & 1 \end{pmatrix} \tag{6.2.24}$$

According to Eqs. (6.2.22), (6.2.24) and the relation $(T') = (\alpha)\,(T)$, we have

$$T_1' = T_{X'X'} = \frac{1}{2}T_1, T_2' = T_{Y'Y'} = \frac{1}{2}T_1, T_3' = T_4' = T_5' = 0, T_6' = T_{X'Y'} = -\frac{1}{2}T_1 \tag{6.2.25}$$

Note that T_1' and T_2' are a half T_1 and a shearing stress component, T_6', appears.

A common mistake often made by a beginner is to misuse the transformation method of force (a vector) for second rank tensors. If this mistake were made, the stress components in the new coordinate system would be

$$T_{X'X'} = \frac{\sqrt{2}}{2}T_{XX}; T_{Y'Y'} = -\frac{\sqrt{2}}{2}T_{XX}$$

with no shearing stress component appearing.

The results shown in Eq. (6.2.25) can also be justified by a straightforward derivation. Suppose that a normal stress component $T_{XX} = T_1$ in the Cartesian coordinate system O–XYZ is shown in Fig. 6.2.3(a) in a two dimensional version. Square *abcd* represents a unit cube with surfaces parallel to the coordinate planes. The force applied on the surface *ad* or *bc* is $T_1 \cdot 1$.

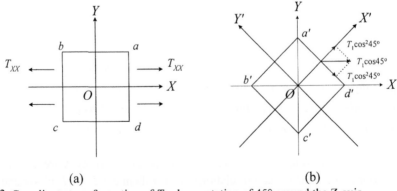

(a)   (b)

Fig. 6.2.3. Coordinate transformation of T_{XX} by a rotation of 45° around the Z-axis

If the coordinate system with the unit cube is rotated around the Z-axis by an angle of 45°, as shown in Fig. 6.2.3(b), the force applied on the unit surface $a'd'$ is $T_1 \cdot \cos 45°$ in the x-direction as the unit cube has been rotated with the coordinate system. This force can be resolved into two equal-valued force components: one in the x'-direction and the other in the y'-direction. The values for the two components are both $T_1 \cdot \cos^2 45° = T_1 / 2$. The corresponding stress components in the O–X'Y'Z' are $T_{X'X'} = T_1 / 2$ and $T_{X'Y'} = -T_1 / 2$. By a similar discussion on the force on plane $a'b'$ or $c'd'$, we find the stress components: $T_{Y'Y'} = T_1 / 2$ and $T_{Y'X'} = -T_1 / 2$. The results agree with those given in Eq. (6.2.25).

§6.3. Coordinate Transformation of Piezoresistive Coefficient

§6.3.1. Coordinate Transformation of Forth Rank Tensors

As discussed in the above sections, the piezoresistive coefficient is a forth rank tensor relating two tensors of the second rank, the resistivity tensor ρ and the stress tensor T. In the coordinate system O–XYZ, the relation is $\Delta = \pi T$ as shown by Eq. (6.1.5). The same expression in the O–X'Y'Z' coordinate system is [2]

$$\Delta' = \pi' T' \tag{6.3.1}$$

As Δ and T are second rank tensors, we have $\alpha \Delta = \pi' \alpha T$. Therefore, we obtain

$$\Delta = \alpha^{-1} \pi' \alpha T \tag{6.3.2}$$

When Eq. (6.3.2) is compared with Eq. (6.1.5), we have the relation of coordinate transformation for the piezoresistive coefficient tensor $\pi = \alpha^{-1} \pi' \alpha$, or

$$\pi' = \alpha \pi \alpha^{-1} \tag{6.3.3}$$

This means that the components of π' can be expressed by the components of π by

$$\pi'_{ij} = \sum_{k,l=1}^{6} \alpha_{ik} \pi_{kl} \alpha_{lj}^{-1} \tag{6.3.4}$$

§6.3.2. Simplification by Symmetric Operations

For a silicon crystal, there are 32 symmetric operations in O_h group that do not change any physical property of the crystal. This means that all physical parameters (including the piezoresistive coefficient tensor) remain unchanged by any of the symmetric operations. By making use of this property, some very useful qualitative information can be obtained. For example, one of the symmetric operations is rotating the crystal around the Z-axis of the crystallographic coordinate system by an angle of π. This operation can be expressed by

$$(R) = \begin{pmatrix} -1 & 0 & 0 \\ 0 & -1 & 0 \\ 0 & 0 & 1 \end{pmatrix}$$

Therefore, we have

$$(\alpha) = \begin{pmatrix} 1 & 0 & 0 & 0 & 0 & 0 \\ 0 & 1 & 0 & 0 & 0 & 0 \\ 0 & 0 & 1 & 0 & 0 & 0 \\ 0 & 0 & 0 & -1 & 0 & 0 \\ 0 & 0 & 0 & 0 & -1 & 0 \\ 0 & 0 & 0 & 0 & 0 & 1 \end{pmatrix}$$

and

$$(\alpha^{-1}) = \begin{pmatrix} 1 & 0 & 0 & 0 & 0 & 0 \\ 0 & 1 & 0 & 0 & 0 & 0 \\ 0 & 0 & 1 & 0 & 0 & 0 \\ 0 & 0 & 0 & -1 & 0 & 0 \\ 0 & 0 & 0 & 0 & -1 & 0 \\ 0 & 0 & 0 & 0 & 0 & 1 \end{pmatrix}$$

Starting from a general form of (π) shown in Eq. (6.1.5), (π') in the new coordinate system is found to be

$$(\pi') = \alpha\pi\alpha^{-1} = \begin{pmatrix} \pi_{11} & \pi_{12} & \pi_{13} & -\pi_{14} & -\pi_{15} & \pi_{16} \\ \pi_{21} & \pi_{22} & \pi_{23} & -\pi_{24} & -\pi_{25} & \pi_{26} \\ \pi_{31} & \pi_{32} & \pi_{33} & -\pi_{34} & -\pi_{35} & \pi_{36} \\ -\pi_{41} & -\pi_{42} & -\pi_{43} & \pi_{44} & \pi_{45} & -\pi_{46} \\ -\pi_{51} & -\pi_{52} & -\pi_{53} & \pi_{54} & \pi_{55} & -\pi_{56} \\ \pi_{61} & \pi_{62} & \pi_{63} & -\pi_{64} & -\pi_{65} & \pi_{66} \end{pmatrix} \qquad (6.3.5)$$

For a symmetric operation, a physical parameter remains unchanged after the coordinate transformation, i.e. $\pi'=\pi$. From Eq. (6.3.5) and Eq. (6.1.5), we conclude that the components π_{14}, π_{15}, π_{24}, π_{25}, π_{34}, π_{35}, π_{41}, π_{42}, π_{43}, π_{46}, π_{51}, π_{52}, π_{53}, π_{56}, π_{64}, π_{65} are all zero.

By using the symmetric operations of the silicon crystal one by one, we can verify that the piezoresistive coefficient tensor of silicon has only three independent non-zero components: $\pi_{11} = \pi_{22} = \pi_{33}$, $\pi_{44} = \pi_{55} = \pi_{66}$ and $\pi_{12} = \pi_{21} = \pi_{13} = \pi_{31} = \pi_{23} = \pi_{32}$ as shown in Eq. (6.1.6)

§6.3.3. Piezoresistance in an Arbitrary Coordinate System

Using Eq. (6.3.4), all components of the piezoresistive coefficient tensor in any coordinate system can be expressed by the components in a crystallographic coordinate system. For example, π'_{11} can be found by Eq. (6.3.4)

$$\pi'_{11} = \sum_{i,j=1}^{3} \alpha_{1i} \pi_{ij} \alpha_{j1}^{-1} \tag{6.3.6}$$

As π has only 12 non-zero components and each equals one of the three independent components, π_{11}, π_{22} and π_{44}, as shown in Eq. (6.1.6), the relationships can be significantly simplified.

For example, from Eq. (5.3.6), π'_{11} is simplified as

$$\begin{aligned}
\pi'_{11} = \ & \pi_{11}\left(\alpha_{11}\alpha_{11}^{-1} + \alpha_{12}\alpha_{21}^{-1} + \alpha_{13}\alpha_{31}^{-1}\right) \\
& + \pi_{12}\left(\alpha_{11}\alpha_{21}^{-1} + \alpha_{11}\alpha_{31}^{-1} + \alpha_{12}\alpha_{11}^{-1} + \alpha_{12}\alpha_{31}^{-1} + \alpha_{13}\alpha_{11}^{-1} + \alpha_{13}\alpha_{21}^{-1}\right) \\
& + \pi_{44}\left(\alpha_{14}\alpha_{41}^{-1} + \alpha_{15}\alpha_{51}^{-1} + \alpha_{16}\alpha_{61}^{-1}\right)
\end{aligned} \tag{6.3.7}$$

The components of α and α^{-1} can be found from Eqs. (6.2.22) and (6.2.23). Therefore, we have

$$\begin{aligned}
\pi'_{11} = \ & \pi_{11}\left(l_1^2 l_1^2 + m_1^2 m_1^2 + n_1^2 n_1^2\right) + \pi_{12}\left(l_1^2(m_1^2 + n_1^2) + m_1^2(l_1^2 + n_1^2) + n_1^2(l_1^2 + m_1^2)\right) \\
& + \pi_{44}\left(2m_1^2 n_1^2 + 2n_1^2 l_1^2 + 2l_1^2 m_1^2\right)
\end{aligned}$$

By using the relations between $l's$, $m's$ and $n's$ given in Eq. (6.2.8), we finally have [7]

$$\pi'_{11} = \pi_{11} + 2(\pi_{44} + \pi_{12} - \pi_{11}) \cdot \left(l_1^2 m_1^2 + l_1^2 n_1^2 + m_1^2 n_1^2\right)$$

Using the notation of $\pi_o = \pi_{11} - \pi_{12} - \pi_{44}$, we have

$$\pi'_{11} = \pi_{11} - 2\pi_o\left(l_1^2 m_1^2 + l_1^2 n_1^2 + m_1^2 n_1^2\right)$$

All other components of π' matrix are listed in Table 6.3.1. The same results have been derived in a batch by using matrix algebra method by M. Bao, et al [8] recently. Table 6.3.1 is also applicable to any other forth rank tensors.

When the relations are used for compliance tensor of a homogeneous material (which is of the highest symmetry), we will have $\Sigma_o = \Sigma_{11} - \Sigma_{12} - \Sigma_{44} = 0$, or, $\Sigma_{11} - \Sigma_{12} = \Sigma_{44}$. When this relation is considered with Eq. (2.1.19), the relation given by Eq. (2.1.16) is verified.

Table 6.3.1. Components of the piezoresistive coefficient tensor in an arbitrary coordinate system (Note: $\pi_o = \pi_{11} - \pi_{12} - \pi_{44}$)

$\pi'_{11} = \pi_{11} - 2\pi_o\left(l_1^2 m_1^2 + l_1^2 n_1^2 + m_1^2 n_1^2\right)$	$\pi'_{14} = 2\pi'_{41}$
$\pi'_{21} = \pi_{12} + \pi_o\left(l_1^2 l_2^2 + m_1^2 m_2^2 + n_1^2 n_2^2\right)$	$\pi'_{24} = 2\pi'_{42}$
$\pi'_{31} = \pi_{12} + \pi_o\left(l_1^2 l_3^2 + m_1^2 m_3^2 + n_1^2 n_3^2\right)$	$\pi'_{34} = 2\pi'_{43}$
$\pi'_{41} = \pi_o\left(l_1^2 l_2 l_3 + m_1^2 m_2 m_3 + n_1^2 n_2 n_3\right)$	$\pi'_{44} = \pi_{44} + 2\pi_o\left(l_2^2 l_3^2 + m_2^2 m_3^2 + n_2^2 n_3^2\right)$
$\pi'_{51} = \pi_o\left(l_1^3 l_3 + m_1^3 m_3 + n_1^3 n_3\right)$	$\pi'_{54} = 2\pi_o\left(l_1 l_2 l_3^2 + m_1 m_2 m_3^2 + n_1 n_2 n_3^2\right.$
$\pi'_{61} = \pi_o\left(l_1^3 l_2 + m_1^3 m_2 + n_1^3 n_2\right)$	$\pi'_{64} = 2\pi_o\left(l_1 l_2^2 l_3 + m_1 m_2^2 m_3 + n_1 n_2^2 n_3\right)$
$\pi'_{12} = \pi'_{21}$	$\pi'_{15} = 2\pi'_{51}$
$\pi'_{22} = \pi_{11} - 2\pi_o\left(l_2^2 m_2^2 + l_2^2 n_2^2 + m_2^2 n_2^2\right)$	$\pi'_{25} = 2\pi'_{52}$
$\pi'_{32} = \pi_{12} + \pi_o\left(l_2^2 l_3^2 + m_2^2 m_3^2 + n_2^2 n_3^2\right)$	$\pi'_{35} = 2\pi'_{53}$
$\pi'_{42} = \pi_o\left(l_2^3 l_3 + m_2^3 m_3 + n_2^3 n_3\right)$	$\pi'_{45} = 2\pi'_{63} = \pi'_{54}$
$\pi'_{52} = \pi_o\left(l_1 l_2^2 l_3 + m_1 m_2^2 m_3 + n_1 n_2^2 n_3\right)$	$\pi'_{55} = \pi_{44} + 2\pi_o\left(l_1^2 l_3^2 + m_1^2 m_3^2 + n_1^2 n_3^2\right)$
$\pi'_{62} = \pi_o\left(l_1 l_2^3 + m_1 m_2^3 + n_1 n_2^3\right)$	$\pi'_{65} = 2\pi'_{41}$
$\pi'_{13} = \pi'_{31}$	$\pi'_{16} = 2\pi'_{61}$
$\pi'_{23} = \pi'_{32}$	$\pi'_{26} = 2\pi'_{62}$
$\pi'_{33} = \pi_{11} - 2\pi_o\left(l_3^2 m_3^2 + l_3^2 n_3^2 + m_3^2 n_3^2\right)$	$\pi'_{36} = 2\pi'_{63}$
$\pi'_{43} = \pi_o\left(l_2 l_3^3 + m_2 m_3^3 + n_2 n_3^3\right.$	$\pi'_{46} = 2\pi'_{52}$
$\pi'_{53} = \pi_o\left(l_1 l_3^3 + m_1 m_3^3 + n_1 n_3^3\right)$	$\pi'_{56} = 2\pi'_{41}$
$\pi'_{63} = \pi_o\left(l_1 l_2 l_3^2 + m_1 m_2 m_3^2 + n_1 n_2 n_3^2\right)$	$\pi'_{66} = \pi_{44} + 2\pi_o\left(l_1^2 l_2^2 + m_1^2 m_2^2 + n_1^2 n_2^2\right)$

§6.4. Piezoresistive Sensing Elements

§6.4.1. Piezoresistor

Consider a layer of rectangular material of silicon. The thickness, h, of the layer is much smaller than its width, w, and its width is smaller than its length, L (i.e., $L \gg w \gg h$). If electrodes are positioned at the two ends as shown in Fig. 6.4.1, the structure is simply a resistor. For analysis the coordinate system is taken as: x' is in the length direction, y' in the structural plane and in perpendicular to the length direction, and z' normal to the structural plane. Generally, the coordinate system $O–x'y'z'$ is referred to as a resistor coordinate system and is related to the crystallographic coordinate system $O–XYZ$ of silicon crystal by a rotation operation represented by

$$(R) = \begin{pmatrix} l_1 & m_1 & n_1 \\ l_2 & m_2 & n_2 \\ l_3 & m_3 & n_3 \end{pmatrix}$$

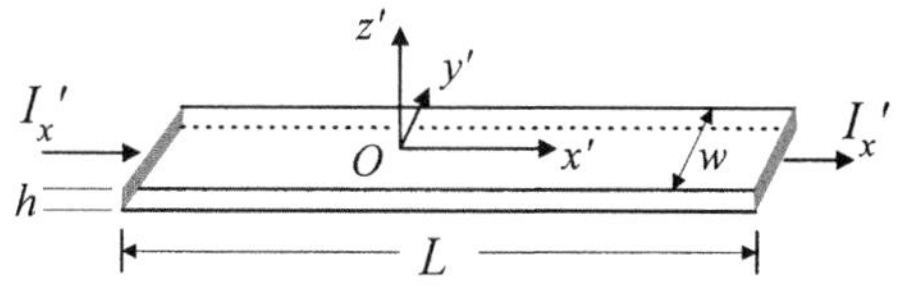

Fig. 6.4.1. A piezoresistor and its coordinate system

If the material is free of stress, the resistivity of the silicon material is a scalar, ρ_o, and the resistance between the two electrodes is $R_o = L\rho_o/wh$. The $I–V$ relation of the resistor is the well-known Ohms' law for isotropic material, $V = IR_o$. When the material of the resistor is stressed, the resistivity of the material, ρ', is a second rank tensor relating the electric field tensor and the current density tensor

$$\begin{pmatrix} E'_x \\ E'_y \\ E'_z \end{pmatrix} = \begin{pmatrix} \rho'_1 & \rho'_6 & \rho'_5 \\ \rho'_6 & \rho'_2 & \rho'_4 \\ \rho'_5 & \rho'_4 & \rho'_3 \end{pmatrix} \begin{pmatrix} J'_x \\ J'_y \\ J'_z \end{pmatrix} \tag{6.4.1}$$

Because the electric field and the electric current in the normal direction of the layer are negligible, Eq. (6.4.1) can be simplified to a two-dimensional form

$$\begin{pmatrix} E'_x \\ E'_y \end{pmatrix} = \begin{pmatrix} \rho'_1 & \rho'_6 \\ \rho'_6 & \rho'_2 \end{pmatrix} \begin{pmatrix} J'_x \\ J'_y \end{pmatrix} \tag{6.4.2}$$

As the length, L, is much larger than the width, w, we have $J_y' = 0$ except for the areas near the electrodes of the resistor. Thus, Eq. (6.4.2) can be further simplified to

$$E'_x = \rho'_1 J'_x$$
$$E'_y = \rho'_6 J'_x \tag{6.4.3}$$

Since $E_x'=V_S/L$, where V_S is the voltage difference between the two electrodes, the current passing through the resistor is

$$I_x' = J_x' wh = \frac{wh}{L}\frac{V_S}{\rho_1'}$$

If this relation is compared with Ohms' law for isotropic material, the resistance is

$$R = \frac{L}{wh}\rho_1' \tag{6.4.4}$$

The resistance is stress dependent as ρ_1'. When Eq. (6.4.4) is compared with the original resistance, $R_o=L\rho_o/wh$, the relative change of the resistance is

$$\frac{\Delta R}{R_o} = \frac{\rho_1' - \rho_o}{\rho_o} = \Delta_1' \tag{6.4.5}$$

where

$$\Delta' = \pi_{11}'T_1' + \pi_{12}'T_2' + \pi_{13}'T_3' + \pi_{14}'T_4' + \pi_{15}'T_5' + \pi_{16}'T_6' \tag{6.4.6}$$

For most applications in piezoresistive pressure transducers and accelerometers, the resistor is located on the surface of a diaphragm or beam. Therefore, the material is stressed in two dimensions at the surface plane. In this case, we have $T_3'= T_4'= T_5'=0$ and Eq. (6.4.6) can be simplified to

$$\frac{\Delta R}{R} = \pi_{11}'T_1' + \pi_{12}'T_2' + \pi_{16}'T_6' \tag{6.4.7}$$

So we can conclude that the resistance of the two-terminal silicon resistor shown in Fig. 6.4.1 is sensitive to the stress or stain in the material. Therefore, the resistor can be used as a sensing element for stress or strain.

In literatures, Eq. (6.4.7) is often written as

$$\frac{\Delta R}{R} = \pi_l T_l + \pi_t T_t + \pi_s T_s \tag{6.4.8}$$

where $\pi_l=\pi_{11}'$ is referred to as the longitudinal piezoresistive coefficient, $\pi_t=\pi_{12}'$ the transversal piezoresistive coefficient and $\pi_s=\pi_{16}'$ the shearing piezoresistive coefficient. As mentioned in Chapter 2, the maximum strain applicable in a silicon structure is usually of the order of 10^{-4}, i.e., the maximum stress is of the order of 10^7 Pa. As the piezoresistive coefficient of silicon is of the order of 10^{-9}/Pa, according to Eqs. (6.4.7) or (6.4.8), the maximum piezoresistance caused by stress in silicon is of the order of 10^{-2}.

However, for a typical silicon resistor, the temperature coefficient of resistance is about 0.2%/°C. This means that the piezoresistance effect of a resistor can easily be over-shadowed by its temperature drift. Therefore, a resistor is rarely used alone. The piezoresistors are usually used in the form of Wheatstone bridge as shown in Fig. 6.4.2 [9]. Fig. 6.4.2(a) gives a schematic view of a piezoresistive accelerometer with a beam-mass structure. The beam is along the <110> direction and four p-type resistors, R_1~R_4, are formed by diffusion or ion-implantation on the surface of the n-Si beam [9]. Two of the resistors (R_2 and R_3) are parallel to the beam direction and the other two (i.e., R_1 and R_4) are perpendicular to the beam direction. The four resistors are connected through metallization to form a Wheatstone bridge as shown in Fig. 6.4.2 (b). In the following discussion, for simplicity, we assume that the four resistors are close together so that they experience the same stress with a stress component T_1 in the beam direction.

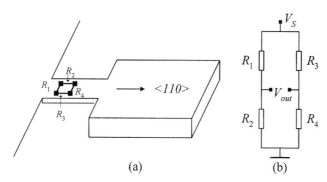

(a)   (b)

Fig. 6.4.2. Piezoresistive sensor on a beam

For resistors R_2 and R_3, the resistor's coordinate system is taken as: x' along the direction of the resistors (the beam direction, i.e., a <110> direction), y' in the surface plane and perpendicular to the beam direction (another <110> direction) and z' normal to the surface plane. When referred to the crystallographic coordinate system of the silicon crystal, the coordinate transformation of the resistor coordinate system is

$$(R) = \begin{pmatrix} \sqrt{2}/2 & \sqrt{2}/2 & 0 \\ -\sqrt{2}/2 & \sqrt{2}/2 & 0 \\ 0 & 0 & 0 \end{pmatrix} \equiv \begin{pmatrix} l_1 & m_1 & n_1 \\ l_2 & m_2 & n_2 \\ l_3 & m_3 & n_3 \end{pmatrix} \tag{6.4.9}$$

According to Table 6.3.1 and Eq. (6.4.9), we have $\pi_{11}'=\pi_{44}/2$ and $\pi_{12}'=-\pi_{44}/2$. Therefore, for R_2 and R_3, the relative change in piezoresistance is (note that $T_1=T_1$ and $T_1=T_s=0$)

$$\frac{\Delta R}{R} = \frac{\pi_{44}}{2} T_1 \tag{6.4.10}$$

For resistors R_1 and R_4, x' is perpendicular to the beam direction (also in the direction of the resistors). Thus, we have $T_t = T_1$ and $T_l = T_s = 0$. The change in piezoresistance for resistors R_1 and R_4 is

$$\frac{\Delta R}{R} = -\frac{\pi_{44}}{2} T_1 \tag{6.4.11}$$

Therefore, if the Wheatstone bridge shown in Fig. 6.4.2 has a supply voltage of V_S, the output is $V_{out} = V_S |\Delta R / R|$. If $V_S = 5V$ and $|\Delta R / R| = 2\%$, the output is 100 mV. As the temperature coefficients of the four resistors are identical, the effect of the temperature coefficient can be canceled out by the balanced configuration.

Piezoresistive Wheatstone bridge can also be composed of four resistors all subjected to lateral stresses, but with two opposite signs. A typical design is the twin-island structure pressure transducer as shown in Fig. 6.4.3 [10]. With applied pressure, the stresses on R_1 and R_4 are tensile and the stresses on R_2 and R_3 are compressive, or vice verse.

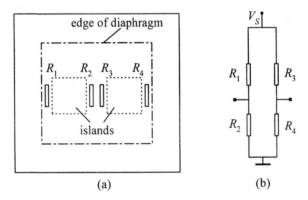

(a)   (b)

Fig. 6.4.3. Pressure sensor with four resistors all subjected to lateral stress

§6.4.2. Four-terminal Sensing Element

(1) Working Principle

For a rectangular resistor as shown in Fig. 6.4.1, Eq. (6.4.3) predicts that lateral electric field E_y' will be caused by a stress and a longitudinal current density, J_x', though the lateral electric field was not considered there due to the small width of the resistors. If the finite width of the resistor is considered, a transverse (or a lateral) voltage, V_T, is established across the resistor. According to Eq. (6.4.3), the lateral voltage is

$$V_T = E_y' w = \frac{w}{L} \frac{\rho_6'}{\rho_1'} V_S \tag{6.4.12}$$

If ρ_1' and ρ_6' are expressed by π_{ij}'s and $T's$, Eq. (6.4.12) can be written as

$$V_T = \frac{wV_S}{L}\left(\frac{\pi'_{61}T'_1 + \pi'_{62}T'_2 + \pi'_{66}T'_6}{1 + \pi'_{11}T'_1 + \pi'_{12}T'_2 + \pi'_{66}T'_6}\right) \tag{6.4.13}$$

As the relative change of resistance caused by the piezoresistance effect is much smaller than unity, Eq. (6.4.13) can be approximated as

$$V_T = \frac{wV_S}{L}\left(\pi_{61}\,T'_1 + \pi_{62}\,T'_2 + \pi_{66}\,T'_6\right) \tag{6.4.14}$$

By making use of the lateral voltage caused by a stress, a four-terminal symmetric structure as shown in Fig. 6.4.4 can be used as a sensing element for stress or strain. When a voltage V_S is supplied between two input electrodes, 1 and 2, a voltage output V_T between the output electrodes, 3 and 4, will be caused by the stress in the material. As the four-terminal sensing element has a symmetric geometry, the original output of the sensing element should be close to zero and the output is insensitive to the temperature coefficient of resistivity. Therefore, the sensing element can be used alone. As the sensing element shown in Fig. 6.4.4 looks like a Hall effect element for a magnetic field sensor, it is also called a "Hall-like" sensing element [11].

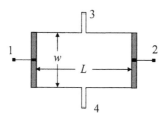

Fig. 6.4.4. Schematic drawing of a four-terminal (Hall-like) piezoresistive sensing element

Now let us consider the orientation of the element for a maximum sensitivity for a p-type four-terminal sensing element on a beam-mass structure. Supposing that the element is oriented in a direction with an angle β with the beam direction, as shown in Fig. 6.4.5, and the only stress component is a normal stress in the beam direction, T_1, according to the relation $T' = \alpha T$ and Eq. (6.2.22) for α, we have

$$T_1'= T_1\cos^2\beta;\ T_2'= T_1\sin^2\beta;\ T_6'= -T_1\sin\beta\cos\beta = -\tfrac{1}{2}T_1\sin(2\beta) \tag{6.4.15}$$

As the coordinate transformation for the piezoresistive coefficient is referred to the crystallographic system (X=[100], etc., as shown in Fig. 6.4.5), we find

$$\begin{aligned}
\pi'_{61} &= -\pi_{44}\left(-\cos^3(\beta+45°)\sin(\beta+45°)+\sin^3(\beta+45°)\cos(\beta+45°)\right) \\
\pi'_{62} &= -\pi_{44}\left(-\cos(\beta+45°)\sin^3(\beta+45°)+\sin(\beta+45°)\cos^3(\beta+45°)\right) \\
\pi'_{66} &= +\pi_{44}\left(1-4\sin^2(\beta+45°)\cos^2(\beta+45°)\right)
\end{aligned} \tag{6.4.16}$$

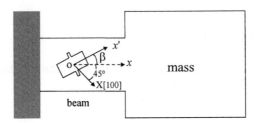

Fig. 6.4.5. Design of a four-terminal sensing element on a beam-mass structure accelerometer

By substituting Eqs. (6.4.15) and (6.4.16) into Eq. (6.4.14), we finally have

$$V_T = \frac{wV_S}{L}\left(-\frac{1}{2}\pi_{44}T_1\sin 2\beta\right) \tag{6.4.17}$$

Obviously, the β for maximum sensitivity is $\pm 45°$[12]. As the four terminal sensing element looks like an "X", it is trade-marked as "X-ducer" by the manufacturer.

According to Eqs. From (6.4.12) to (6.4.14), the sensitivity of a four-terminal sensing element is proportional to the ratio of w/L. It seems that for a "fat" design with a high w/L ratio, the sensitivity could be very high. This is in fact not true as the equations apply only when w/L is much smaller than unity so that $J_y'=0$ is true for most of the element. As a matter of fact, J_y' has a finite value near the electrodes due to the short-circuit effect of the electrodes. This will be discussed as follows.

(2) Short-circuit Effect of Four-terminal Sensing Element

Generally, the potential distribution in a single element sensor with a finite w/L can be found by solving a differential equation. The output voltage, V_T, can then be found according to the potential distribution.

For a rectangular four-terminal element as shown in Fig. 6.4.6, Eq. (6.4.2) is written in an inverse form (the prime signs have been removed for simplicity)

$$\begin{aligned} J_x &= \sigma_1 E_x + \sigma_6 E_y \\ J_y &= \sigma_6 E_x + \sigma_2 E_y \end{aligned} \tag{6.4.18}$$

By defining $\sigma_o \equiv 1/\rho_o$ and $B \equiv (\pi_{61}T_1' + \pi_{62}T_2' + \pi_{66}T_6')$, σ_1, σ_2 and σ_6 can be approximated, respectively, as

$$\sigma_1 = \frac{\rho_2}{\rho_1\rho_2 - \rho_6^2} \cong \frac{1}{\rho_o} = \sigma_o; \sigma_2 = \frac{\rho_1}{\rho_1\rho_2 - \rho_6^2} \cong \frac{1}{\rho_o} = \sigma_o; \sigma_6 = -\frac{B\rho_o}{\rho_1\rho_2 - \rho_6^2} \cong -\frac{B}{\rho_o} = -B\sigma_o \tag{6.4.19}$$

As there is no further charge accumulation once the current is stabilized, the electric current should satisfy the condition of $\nabla \vec{J} = 0$, i.e.,

$$\frac{\partial J_x}{\partial x} + \frac{\partial J_y}{\partial y} = 0 \tag{6.4.20}$$

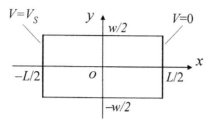

Fig. 6.4.6. The geometries of a single-element sensing element

By substituting Eqs. (6.4.18) and (6.4.19) into Eq. (6.4.20) and making use of the relation between the electric field, E, and the potential distribution $V(x,y)$,

$$E_x = -\frac{\partial V(x,y)}{\partial x}, E_y = -\frac{\partial V(x,y)}{\partial y}$$

the differential equation of potential distribution is

$$\frac{\partial^2 V}{\partial x^2} - 2B\frac{\partial^2 V}{\partial x \partial y} + \frac{\partial^2 V}{\partial y^2} = 0 \tag{6.4.21}$$

The boundary conditions for the equation are

$$V\left(-\frac{L}{2}, y\right) = V_S, V\left(\frac{L}{2}, y\right) = 0, \frac{\partial V}{\partial y}\left(x, \pm\frac{w}{2}\right) = 0 \tag{6.4.22}$$

From Eqs. (6.4.21) and (6.4.22), $V(x,y)$ can be found by numerical analysis. Once $V(x,y)$ is found, the output $V_T = V(0,w/2) - V(0,-w/2)$ can be obtained. According to [13], V_T can be approximated as $BV_S w/L$ for small w/L. However, when w/L is large enough, V_T levels off, approaching BV_S. The dependence of V_T on w/L is shown in Fig. 6.4.7.

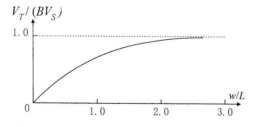

Fig. 6.4.7. The dependence of output V_T on w/L

Often, a correction factor is added to Eq. (6.4.12) and the output is

$$V_T = f\left(\frac{w}{L}\right) \cdot \frac{w}{L}\frac{\rho_6{'}}{\rho_1{'}} V_S \tag{6.4.23}$$

where $f(w/L)$ is referred to as the correction factor of short circuit effect of the electrodes. Obviously, $f(w/L)$ equals 1 for small w/L, and decreases which w/L monotonously, as shown in Fig. 6.4.8.

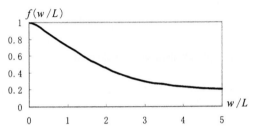

Fig. 6.4.8. The dependence of the correction factor for short circuit effect on w/L

As the power consumption in the element increases with the square of w/L, the ratio of w/L should not be too large. Usually, the w/L ratio used is around 0.6. In this case, we have

$$V_T \doteq 0.5BV_S = 0.5 \cdot \left(\pi_{61}\, T_1' + \pi_{62}\, T_2' + \pi_{66}\, T_6'\right) \cdot V_S$$

For a design of $\beta=45°$ as shown in Fig. 6.4.5, we have $B=-\pi_{44}T_1/2$ and the output voltage is

$$V_T \doteq -0.25\pi_{44}T_1V_S$$

The output is about a half as large as that of a Wheatstone bridge configuration.

§6.4.3. Sensing Elements in a Diffusion Layer
In practical applications, piezoresistive sensing elements are usually formed by a diffusion or ion-implantation layer. In either case, the impurity concentration in the layer is laterally uniform (in x'-y' direction), but not uniform in the depth direction. There is usually a concentration peak at or near the surface and the concentration decreases away from the peak in an exponential or a Gausian distribution with distance. As the resistivity and the piezoresistive coefficient of the material are both functions of doping concentration, the effect of non-uniform impurity concentration on sensitivity should be considered.
Consider a rectangular resistor made of a diffusion layer, as shown in Fig. 6.4.9. At any point in the resistor, we have the relation (the prime signs is omitted for simplicity)

$$E_x = \rho_1 j_x + \rho_6 j_y$$
$$E_y = \rho_6 j_x + \rho_2 j_y \tag{6.4.24}$$

where $\rho_i s$ and $j_i s$ are functions of z. As $j_y=0$, we have $E_x=\rho_1 j_x$, or

$$j_x = \frac{E_x}{\rho_1} = \frac{E_x}{\rho_o(1+\pi_{11}T_1+\pi_{12}T_2+\pi_{16}T_6)}$$

where ρ_0 and $\pi_{ij}s$ are functions of z, but the dependence of $T_i s$ on z is negligible. As the variation of resistivity is much smaller than unity, we have the approximation

$$j_x = \sigma_o E_x (1 - A) \tag{6.4.25}$$

where $\sigma_o = 1/\rho_o$ and $A \equiv \pi_{11}T_1 + \pi_{12}T_2 + \pi_{16}T_6$. Both σ_o and A are functions of z.

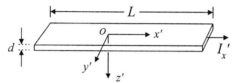

Fig. 6.4.9. A piezoresistor formed by a diffusion layer and its coordinate system

The current flowing through the resistor is

$$I_x = w \int_0^d j_x(z)dz = E_x w \int_0^d \sigma_o(1-A)dz \tag{6.4.26}$$

If the voltage applied between the two electrodes is V_S, the current along the resistor is

$$I_x = \frac{V_S}{L} w \int_0^d \sigma_o(z)(1-A(z))dz \tag{6.4.27}$$

Therefore, the resistance can be written as

$$R = \frac{L}{w} \frac{1}{\int_0^d \sigma_o(z)(1-A(z))dz} \tag{6.4.28}$$

As $A(z)$ is much smaller than 1, we can write

$$R \doteq \frac{L}{w} \cdot \frac{1 + \int_0^d \sigma_o(z)A(z)dz \Big/ \int_0^d \sigma_o(z)dz}{\int_0^d \sigma_o(z)dz} \tag{6.4.29}$$

From $R_o = L \Big/ w \int_0^d \sigma_o(z)dz$, the relative change in resistance is

$$\frac{\Delta R}{R_o} = \frac{\int_0^d \sigma_o(z)(\pi_{11}T_1 + \pi_{12}T_2 + \pi_{16}T_6)dz}{\int_0^d \sigma_o(z)dz} \tag{6.4.30}$$

This result implies that the piezoresistance effect is an average weighted by conductivity $\sigma_o(z)$, which is a function of doping profile.

If the concentration distribution along the z-axis is $N(z)$, we have $\sigma_o(z) = N(z)e\mu(z)$. As the dependence of $\sigma_o(z)$ on z is much stronger than that of $\pi_{ij}(z)$, Eq. (6.4.30) can be approximated by

$$\frac{\Delta R}{R_o} = \pi_{11}(z_m)T_1 + \pi_{12}(z_m)T_2 + \pi_{16}(z_m)T_6 \tag{6.4.31}$$

where z_m is the position where $N(z)$ has a peak. In many cases, Eq. (6.4.31) is written as

$$\frac{\Delta R}{R_o} = \pi_{11}(N_s)T_1 + \pi_{12}(N_s)T_2 + \pi_{16}(N_s)T_6 \tag{6.4.32}$$

where N_s is the maximum concentration of the layer and $\pi_{ij}(N_S)$ is the value of π_{ij} for the doping concentration N_s [14].

§6.5. Polysilicon Piezoresistive Sensing Elements

A major problem for a diffused or ion-implanted piezoresistive sensing element is that significant drift might be caused by the leakage current of the p-n junctions used to isolate the elements electrically from the silicon substrate. As the leakage current increases exponentially with temperature, a piezoresistive sensor may even fail to work when the temperature exceeds 100°C. Therefore, for stable operation at a temperature from 100°C up to 300°C, materials with a layer of crystalline silicon on an insulator substrate (SOI) have been developed for piezoresistive sensing elements. The most noted SOI materials are "Silicon on Sapphire" (SOS), "Separation by Ion Implanted Oxygen" (SIMOX), etc. As piezoresistors are sculptured out of the silicon layer on the insulator, these elements are isolated with each other by air or the dielectric material. This eliminates the leakage problem completely. The design of an SOI piezoresistive sensing element is basically the same as the design of a normal piezoresistive sensing element, but the cost for such a device is usually high due to the material used and the special process needed.

A low cost alternative to single crystalline silicon SOI material is a polysilicon layer deposited on a SiO$_2$/Si substrate. As the characteristics of piezoresistance in polysilicon are quite different from those of a single crystalline silicon material, the design principles of a polysilicon piezoresistive sensing element is discussed in this section.

§6.5.1. Piezoresistance Effect in Polysilicon

A polysilicon layer consists of a large amount of crystalline silicon grains and some non-crystalline regions between the grains (the boundary regions). The grain sizes are usually from a few tens nanometer to a few microns, depending on the deposition conditions and the thickness of the layer. The orientation of the grains can be totally random, or, they may have one or a few preferential growth orientations in the normal direction (z'-direction) of the layer. However, they are always random in the layer plane (the x'-y' planes) for an as-deposited material without any post-deposit thermal treatment (such as laser annealing).

The boundary regions contain a large amount of trap centers. As there are very few free charge carriers in the region, it is virtually "non-conductive". The trap centers in the boundary regions cause a potential barrier between neighboring grains and cause depletion layers in the surface region of the grains. However, electric carriers can still pass through the boundary regions and the surface depletion layers by tunneling.

For simplicity, a simplified model of polysilicon material is used here. The grains in the polysilicon layer are aligned regularly in a resistor as shown in Fig. 6.5.1.

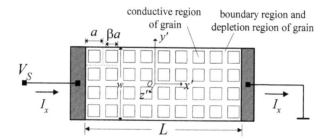

Fig. 6.5.1. A simplified model for a polysilicon material

The grains are assumed to be square in shape of the same size much smaller than the width of the resistor, but the grains are as high as the layer thickness. Also assumed is that the area occupied by a grain (including the boundary region) is a square with a dimension of a, and the conductive area of the grain has a dimension of βa ($\beta < 1$). By using this simplified model, the piezoresistive effect of polysilicon piezo-resistors and four-terminal sensing elements are discussed below.

(1) Polysilicon Piezoresistor

The resistance of a polysilicon piezoresistor, R, consists of two parts: one from the conductive region of the grains, R_g, and the other from the boundary regions and the depletion layers of the grains, R_l. The total resistance is

$$R = R_g + R_l \tag{6.5.1}$$

The relative change in resistance is

$$\frac{\Delta R}{R_o} = \frac{\Delta R_g + \Delta R_I}{R_{go} + R_{Io}} \tag{6.5.2}$$

where subscript "*o*" denotes the original value when the material is not stressed. As R_I is dominated by tunneling mechanism, it depends on the distance between the conductive regions.

As, for most cases, the piezoresistance effect is the dominant mechanism, Eq. (6.5.5) can be approximated as

$$\frac{\Delta R}{R_o} = \frac{\Delta R_g}{R_{go} + R_{Io}} \tag{6.5.3}$$

As the orientations of grain are random, at least in two dimensions, the piezoresistance of the grains has to be found by taking an average over the directions, i.e.,

$$\Delta R_g = R_{go}(\overline{\pi}_l\, T_l + \overline{\pi}_t\, T_t + \overline{\pi}_s\, T_s) \tag{6.5.4}$$

Using Eqs. (6.5.3) and (6.5.4), we have

$$\frac{\Delta R}{R_o} = \frac{\overline{\pi}_l\, T_l + \overline{\pi}_t\, T_t + \overline{\pi}_s\, T_s}{R_{go} + R_{Io}}\, R_{go} \tag{6.5.5}$$

(2) Four-terminal Sensing Element

For a polysilicon four-terminal sensing element, the lateral electric field inside a grain is

$$E_Y = (\overline{\pi}_{61}\, T_l + \overline{\pi}_{62}\, T_t + \overline{\pi}_{66}\, T_s)E_X\,.$$

According to the model shown in Fig. 6.5.1, the electric filed in the *x*-direction is

$$E_X = \frac{R_{go}V_S}{R_{go} + R_{Io}}\frac{1}{\beta L}$$

where V_S is the supply voltage and L is the length of the resistor. Thus, we have

$$E_Y = \frac{V_S}{\beta L}\frac{R_{go}}{R_{go} + R_{Io}}(\overline{\pi}_{61}\, T_l + \overline{\pi}_{62}\, T_t + \overline{\pi}_{66}\, T_s) \tag{6.5.6}$$

As the orientations of grain are random and the lateral electric field is induced only in the conductive regions of a grain, the output voltage is

$$V_T = \beta w E_Y = V_S \frac{w}{L} \frac{R_{go}}{R_{go} + R_{Io}} (\bar{\pi}_{61} T_l + \bar{\pi}_{62} T_t + \bar{\pi}_{66} T_s) \qquad (6.5.7)$$

where w is the width of the resistor. If the width of the element is not small when compared with L, the short-circuit effect of the electrodes has to be considered and a correction factor $f(w/L)$ has to be added. In this case, the transverse output voltage is

$$V_T = V_S \frac{w}{L} f(w) \frac{R_{go}}{R_{go} + R_{Io}} (\bar{\pi}_{61} T_l + \bar{\pi}_{62} T_t + \bar{\pi}_{66} T_s) \qquad (6.5.8)$$

In next section, the components of the average piezoresistive coefficient, $\bar{\pi}_{ij}s$, in Eqs. (6.5.6) to (6.5.8) will be analyzed under two conditions:
(i) The grains have a specific growth orientation in their normal direction.
(ii) The grains are completely random in orientation.

§6.5.2. Average Piezoresistive Coefficient

The grains in the polysilicon layer may have one or a few preferential growth orientations in the normal direction, but they are always random in the layer plane. Therefore, the piezoresistance of a polysilicon sensing element is determined by the average piezoresistive coefficient of the grains. In this section, we will first be analyzed the average piezoresistive coefficient of polysilicon layer for two important series of growth orientations and then the average piezoresistive coefficient of polysilicon layer with completely random growth orientation [15].

(1) Average for Specific Orientations

If the Miller index of a specific growth orientation is [k l m], the normal direction of the layer is one of the <k l m> orientations for each grain. The orientation can be described by two Euler's angles, ϕ and θ, with reference to the crystallographic coordinate system of the grain. All the grains in the layer have the same Euler's angles ϕ and θ, but the third Euler's angle, ψ, is random. Therefore, the average of piezoresistive coefficient tensor component, $\bar{\pi}_{ij}$, can be found by taking the average for ψ in the range of 0°~360°. Obviously, the results are a function of ϕ and θ. Analyses on two important series of growth orientations are given below.

(a) Growth Orientations of [k, 0, m] Plane

If the growth orientation of grains is in a (010) plane, the series of orientations are in the form of [k 0 m]. Some important directions in the series are: [001], [103], [102], [101], [201], [301] and [100] as shown in Fig. 6.5.2. For these directions, the first two Euler's angles are $\phi=0$ and $\theta=\tan^{-1}(k/m)$ but the third Euler's angle, ψ, is random for the grains.

According to the relations given by Eq. (6.2.15), the direction cosines of the resistor's coordinate system with respect to the crystallographic coordinate system of a grain are given by

$$\begin{pmatrix} l_1 & m_1 & n_1 \\ l_2 & m_2 & n_2 \\ l_3 & m_3 & n_3 \end{pmatrix} = \begin{pmatrix} \cos\theta\cos\psi & \sin\psi & -\sin\theta\cos\psi \\ -\cos\theta\sin\psi & \cos\psi & \sin\theta\sin\psi \\ \sin\theta & 0 & \cos\theta \end{pmatrix} \tag{6.5.9}$$

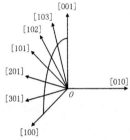

Fig. 6.5.2. Some main directions in (010) plane

According to Table 6.3.1 and Eq. (6.5.9), we have [15]

$$\overline{\pi}'_{11} = \pi_{11} - 2\pi_o \frac{1}{2\pi} \int_0^{2\pi} \left(l_1^2 m_1^2 + l_1^2 n_1^2 + m_1^2 n_1^2 \right) d\psi = \pi_{11} - \frac{1}{4}\left(1 + 3\cos^2\theta\sin^2\theta\right)\pi_o \tag{6.5.10}$$

Other useful components found are

$$\overline{\pi}'_{12} = \pi_{12} + \frac{1}{8}\left(1 + \cos^4\theta + \sin^4\theta\right)\pi_o$$

$$\overline{\pi}'_{16} = \overline{\pi}'_{61} = \overline{\pi}'_{62} = 0 \tag{6.5.11}$$

$$\overline{\pi}'_{66} = \pi_{44} + \frac{1}{4}\left(1 + \cos^4\theta + \sin^4\theta\right)\pi_o$$

If the data of π_{11}, π_{12}, π_{44} in Table 6.1.1 are used, the results for some main directions are listed in Table 6.5.1. The curves showing the dependence of $\overline{\pi}'_{11}$, $\overline{\pi}'_{12}$ and $\overline{\pi}'_{66}$ for p-Si on angle θ are given in Fig. 6.5.3.

Table 6.5.1. $\overline{\pi}'_{11}$, $\overline{\pi}'_{12}$ and $\overline{\pi}'_{66}$ of p-polysilicon on some main directions in (010) plane

directions	θ	n-Si			p-Si		
		$\overline{\pi}'_{11}$	$\overline{\pi}'_{12}$	$\overline{\pi}'_{66}$	$\overline{\pi}'_{11}$	$\overline{\pi}'_{12}$	$\overline{\pi}'_{66}$
[001]	0°	-66.7	18.0	-89.6	39.2	-33.7	72.9
[103]	18.35°	-57.1	21.1	-78.2	48.0	-30.8	78.8
[102]	26.57°	-49.7	23.7	-73.2	54.9	-28.5	83.3
[101]	45°	-40.1	26.8	-66.9	63.7	-25.6	89.2
[201]	63.43°	-49.7	23.7	-73.2	54.9	-28.5	83.3
[301]	71.57°	-57.1	21.1	-78.2	48.0	-30.8	78.8
[100]	90°	-66.7	18.0	-89.6	39.2	-33.7	72.9

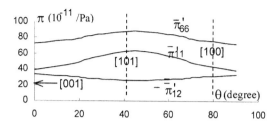

Fig. 6.5.3. The dependence of $\overline{\pi}'_{11}, \overline{\pi}'_{12}$ and $\overline{\pi}'_{66}$ of p-polysilicon on orientations in (010) plane

(b) Growth Orientations of [k k n] Plane

If the growth orientations of the grains are in the $(10\overline{1})$ plane, the Miller index of the orientation is in the form of $[k\ k\ n]$. Some of the main directions are [001], [113], [112], [111], [221], [331] and [110], as shown in Fig. 6.5.4. For this series of directions, the first two Euler's angles are $\phi = 45°$ and $\theta = \tan^{-1}(\sqrt{2}k/n)$, but the third Euler's angle ψ is random for the grains.

Therefore, the direction cosines of the resistor's coordinate system with respect to the crystallographic coordinate system of a grain are given by

$$
\begin{pmatrix} l_1 & m_1 & n_1 \\ l_2 & m_2 & n_2 \\ l_3 & m_3 & n_3 \end{pmatrix} = \begin{pmatrix} \dfrac{\cos\theta\cos\psi - \sin\psi}{\sqrt{2}} & \dfrac{\cos\theta\cos\psi + \sin\psi}{\sqrt{2}} & -\sin\theta\cos\psi \\ \dfrac{\cos\psi - \cos\theta\sin\psi}{\sqrt{2}} & \dfrac{\cos\psi - \cos\theta\sin\psi}{\sqrt{2}} & \sin\theta\sin\psi \\ \sin\theta/\sqrt{2} & \sin\theta/\sqrt{2} & \cos\theta \end{pmatrix} \quad (6.5.12)
$$

According to Table 6.3.1 and Eq. (6.5.12), we find

$$
\overline{\pi}'_{11} = \pi_{11} - \frac{1}{16}\pi_o\left(3 + 3\cos^4\theta - 2\cos^2\theta + 12\cos^2\theta\sin^2\theta + 4\sin^2\theta\right)
$$

$$
\overline{\pi}'_{12} = \pi_{12} + \frac{1}{16}\pi_o\left(1 + \cos^4\theta + 2\cos^2\theta + 2\sin^4\theta\right)
$$

$$
\overline{\pi}'_{66} = \pi_{44} + \frac{1}{8}\pi_o\left(1 + \cos^4\theta + 2\cos^2\theta + 2\sin^4\theta\right) \quad (6.5.13)
$$

$$
\overline{\pi}'_{16} = \overline{\pi}'_{61} = \overline{\pi}'_{62} = 0
$$

If the data of π_{11}, π_{12}, π_{44} in Table 6.1.1 are used, the non-zero components in Eq. (6.5.13) are found and listed in Table 6.5.2 and the curves showing the dependence of $\overline{\pi}'_{11}, \overline{\pi}'_{12}$ and $\overline{\pi}'_{66}$ on angle θ for p-Si are given in Fig. 6.5.5.

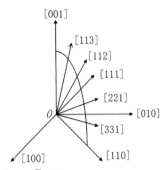

Fig. 6.5.4. Some main directions in $(10\overline{1})$ plane

Table 6.5.2. $\overline{\pi}'_{11}, \overline{\pi}'_{12}$ and $\overline{\pi}'_{66}$ on some main directions in $(10\overline{1})$ plane

directions	θ	n-Si			p-Si		
		$\overline{\pi}'_{11}$	$\overline{\pi}'_{12}$	$\overline{\pi}'_{66}$	$\overline{\pi}'_{11}$	$\overline{\pi}'_{12}$	$\overline{\pi}'_{66}$
[001]	0°	-66.7	17.9	-84.6	39.2	-33.7	72.9
[113]	25.24°	-50.0	23.5	-73.5	54.6	-38.6	83.1
[112]	35.26°	-40.1	26.8	-66.6	63.7	-25.5	98.1
[111]	54.74°	-31.2	29.7	-60.9	71.8	-22.6	94.6
[221]	70.53°	-35.1	28.4	-63.6	68.2	-24.0	92.2
[331]	76.74°	-37.5	27.6	-65.1	66.0	-24.7.8	90.7
[110]	90°	-40.1	26.8	-66.9	63.7	-25.5	89.1

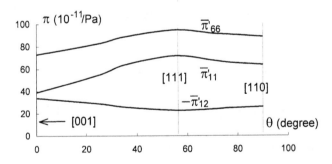

Fig. 6.5.5. Dependence of $\overline{\pi}'_{11}, \overline{\pi}'_{12}$ and $\overline{\pi}'_{66}$ of p-polysilicon on growth orientations in $(10\overline{1})$ plane

(2) Completely Random Distribution

In many cases, there are too many preferential growth orientations for a polysilicon layer so that none of them dominates, or, the preferential growth is not significant. In these cases, it would be convenient to use just the average results for a completely random distribution for design considerations. The average of the components of piezoresistive coefficient for a completely random distribution can be found by taking the average over the whole space angle for three Euler's angles

$$\overline{\pi}_{ij} = \frac{1}{4\pi} \int\limits_{0}^{2\pi} d\phi \int\limits_{0}^{\pi} \sin\theta d\theta \left(\frac{1}{2\pi} \int\limits_{0}^{2\pi} \pi_{ij}(\phi,\theta,\psi)d\psi \right)$$

The non-zero components are

$$\overline{\pi}'_{11} = \pi_{11} - \frac{2}{5}\pi_o, \ \overline{\pi}'_{12} = \pi_{12} + \frac{1}{5}\pi_o, \ \overline{\pi}'_{66} = \pi_{44} + \frac{2}{5}\pi_o \qquad (6.5.14)$$

If the data of π_{11}, π_{12} and π_{44} in Table 6.1.1 are used, the averages are given in Table 6.5.3.

Table 6.5.3. The average of $\overline{\pi}_{11}$, $\overline{\pi}_{12}$, $\overline{\pi}_{66}$ for a random distribution $(\times 10^{-11}/Pa)$

	$\overline{\pi}'_{11}$	$\overline{\pi}'_{12}$	$\overline{\pi}'_{66}$
n-Si	−45.4	25.0	−70.4
p-Si	58.8	−27.2	85.9

§6.5.3. Design of Polysilicon Piezoresistive Sensors

(1) Factors Affecting Sensitivity of a Polysilicon Sensor

(a) The Effect of the Piezoresistive Coefficient

According to the analysis given above, the effect of piezoresistance does not cancel out by the random distribution of grain orientation. As shown by Tables 6.5.1 to 6.5.3, the average piezoresistance effect of a polysilicon layer is still quite significant. For a p-type polysilicon layer with a completely random distribution of grain orientation, we have

$$\overline{\pi}'_{11} = 58.8 \times 10^{-11}/Pa = 0.425\pi_{44}$$
$$\overline{\pi}'_{12} = -27.2 \times 10^{-11}/Pa = -0.2\pi_{44} \qquad (6.5.15)$$
$$\overline{\pi}'_{66} = 85.9 \times 10^{-11}/Pa = 0.62\pi_{44}$$

Note that, for the design of a p-Si piezoresistor along the <110> direction in a (001) plane, the components of piezoresistive coefficient are $\pi_{11}=0.5\pi_{44}$, $\pi_{12}=-0.5\pi_{44}$ and $\pi_{66}=\pi_{44}$. On average, the piezoresistive effect of polysilicon is lower than that of single crystalline silicon by a factor of about 40%.

(b) The Effect of the Boundary Region

The ratio of $R_g/(R_g+R_l)$ is another major factor that reduces the sensitivity of polysilicon piezoresistive sensors. The ratio depends mainly on the grain size and the doping level of the material. The experimental results given by Lu, et al, [16] show that for the most commonly used conditions, i.e., a grain size of about 100 nm and a doping level of $1\times10^{18}/cm^3$ to $1\times10^{19}/cm^3$, the factor is around 0.6.

(c) The Effect of Stress Transformation

As the thickness of the polysilicon layer and the insulator layer are much smaller than that of the substrate, the strain in the polysilicon, ε_{poly}, is the same as that in the adjacent region of the silicon substrate, ε_{Si}. Therefore, the stress in the polysilicon layer is

$$T_{poly} = \frac{E_{poly}}{E_{Si}} T_{Si} \tag{6.5.16}$$

where T_{Si} is the stress in the silicon substrate and E_{poly} and E_{Si} are the Young's modulus of polysilicon and silicon, respectively.

Though the ratio of E_{poly}/E_{Si} is usually considered to be lower than 1.0, the experimental data are quite diverse, from close 1.0 to under 0.5. We assume here that the value is about 0.7.

According to the three factors discussed above, the sensitivity of a polysilicon piezoresistive sensor is about one forth of the sensitivity of a single crystalline piezoresistive sensor for similar conditions. Or, the gauge factor of a polysilicon piezoresistor is about 25. It is still more than one order of magnitude higher than that of a conventional metal strain gauge.

(2) Design Considerations on Polysilicon Sensors

(a) Piezoresistor

According to §6.5.2, $\overline{\pi}'_{16}$ is zero. Therefore, from Eq. (6.5.5), we find

$$\frac{\Delta R}{R_o} = \left(\overline{\pi}'_l T_l + \overline{\pi}'_t T_t \right) \frac{R_{go}}{R_{go} + R_{lo}} \tag{6.5.17}$$

Note from Eq. (6.5.15) that $\overline{\pi}'_l$ is more than twice as large as $|\pi'_t|$. Therefore, the design rule for a polysilicon piezoresistor is to make full use of the longitudinal stress, T_l, instead of both T_l and T_t. A typical design of polysilicon pressure transducer using twin-island structure is shown in Fig. 6.5.6.

(b) Four-terminal Sensing Element

According to §6.5.2, $\overline{\pi}'_{61}$ and $\overline{\pi}'_{62}$ are zero for polysilicon layer. Therefore, from Eq. (6.5.8), the output voltage of a four-terminal sensing element is

$$V_T = \frac{R_{go}}{R_{go} + R_{lo}} f\left(\frac{w}{L}\right) \cdot \frac{w}{L} V_s \overline{\pi}'_{66} T_s \tag{6.5.18}$$

Thus, we conclude that the design of a four-terminal sensing element of polysilicon is the same as that of single crystal silicon material.

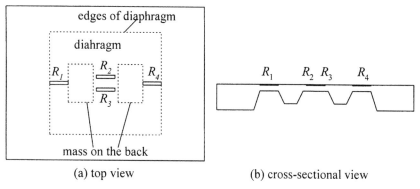

(a) top view   (b) cross-sectional view

Fig. 6.5.6. The design of a typical polysilicon pressure transducer

§6.6. Analyzing Piezoresistive Bridge

§6.6.1. Offset Voltage and Temperature Coefficient of Offset

(1) Offset Voltage

If the circuit of a Wheatstone bridge (or the structure geometries of the four-terminal sensing element) is ideally symmetric, the output of the bridge is zero when the resistors are not stressed. And, the output remains zero for any temperature since the four resistors have the same temperature coefficient of resistance.

However, there are always some non-ideal factors that cause a non-zero output voltage for a Wheatstone bridge. This original non-zero output is referred to as the offset voltage of the bridge.

There are two main factors that cause offset voltage. The first factor is the geometric deviation of the resistors from their nominal value. The second one is the stresses in the chip caused by the mismatch of thermal expansion coefficient between the silicon substrate and the deposited films on surface or the packaging materials. In this section, the offset voltage and the temperature coefficient of offset (*TCO*) will be studied on the assumption that they are caused only by the deviations in resistance from the nominal value.

Consider a Wheatstone bridge consisting of four resistors with the same nominal resistance R_B, i.e., $R_1=R_2=R_3=R_4=R_B$. If each resistor has a specific relative deviation from its nominal value, i.e., $R_1=R_B(1+\beta_1)$, $R_2=R_B(1+\beta_2)$, $R_3=R_B(1+\beta_3)$ and $R_4=R_B(1+\beta_4)$, as shown in Fig. 6.6.1 (a), the offset voltage of the bridge is

$$V_{OS} = V_{o1} - V_{o2} = V_S \frac{R_2 R_3 - R_1 R_4}{(R_1 + R_2) \cdot (R_3 + R_4)} \doteq V_S \left(\frac{\beta_2}{4} + \frac{\beta_3}{4} - \frac{\beta_1}{4} - \frac{\beta_4}{4} \right) \qquad (6.6.1)$$

where V_S is the supply voltage. Note that the contribution by the deviation of a specific resistor R_i is one fourth of its relative deviation β_i, and the effects of R_2, R_3 and R_1, R_4 are in the opposite directions. Thus, For the convenience of further discussion, the circuit shown in Fig. 6.6.1(a) is simplified as the circuit shown in Fig. 6.6.1(b), where

$$\beta = (\beta_2 + \beta_3 - \beta_1 - \beta_4) \tag{6.6.2}$$

Therefore, the offset voltage is $V_{OS} = \beta V_S/4$. For the simplicity of discussion, β is assumed to be positive in value.

Note that the offset voltage shown in Eq. (6.6.1) is not temperature dependent if all the resistors have the same temperature coefficient. This is indeed the case, as the resistors are made in a batch by diffusion or ion implantation. However, as part of the offset voltage is caused by the thermal stress, the offset voltage is often temperature dependent.

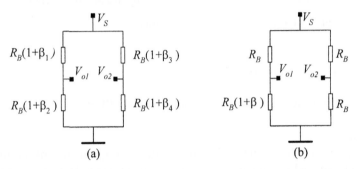

Fig. 6.6.1. A piezoresistive bridge of resistors with deviations from nominal resistance

(2) Compensation of Offset Voltage

For practical applications, the offset voltage of a piezoresistive bridge has to be compensated for by some means before further signal conditioning. The compensation of the offset voltage is usually made using external resistors by a parallel or/and series connection. The temperature coefficient of the compensation resistors is usually much smaller than that of the silicon resistors.

(a) Parallel Compensation of Offset Voltage

For parallel compensation, a resistor R_P is connected in parallel with the piezoresistor R_2 as shown in Fig. 6.6.2(a). The condition for compensation is

$$\frac{R_P \cdot R_B(1+\beta)}{R_P + R_B(1+\beta)} = R_B \tag{6.6.3}$$

For small β, the resistance of R_P is

$$R_P \cong \frac{1}{\beta} R_B \tag{6.6.4}$$

For example, if $R_B = 5k\Omega$ and $\beta = 0.02$, we have $R_P = 250k\Omega$.

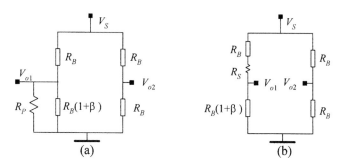

Fig. 6.6.2. Compensation for offset voltage (a) parallel compensation; (b) series compensation

(b) Series Compensation of Offset Voltage

Compensation can also be made using series connection. For series compensation, a small resistors $R_S=R_B$ is connected in series with resistor R_1, as shown in Fig. 6.6.2(b). If $R_B=5k\Omega$ and $\beta=0.02$, the resistance of the series resistor is: $R_B=100\Omega$.

For series compensation, the bridge has to be broken so that the compensation resistor can be inserted. For the convenience of offset compensation, the commercially available piezoresistive transducers are often provided with a five-terminal version or a dual half-bridge version as shown in Fig. 6.6.3(a) and Fig. 6.6.3(b), respectively.

However, either the parallel or the series scheme can only null the offset voltage at a specific temperature. The offset appears again once the temperature changes due to the difference in temperature coefficient between the bridge resistors and the compensation resistors. The offset voltage might be large for large temperature variation.

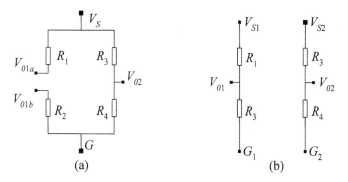

Fig. 6.6.3. Typical piezoresistive sensors (a) five-terminal version; (b) dual half-bridge version

(c) *TCO* Caused by Parallel and Series Compensation

Assume that the temperature coefficient of the bridge resistors (TCR_B) is α_b and the temperature coefficient of the compensation resistors is α_c. α_b is between +0.1% to +0.3% according to the doping level of the resistors and α_c is usually much smaller than α_b.

Due to the compensation resistor in parallel with R_2, a *TCO* is induced

$$TCO = \frac{d}{dT}\left[\left(\frac{R_B R_p}{R_B + R_p}\right)\bigg/\left(\frac{R_B R_p}{R_B + R_p} + R_B\right)\right]$$

Under the condition that $R_p >> R_B$, the temperature coefficient of offset of the circuit with a parallel compensation resistor is

$$TCO_p = -\frac{1}{4}(\alpha_b - \alpha_c)\cdot\frac{R_B}{R_p}V_S \qquad (6.6.5)$$

If $\alpha_b = +0.2\%$, $\alpha_c = 0$, $\beta = 0.04$, $V_S = 5V$, we have $TCO_p = -100\mu V/°C$. For a temperature variation of $100°C$, the offset voltage is $V_{OS} = -10mV$, a significant value for practical applications. For a compensation resistor R_s in series with R_1, the TCO caused is

$$TCO_s = \frac{1}{4}(\alpha_b - \alpha_c)\cdot\frac{R_S}{R_p}V_S \qquad (6.6.6)$$

Note that TCO_p and TCO_s have the opposite signs. This reminds us that the offset voltage and the TCO of a Wheatstone bridge might be compensated for simultaneously if the compensation is made partially parallel and partially series. This will be discussed below.

(3) Simultaneous Compensation of Offset Voltage and *TCO*

For many applications, the TCO and the offset voltage have to be compensated at the same time so that the offset voltage remains small in a large temperature range. As discussed above, this can be done by a series-parallel compensation scheme as shown in Fig. 6.6.4.

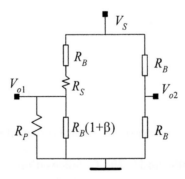

Fig. 6.6.4. Series-parallel compensation scheme for simultaneous compensation of offset and TCO

For the series-parallel compensation scheme shown in Fig. 6.6.4, the conditions for the compensation of offset voltage is

$$R_S + R_B = \frac{(1+\beta)R_B R_P}{R_P + (1+\beta)R_B}$$

(6.6.7)

The condition for the compensation of TCO is

$$\frac{1}{4}\frac{R_S}{R_B}(\alpha_b - \alpha_c) = \frac{1}{4}\frac{R_B}{R_P}(\alpha_b - \alpha_c)$$

(6.6.8)

From Eqs. (6.6.7) and (6.6.8), we find

$$R_S \cong \frac{\beta}{2}R_B \, ; R_P \cong \frac{2R_B}{\beta}$$

(6.6.9)

As mentioned before, the offset voltage may also be caused by the residual stress arising from the mismatch of thermal expansion between silicon and the packaging materials. This factor also affects the *TCO*. If the residual stress plays an important role in *TCO*, the results given in Eqs. (6.6.9) are no longer valid. Another problem is that *TCO* is zero only at the temperature the compensation is made. It reappears again when temperature changes. Therefore, the discussion above only gives a useful guide for compensation: *TCO* goes in the negative direction if the parallel resistance is increased (or, if the series resistance is reduced) and vice versa.

As the compensation of offset voltage and *TCO* is very important in sensor applications, many sophisticated schemes have been developed. For example, a practical compensation scheme is to null the offset voltage at two or more separate temperatures by a network, including resistors, thermo-resistors, diodes or transistors. We will not go more detail on this regard.

§6.6.2. Temperature Coefficient of Sensitivity

For a typical piezoresistive sensor, the output voltage is proportional to π_{44} of the resistor material. According to §6.1, π_{44} is a function of doping level. π_{44} and the temperature coefficient of π_{44} (i.e., the temperature coefficient of piezoresistance, $TC\pi$) are also a function of the doping level. $TC\pi$ is usually negative in sign.

For example, for a pressure transducer with voltage supply V_S, the output is

$$V_{out} = \frac{1}{2}\pi_{44}cpV_S$$

(6.6.10)

where c is a constant related to the diaphragm structure of the sensor. Therefore, for a constant supply voltage, the temperature coefficient of sensitivity (*TCS*) of the pressure transducer is

$$TCS = TC\pi$$

(6.6.11)

For example, if $TC\pi= -0.2\%$, the sensitivity of the pressure transducer will be reduced by about 10% if the temperature is raised by 50°C. As the *TCS* is significant, the compensation for *TCS* has to be considered for many applications.

A widely used compensation scheme for *TCS* is to use a constant current supply instead of a constant voltage supply. If the supply current for a piezoresistive pressure transducer is I_S, the output voltage becomes

$$V_{out} = \frac{1}{2}\pi_{44}cpI_S R_B \qquad (6.6.12)$$

where R_B is the bridge resistance. From Eq. (6.6.12), we have

$$TCS=TCR+TC\pi \qquad (6.6.13)$$

where the temperature coefficient of resistance (*TCR*) is also a function of doping level. As *TCR* has a positive temperature coefficient (i.e., *TCR*>0), the effect of *TCR* is opposite to that of *TCπ*. The curves in Fig. 6.6.5 show *TCπ* and *TCR* as functions of doping level.

From Fig. 6.6.5, *TCπ* and *TCR* have the same value (but with opposite signs) at two critical doping levels: $N_{C1}{\cong}2{\times}10^{18}/cm^3$ and $N_{C2}{\cong}5{\times}10^{20}/cm^3$. Therefore, if the doping level of the resistors is controlled to be equal to one of the two critical doping levels, the *TCS* of the pressure transducer will be zero due to the cancellation of *TCR* and *TCπ*.

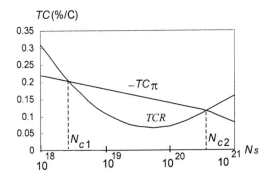

Fig. 6.6.5. *TCR* and *TCπ* as a function of doping concentration N_S

As exact cancellation of *TCR* and *TCπ* is not readily attainable, an adjustable approach is preferred for practical applications. The main feature of this approach is to control the doping level in the region $N_S<N_{C1}$ or $N_S>N_{C2}$ so that $TCR>|TC\pi|$. The *TCS* of the pressure transducer can thus be adjusted to zero by trimming the parallel compensation resistor, R_P, as shown in Fig. 6.6.6.

According to Fig. 6.6.6, the output of the pressure transducer is

$$V_{out} = \frac{1}{2}\pi_{44}cpI_S \frac{R_P R_B}{R_P + R_B}$$

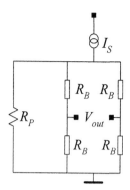

Fig. 6.6.6. Parallel resistor compensation for TCS

Thus, the temperature coefficient of sensitivity is

$$TCS = TC\pi + \frac{R_P TCR + R_B \alpha_c}{R_B + R_P} \tag{6.6.14}$$

where α_c is the temperature coefficient of R_P. The condition for zero TCS is found to be

$$R_P = -\frac{R_B(TC\pi + \alpha_c)}{TCR + TC\pi} \tag{6.6.15}$$

As $TCR>0$ and $TC\pi<0$, Eq. (6.6.15) can be written as

$$R_P = \frac{R_B(|TC\pi| - \alpha_c)}{TCR - |TC\pi|} \tag{6.6.16}$$

Under the condition that $TCR>|TC\pi|>>\alpha_c$, a positive-valued R_P exists for zero TCS.

In addition to the compensation schemes described above, there are many other schemes using circuit with temperature sensitive components (such as thermo-resistors, diodes, transistors) for TCS compensation. We will not go more detail on this regard.

§6.6.3. Nonlinearity of Piezoresistive Transducers

(1) Definitions of Nonlinearity

The piezoresistive transducer is categorized as a "linear transducer", i.e., the output response of the transducer is in direct proportion to the input measurand. This is true for most piezoresistive transducers within a tolerance of about one percent of the operation range (i.e., the full scale output, FSO). When the output-input relation of a piezoresistive transducer is calibrated with higher accuracy, the relationship is a curve instead of a straight line. Therefore, the calibrated output-input relation of a transducer is often referred to as the calibration curve of the transducer. For linear transducers, the calibration curve of transducer is often compared to a specified straight-line.

The deviation of the specified straight line from the calibration curve of the transducer is characterized by a parameter called the nonlinearity. For each calibration point, there is a specific deviation. The nonlinearity error of a specific calibration point is defined as the deviation at this calibration point and is generally expressed as a percentage of *FSO*. The nonlinearity of a transducer is defined as the maximum deviation of all the calibration points, also expressed as a percentage of *FSO*. The nonlinearity of a piezoresistive transducer is typically in the range of 0.5%~0.05%.

There are quite a few methods for defining the specified straight line according to the calibration curve of a transducer. The straight line can be defined as the line connecting the two end points (at 0 and 100% operation range). This line is called an "end-point straight-line" or a "terminal-based straight-line" [17]. The end-point line can be shifted in parallel to a certain extent to equalize the maximum deviations on both sides of the line so that the maximum value of deviation, i.e., the nonlinearity is minimized. Then, the line is called the best-fit line. Manufacturers like to use the best-fit lines as they give better-looking data. Sometimes, the best-fit line is obtained by the least squares method. As the end-point straight-line method is the most straightforward, the most convenient and the most widely used method in practical applications, it is exclusively used in the discussion in this book. Also for the convenience of description, a piezoresistive pressure sensor is used as an example in the following discussions. Thus, the input measurand is pressure in Pa and the output of the transducer is voltage in mV.

As, in most practical applications, the pressure to be measured is directly read from the output of the pressure transducer according to the specified straight line instead of the calibration curve, the accuracy of the pressure measurement is directly related to the nonlinearity of the transducer. Therefore, nonlinearity is one of the most important parameters for the transducer, in addition to *TCO* and *TCS* discussed in §6.6.1 and §6.6.2.

For example, the calibration curve of a pressure transducer is shown by the solid curve and the end-point line is shown by the dotted line in Fig. 6.6.7, where p_m designates the maximum pressure input and the corresponding output $V_o(p_m)$ is *FSO*. Both the calibration curve and the end-point line start from zero pressure input (or, the lower limit of operation range) according to the definition.

According to the definition, the nonlinearity at a specific test pressure p_i is

$$NL_i = \frac{V_o(p_i) - \dfrac{V_o(p_m)}{p_m} p_i}{V_o(p_m)} \times 100\% \qquad (6.6.17)$$

According to definition, the nonlinearity can be either positive or negative for a calibration point. The nonlinearity of the pressure transducer is the maximum value of NL_i.

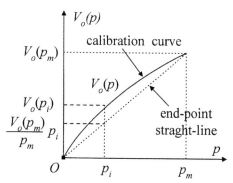

Fig. 6.6.7. Calibration curve and corresponding end-point straight line

Mathematically, the calibration curve $V_o(p)$ can be developed into a series in power of p. Some typical conditions can be discussed as follows:

(i) $V_o(p)=ap$, where a is the sensitivity of the pressure transducer. As the calibration curve is a straight line, the nonlinearity is zero.

(ii) $V_o(p)=ap+bp^2$ (often $a \gg bp$). According to Eq. (6.6.17), we find

$$NL(p) = \frac{ap+bp^2-(a+bp_m)p}{ap_m+bp_m^2} \cong \frac{b(p-p_m)p}{ap_m} \tag{6.6.18}$$

The maximum nonlinearity value for the whole operation range appears at $p = p_m/2$ and the value is

$$NL = NL\left(\frac{p_m}{2}\right) = -\frac{bp_m}{4a} \tag{6.6.19}$$

If $(bp_m)/a>0$, the nonlinearity is negative, i.e., $NL < 0$, and, if $(bp_m)/a<0$, the nonlinearity is positive, i.e., $NL > 0$. The two situations are shown in Fig. 6.6.8.

(iii) $V(p)=ap+bp^2+cp^3$. The nonlinearity of the pressure transducer is

$$NL(p) \cong \frac{b}{ap_m}(p-p_m)p + \frac{c}{ap_m}(p^2-p_m^2)p \tag{6.6.20}$$

The maximum of $NL(p)$ appears at $p_1 = \dfrac{-2b+\sqrt{4b^2+12c(bp_m+cp_m^2)}}{6c}$.

Therefore, the nonlinearity of the transducer is

$$NL = -b\frac{p-p_m}{ap_m}p_1 - c\frac{p_m^2-p_1^2}{ap_m}p_1 \tag{6.6.21}$$

In case the quadratic term is missing, i.e., $b=0$, we have $p_1 = p_m / \sqrt{3} = 0.577 p_m$ and

$$NL = -\frac{2}{3\sqrt{3}} \frac{c}{a} p_m^2 \cong -0.385 \frac{c}{a} p_m^2 \tag{6.6.22}$$

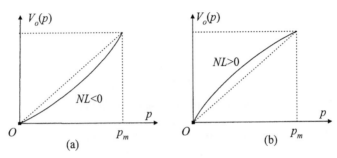

Fig. 6.6.8. Calibration curves for (a) negative nonlinearity and (b) positive nonlinearity

(2) Nonlinearity of a Piezoresistive Bridge

Now let us consider a piezoresistive pressure transducer with a Wheatstone bridge on a silicon diaphragm. The output of the bridge is

$$V_o = \frac{R_2 R_3 - R_1 R_4}{(R_1 + R_2) \cdot (R_3 + R_4)} V_S \tag{6.6.23}$$

where V_S is the supply voltage. Obviously, the nonlinearity of the pressure transducer is determined by the nonlinearities of the four individual resistors.

According to § 6.4, each resistance R_i is a function of the stress applied

$$R_i = R_{io} + R_{io}(\pi_l T_l + \pi_t T_t + \pi_s T_s) \tag{6.6.24}$$

Therefore, the nonlinearity of a piezoresistive Wheatstone bridge is determined by three factors:

(i) The nonlinear relation between stress, T, and the applied pressure p

(ii) The nonlinear relation between the piezoresistive coefficient π and stress T.

(iii) The difference in sensitivity among the four piezoresistors

Firstly, we discuss the last factor on its effect on the nonlinearity of a Wheatstone bridge. The discussion on the other two factors will follow. In case the four resistors have different sensitivities to the pressure, the $R \sim p$ relation is generally expressed as

$$R_i = R_{io}(1 + a_i p + b_i p^2) \tag{6.6.25}$$

where subscript, i, designates an individual resistor of the bridge. According to Eqs. (6.6.25) and (6.6.19), the nonlinearity of $R_i(p)$ is

$$NL_i = -\frac{b_i}{4a_i} p_m \tag{6.6.26}$$

By substituting Eq. (6.6.25) into Eq. (6.6.23) and assuming that $R_{2o}R_{3o} = R_{1o}R_{4o}$ (i.e., the offset voltage of the bridge is zero), we have

$$V_o = V_S \frac{(1+a_2p+b_2p^2)(1+a_3p+b_3p^2)-(1+a_1p+b_1p^2)(1+a_4p+b_4p^2)}{[2+(a_1+a_2)p+(b_1+b_2)p^2]\cdot[2+(a_3+a_4)p+(b_3+b_4)p^2]} \tag{6.6.27}$$

To the second power of p, Eq. (6.6.27) can be approximated as

$$V_o = \frac{(a_2+a_3-a_1-a_4)}{4}\left\{p+\left[\frac{b_2+b_3-b_1-b_4}{a_2+a_3-a_1-a_4}-\frac{a_2^2+a_3^2-a_1^2-a_4^2}{2(a_2+a_3-a_1-a_4)}\right]p^2\right\}V_S \tag{6.6.28}$$

From Eq. (6.6.28), the sensitivity of the pressure transducer is

$$S = \frac{1}{4}(a_2+a_3-a_1-a_4)\cdot V_S \tag{6.6.29}$$

For most designs, a_1 and a_4 have opposite signs to a_2 and a_3. Therefore, we have

$$S = \frac{1}{4}(a_2+a_3+|a_1|+|a_4|)\cdot V_S \tag{6.6.30}$$

This means that the sensitivity of the pressure transducer is the average sensitivity of the four individual piezoresistors.

According to Eq. (6.6.28), the nonlinearity of the pressure transducer is

$$NL = -\frac{1}{4}\frac{b_2+b_3-b_1-b_4}{a_2+a_3-a_1-a_4}p_m + \frac{a_2^2+a_3^2-a_1^2-a_4^2}{8(a_2+a_3-a_1-a_4)}p_m \tag{6.6.31}$$

Assuming that a_2 and a_3 are positive while a_1 and a_4 are negative and defining $\bar{a} \equiv (a_2+a_3-a_1-a_4)/4$, the nonlinearity of the Wheatstone bridge is

$$NL = \frac{1}{4\bar{a}}\left[a_2(NL_2)+a_3(NL_3)-a_1(NL_1)-a_4(NL_4)\right]+\frac{1}{32\bar{a}^2}(a_2^2+a_3^2-a_1^2-a_4^2)p_m\bar{a} \tag{6.6.32}$$

The first term in Eq. (6.6.32) is the average of the nonlinearities of the four resistors weighted by their respective sensitivities. The factor, $p_m\bar{a}$, in the second term is the average piezoresistance of the full scale pressure, that is usually in the range of 1%~5%. Note that if the four resistors are identical in sensitivity on p the nonlinearity of the bridge is zero.

For example, if we have $a_2=a_3$ and $a_1=a_4=-a_2/2$ due to some design and/or process reasons, and $p_m\bar{a} = 3\%$, the nonlinearity caused by the second term in Eq. (6.6.32) is *NL*=0.25%. This is not negligible for many pressure transducers.

(3) Nonlinearity of Stress

General speaking, there are three main factors that cause the non-linearity of piezoresistive pressure transducers: (i) the nonlinearity caused by the difference in piezoresistive sensitivity between bridge resistors, (ii) the nonlinear relationship between the stress and the pressure applied, and (iii) the nonlinear relationship between the piezoresistive coefficient and the stress. The first factor has been clearly shown in Eq. (6.6.32). The second factor will be discussed below in a qualitative way, and the discussion on the third factor will follow.

The discussion on stress in §2.6 is based on a linear theory, which concluded that the stress in a diaphragm is proportional to the pressure applied. The linear theory assumes that the stress distribution is a result of pure bending, that is, the neutral plane of the diaphragm is not stretched. This assumption requires that the deflection of the diaphragm be small when compared with its thickness. If the deflection of the diaphragm is not small, the neutral plane of the diaphragm will be stretched like a balloon, hence the name of the "Balloon effect ". When the stress caused by the stretch of the neutral plane is considered, the stress in the diaphragm consists of two parts; the first part T_b is caused by the bending of the diaphragm and the second part T_c is caused by the stretch of the neutral plane, i.e.,

$$T = T_b + T_c \tag{6.6.33}$$

When compared with the linear theory, the stress caused by bending, T_b, is reduced in magnitude as the stretch of the diaphragm takes part of the pressure load.

As the Balloon effect is due to the stretch of the diaphragm, it is related to the displacement of the diaphragm. According to §2.6, the displacement of a flat diaphragm is proportional to $(a/h)^4$ and the stress is proportional to $(a/h)^2$. This means that the larger the ratio of a/h, the larger the balloon effect. Thus, the nonlinearity of T_c is negative and accordingly, the non-linearity of T_b is positive as schematically shown in Fig. 6.6.9.

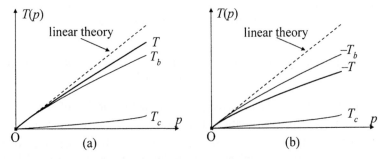

Fig. 6.6.9. "Balloon effect" on the stress (a) pressure on front, (b) pressure on back

Since T_b can either be positive or negative (depending on which side of the neutral plane the pressure is applied) and T_c is always positive, the Balloon effect is different for front pressure and for pressure on back. Let us consider a pressure transducer with resistors at the edge of the diaphragm. The bending stress at the resistors is positive when pressure is applied from the front surface while the bending stress at the resistors is negative when the pressure is applied from the back of the diaphragm, but T_c is always positive. This means that both T_b and T_c are positive when the pressure is applied from the front side while T_b is negative and T_c is positive when the pressure in applied from the back. As a general result, the nonlinearity caused by the Balloon effect is smaller for a front pressure than for a pressure on back.

(4) Nonlinearity of Piezoresistance

So far in this chapter, the piezoresistance effect of silicon was considered to be linear, i.e., the piezoresistive coefficient of silicon was considered to be independent of stress. In fact, this is not true if it is examined with high accuracy. However, it is very difficult to investigate the dependence of piezoresistive coefficient on stress because there are many components of the stress tensor and the measurement of higher order effects requires very high accuracy. Therefore, published data are scarce and incomplete, and the accuracy of data is difficult to evaluate.

According to the experimental results of Matsuda et al [17], for p-resistors in the <110> orientations with a doping level of $2 \times 10^{18}/cm^3$, the dependence of nonlinearity on stress is shown in Fig. 6.6.10. Based on Eq. (6.6.32) and the experimental results, some design considerations on the nonlinearity of pressure transducer can be discussed.

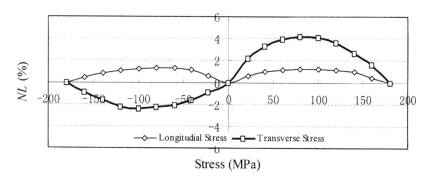

Fig. 6.6.10. Nonlinearity of p-type piezoresistors for <110> stress (doping level: $2 \times 10^{18}/cm^3$)

Let us consider a design with a square flat diaphragm and four resistors at the edge centers as shown in Fig. 6.6.11. Suppose that the pressure is applied on the front. Thus, the stresses on resistors R_1 and R_4 are the same and can be approximated to be pure longitudinal T_l. On the other hand, the stresses on resistors R_2 and R_3 are the same transverse stress, T_t. Also, we assume that $T_l=T_t=T$. This implies that all the a_i's in Eq. (6.6.32) have the same value.

In this case, the nonlinearity of the pressure transducer can be estimated by a simplified equation

$$NL = \frac{1}{4}\left[(NL_2)+(NL_3)+(NL_1)+(NL_4)\right] \tag{6.6.34}$$

According to Fig. 6.6.10, NL_2 and NL_3 are positive as shown by the bold curve for transverse stress on the right-hand side, while NL_1 and NL_4 are also positive as shown by the thin curve for longitudinal stress on the right-hand side. The nonlinearity of the pressure transducer is then the average value of the four NL_is of the resistors

$$NL = \frac{1}{4}\left(|NL_1|+|NL_4|+|NL_2|+|NL_3|\right) \tag{6.6.35}$$

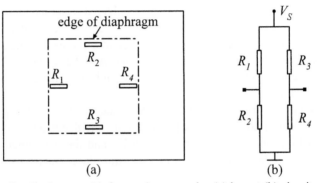

Fig. 6.6.11. Square flat diaphragm with four resistors at edge (a) layout (b) circuit

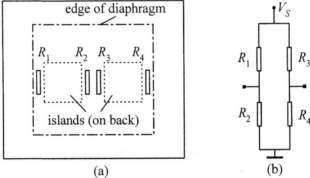

Fig. 6.6.12. Design of four parallel resistors on a twin-island structure (a) layout (b) circuit

Now let us look at the twin-island design shown in Fig. 6.6.12. We assume that the stresses on R_1 and R_4 are $+T$ and the stresses on R_2 and R_3 are $-T$; both are transverse stresses on the resistors. According to Fig. 6.6.10, NL_2 and NL_3 are negative (as shown by the bold curve for transverse stress on the left-hand side) and NL_1 and NL_4 are positive (as shown by the bold curve on the right-hand side). As the nonlinearity of the pressure transducer is the average value of the four *NLs*, we have

$$NL = \frac{1}{4}\left(|NL_1| + |NL_4| - |NL_2| - |NL_3|\right) \tag{6.6.36}$$

This implies that the nonlinearity of the four resistors are canceled out with each other to some degree. Therefore, a pressure transducer using only transverse piezoresistance has the advantage of lower overall nonlinearity. It has been found that the nonlinearity of the pressure transducer using only transverse piezoresistive effect is indeed smaller than those using both transverse and longitudinal piezoresistance effects [10].

§6.6.4. Calibration of Piezoresistive Transducers

For accurate measurement of pressure, the calibration of pressure transducer is extremely important. In fact, the accuracy of a measurement is limited by the calibration accuracy of the transducer. Therefore, the pressure transducer has to be calibrated either by the manufacturer or by the user before it can be used for pressure measurement. As the frequency bandwidth of pressure transducers is usually much higher than the frequency of pressure signals in practical applications, it is usually enough to use static pressure signals for the calibration of pressure transducer.

As the calibration of a pressure transducer needs repeated measurements and is very time-consuming, the calibration cost represents a major part of the total cost of a pressure transducer. A typical calibration procedure for a pressure transducer will be described with reference to Figs. 6.6.13 and 6.6.14, and Tables 6.6.1 and 6.6.2.

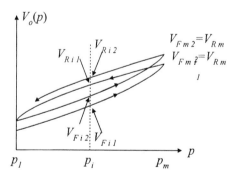

Fig. 6.6.13. Test cycles for the calibration of pressure transducers

The calibration of pressure transducer requires repeated measurements of a few selected pressures (P_i, $i=1,2,,...m$, $m>5$) using the pressure transducer. Each pressure is called a test point. The test points should be roughly uniformly distributed over the full pressure operation range of the pressure transducer, usually including one test point at the lower limit of the operation range and another at the upper limit of operation.

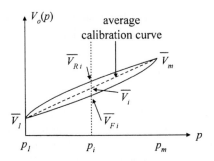

Fig. 6.6.14. Definition of some interim results

The measurements must be done in many cycles (j=1,2,,...n, n>5). Each cycle consists of a forward excursion (k=F, for p_1, p_2,...p_i, ...p_m) and a reverse excursion (k=R, for p_m, p_{m-1},...p_i, ...p_l). The output voltage of the pressure transducer for excursion k (k=F or R), testing point i, and cycle number j, is designated as $V_{k,i,j}$.

Table 6.6.1. Example of test data sheet (nine test points)

$V_{k,i,j}$	Forward excursion Test points i (i=1,2, ... ,8)			End point i=m=9	Reverse excursion Test points i (i=8,7, ... ,1)			
Test cycle j	p_{F1} ···	p_{Fi} ···	p_{F8}	$p_{F9}=p_{R9}$	p_{R8} ···p_{Ri}··· p_{R2}			p_{R1}
j=1	V_{F11}	V_{Fi1}	V_{F81}	$V_{F91}=V_{R91}$	V_{R81}	V_{Ri1}	V_{R21}	$V_{R11}=V_{F12}$
j=2	V_{F12}	V_{Fi2}	V_{F82}	$V_{F92}=V_{R92}$	V_{R82}	V_{Ri2}	V_{R22}	$V_{R12}=V_{F13}$
j=3	V_{F13}	V_{Fi3}	V_{F83}	$V_{F93}=V_{R93}$	V_{R83}	V_{Ri3}	V_{R23}	$V_{R13}=V_{F14}$
...	...			...	...			
j=n	V_{F1n}	V_{Fin}	V_{F8n}	$V_{F9n}=V_{R9n}$	V_{R8n}	V_{Rin}	V_{R2n}	V_{R1n}
Average $\overline{V}_{k,i}$	$\overline{V}_{F,1}$	$\overline{V}_{F,i}$	$\overline{V}_{F,8}$	$\overline{V}_{F,9}=\overline{V}_{R,9}$	$\overline{V}_{R8}$	$\overline{V}_{R,i}$	$\overline{V}_{R,2}$	$\overline{V}_{R,1}$
Standard deviation $s_{k,i}$	$s_{F,1}$	$s_{F,i}$	$s_{F,8}$	$s_{F,9}=s_{R,9}$	$s_{R,8}$	$s_{R,i}$	$s_{R,2}$	$s_{R,1}$

Table 6.6.2. Interim data sheet

Average $\overline{V}_{k,i}$	$\overline{V}_{F,1}$	$\overline{V}_{F,i}$	$\overline{V}_{F,8}$	$\overline{V}_{F,9}=\overline{V}_{R,9}$	$\overline{V}_{R8}$	$\overline{V}_{R,i}$	$\overline{V}_{R,2}$		$\overline{V}_{R,1}$
Standard deviation $s_{k,i}$	$s_{F,1}$ ...	$s_{F,i}$ ...	$s_{F,8}$	$s_{F,9}=s_{R,9}$	$s_{R,8}$	$s_{R,i}$	$s_{R,2}$	...	$s_{R,1}$

The test results for a specific cycle, j, are filled into a row in Table 6.6.1. The test sequence for the first and second cycles is schematically shown in Fig. 6.6.13 (note: the differences of data for the same test pressures have been exaggerated). According to Table 6.6.1, the number of measurements for the calibration is 2(m-1)n+1. If m=9 and n=7, the total test number is 113.

Once the measurement is completed and the data are listed in a data sheet as shown in Table 6.6.1, the experimental data are further processed as follows:

(1) Finding the Average for Each Testing Point
For a test point (k,i), the average output of the pressure transducer is

$$\overline{V}_{ki} = \frac{1}{n}\sum_{j=1}^{n} V_{kij} \quad (k=F,R; i=1,2,\cdots m)$$

(6.6.37)

The standard deviation of measurement for the test point is

$$s_{ki} = \sqrt{\frac{1}{n-1}\sum_{j=1}^{n}\left(V_{kij}-\overline{V}_{ki}\right)^2} \quad (k=F,R; i=1,2,\cdots m)$$

(6.6.38)

The results of Eqs. (6.6.37) and (6.6.38) are filled into an interim data sheet as shown by Table 6.6.2 and are schematically shown by the solid curves in Fig. 6.6.14.

(2) Finding the Average for Each Test Pressure
The averages for each test pressure are found by

$$\overline{V}_i = \frac{1}{2}\left(\overline{V}_{Fi}+\overline{V}_{Ri}\right)$$

(6.6.39)

The curve representing the $\overline{V}_i \sim p_i$ relation is referred to as the calibration curve of the pressure transducer, as shown by the dotted line in Fig. 6.6.14.

(3) Finding the Overall Standard Deviation of the Measurement
The overall standard deviation, s, is calculated by the definition

$$s = \sqrt{\frac{1}{2(m-1)}\sum_{i=1}^{m}\left(s_{Fi}^2+s_{Ri}^2\right)}$$

(6.40)

Based on the results given in the above equations, some important parameters are found as follows:

(1) Full Scale Output (*FSO*)

$$FSO = V_{FS} = \overline{V}_m - \overline{V}_1$$

(6.6.41)

(2) Sensitivity (*S*)

$$S = \frac{V_{FS}}{p_m - p_1}$$

(6.6.42)

As the sensitivity of a piezoresistive pressure transducer is proportional to the supply voltage of the Wheatstone bridge, the sensitivity of a pressure transducer is sometimes defined as

$$S = \frac{V_{FS}}{(p_m - p_1) \cdot V_S}$$

(6.6.43)

where V_S is the supply voltage of the pressure transducer.

(3) Hysteresis Uncertainty (*H*)

For each test pressure, the hysteresis uncertainty is defined as

$$H_i = \frac{\left| \overline{V}_{Ri} - \overline{V}_{Fi} \right|}{V_{FS}} \times 100\%$$

(6.6.44)

The hysteresis uncertainty for a transducer is the maximum of H_is of the transducer.

(4) Repeatability (*R*)

The repeatability uncertainty of the pressure transducers is defined as

$$R = \frac{2s}{V_{FS}} \times 100\%$$

(6.6.45)

for 95% confidence. Or, it is defined as

$$R = \frac{3s}{V_{FS}} \times 100\%$$

(6.6.46)

for 99.73% confidence.

(5) Nonlinearity (*NL*)

If the end-point straight-line scheme is used, the points on the straight-line are

$$V_{io} = \overline{V}_1 + \frac{\overline{V}_m - \overline{V}_1}{p_m - p_1} (p_i - p_1)$$

(6.6.47)

The nonlinearity for each test pressure is

$$NL_i = \frac{\overline{V}_i - V_{io}}{V_{FS}} \times 100\%$$

(6.6.48)

And, the nonlinearity for the pressure transducer is

$$NL = \left(NL_i \right)_{max}$$

(6.6.49)

(6) Nonlinearity and Hysteresis (*NLH*)

The maximum deviation of $\overline{V}_{ki}$ from the corresponding point in the straight line, V_{io}, represents the uncertainty caused by nonlinearity and hysteresis and is referred to as the nonlinearity and hysteresis uncertainty

$$NLH = \frac{\left|\overline{V}_{ki} - V_{io}\right|_{\max}}{V_{FS}} \times 100\%$$ (6.6.50)

(7) Combined Uncertainty (or, Accuracy, δ)

There are two commonly used definitions for the combined uncertainty

(a) Definition I

$$\delta = \pm\sqrt{(NL)^2 + H^2 + R^2}$$ (6.6.51)

(b) Definition II

$$\delta = \pm\left(\left|LH\right| + \left|R\right|\right)$$ (6.6.52)

Obviously, the uncertainty value by definition II is larger than that by definition I.

§6.7. Problems

1. For a silicon beam cutting from a (001) wafer, the length direction of the beam (i.e., the x'-direction as shown in the figure below) makes an angle α with the [100] direction. Find (1) the expressions for Young's modulus $E_{x'x'}$ in the length direction and the Poisson ratio $\nu_{y'x'}$ in the width direction; (2) the values of $E_{x'x'}$ and $\nu_{y'x'}$ for $\alpha=0$ and $\alpha=45°$.

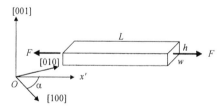

2. A diffused resistor on a (001) Si substrate is shown in the figure below, where α is the included angle between the resistor and the [100] direction. Find (1) the expressions for π_l, π_t, and π_s (in terms of π_{11}, π_{12}, π_{44} and α); and (2) the expressions for π_l, π_t, and π_s, if the resistor is p-type so that $\pi_{11}\approx\pi_{12}\approx0$.

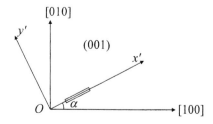

3. A diffused resistor on a (101) Si substrate is shown in the figure below, where α is the included angle between the resistor and the $[10\bar{1}]$ direction. (1) Find the expressions for π_l, π_t, and π_s (in terms of π_{11}, π_{12}, π_{44} and α). And (2) if the resistor is a p-type one (i.e., $\pi_{11} \approx \pi_{12} \approx 0$), verify that the direction for a maximum π_l is the $<111>$ direction.

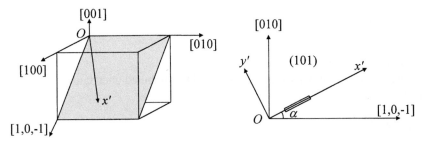

4. A diffused resistor on a (111) Si substrate is shown in the figure below, where α is the included angle between the resistor and the $[\bar{1}10]$ direction. Find the expressions for π_l, π_t, and π_s (in terms of π_{11}, π_{12}, π_{44} and α).

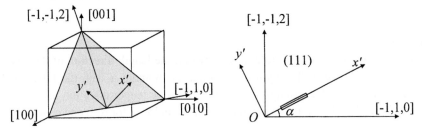

5. A square silicon diaphragm is formed by anisotropic etching as shown in the figure below. The area of the diaphragm is $2a \times 2a$ and the thickness is h. A p-resistor is formed by diffusion on the diaphragm very close to an edge center. If a pressure p is applied on the front surface, find the expression for the relative change of the resistance $\Delta R/R_o$.

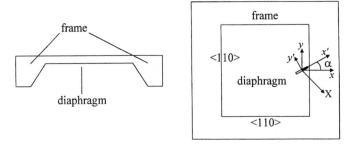

6. A square silicon diaphragm is formed by anisotropic etching as shown in the figure of problem 5. The area of the diaphragm is $2a \times 2a$ and the thickness of the diaphragm is h. A p-type four-terminal sensing element (with its length L much larger than its width w) is

formed by diffusion on the diaphragm very close to the center of an edge. If a pressure p is applied on the front surface, find the expression for the relative output voltage V_T/V_S

7. The calibration data of a pressure transducer is shown in the table below. Find the nonlinearity based on the end-point straight-line definition.

Pressure (kPa)	0	20	40	60	80	100	120
Output (mV)	21	42.92	64.68	86.28	107.72	129.00	150.12

8. A piezoresistive pressure transducer with resistors on a long rectangular diaphragm as shown in the figure below. The sensitivity of the four resistors are: $a_1=a_4=3\times10^{-4}$/kPa and $a_2=a_3=1.5\times10^{-4}$/kPa. Find the nonlinearity caused by the non-uniform sensitivity for a pressure range of $p_m=100$kPa.

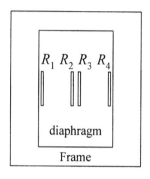 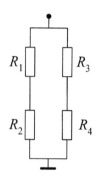

9. If the expression of $e^{\alpha T_1}$ is developed to the second power: $e^{\alpha T_1}=1+\alpha T_1+(\alpha T_1)^2/2$, find the nonlinearity of π_{11} and π_{12} for n-Si based on the end-point straight-line definition for the maximum stress $T_{1max}=10^8$Pa.

10. A parallel compensation scheme for TCS is shown in the figure below. If $R_B=2.0$ kΩ, TCR and TCS of R_B are $TCR_B=0.3\%/°C$ and $TC\pi=-0.2\%/°C$, respectively, and the TCR of R_p is $\alpha_p=TCR_p=-0.05\%/°C$, find the expected value of R_p.

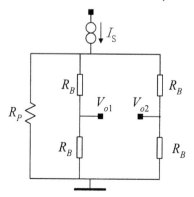

References

[1] C.S. Smith, Piezoresistance effect in germanium and silicon, Phys. Rev., Vol. 94 (1954) 42-49

[2] C.S. Smith, "Macroscopic Symmetry and Properties of Crystals", Solid State Physics, Advance in Research and Applications, Vol. 6 (1958) 175-249, Editors: F. Seitz, D. Turnbull.

[3] O.N. Tufte, E.L. Stelzer, Piezoresistance properties of heavily doped n-type silicon, Phys. Rev., Vol. 133 (1964) A1705-A1716

[4] O.N. Tufte, E.L. Stelzer, Piezoresistance properties of silicon diffused layers, Journal of Applied Physics Vol. 34 (1963) 313-318

[5] W. Pietrenko, Einfluss von temperatur und strörstellenkonzentration auf den piezowiderstandseffekt in n-silizium, Phys. Status Solidi — Section A Appl. Res., Vol. 41 (1977) 197-205

[6] B. kloeck, N. de Rooij, "Mechanical Sensors", Chap. 4 in book "Semiconductor Sensors ", Edited by S. M. Sze, John Wiley & Sons, Inc., 1994

[7] W.G. Pfann, R.N. Thurston, Semiconducting stress transducers utilizing the transverse and shear piezoresistance effects, J. of Appl. Phys. Vol. 32 (1961) 2008-2019

[8] M. Bao, Y. Huang, Batch derivation of poezoresistive coefficient tensor by matrix algebra, Journal Micromechanics and Microengineering, 14 (2004) 332-334

[9] H. Muro, H. Kaneko, S. Kiyota, P. J. French, Stress analysis of SiO_2/Si bi-metal effect in silicon piezoresistive accelerometers, Digest of Technical Papers, The 6th Intl. Conf. on Solid-State Sensors and Actuators, San Francisco, CA, USA June 24-27, 1991, (Transducers'91) 768-771

[10] R.M. Whittier, Basic advantages of anisotropic etched transverse gauge pressure transducer, Endevco Tech Paper, TP277

[11] Y. Kanda, A. Yasukawa, Hall-effect devices as strain and pressure sensors, Sensors and Actuators, 2 (1982) 283-296

[12] M. Bao, Y. Wang, Analysis and design of a four-terminal silicon pressure sensor at center of a diaphragm, Sensors and Actuators, 12 (1987) 49-56

[13] M. Bao, W. Qi, Y. Wang, Geometric design rules of four-terminal gauge for pressure sensors, Sensors and Actuators, 18 (1989) 149-156

[14] D.R. Kerr, A. G. Milnes, Piezoresistance of diffused layers in cubic semiconductors, Journal of Applied Physics, 34 (1963) 727-731

[15] G. Zhao, M. Bao, Statistical analysis for the sensitivity of polysilicon piezoresistance, Chinese Journal of Semiconductors, Vol. 10 (1989) 693-701

[16] N. Lu, L. Gerzberg, C. Lu, J. Meindl, Modeling and optimization of monolithic polycrystalline silicon resistors, IEEE Trans. on Electron Devices, Vol. ED-28 (1981) 818-830

[17] K. Matsuda, Y. Kanda, K. Yamamura, K. Suzuki, Nonlinearity of piezoresistance effect in p- and n-type silicon, Sensors and Actuators A 21-23 (1990) 45-48

Answers to the Problems

Problems in Chapter 2

1. $\Delta w = e_2 w = -0.214\,\mu m$; $\Delta h = e_3 h = -0.0428\,\mu m$.

2. $\Delta L = e_1 L = 2.95\,\mu m$; $\Delta w = e'_2 w = -0.039\,\mu m$; $\Delta h = e'_3 h = -0.0428\,\mu m$.

3. $I_{x'} = I_x + bhr^2$.

4. $z_0 = 1.38\,\mu m$.

5. $I = \dfrac{\pi}{4}a^4 = \dfrac{\pi}{64}d^4$.

6. $I_{x'} = \dfrac{1}{12}b_1 h_1{}^3 + \dfrac{1}{12}b_2 h_2{}^3 + \dfrac{b_1 h_1 b_2 h_2}{4(b_1 h_1 + b_2 h_2)}(h_1 + h_2)^2$.

7. $w_{max} = 0.62\,\mu m$.

8. $L_{max} = 2.51\,mm$.

9. $w(L/2) = \dfrac{5\rho g L^4}{32 E h^2}$.

10. $w(a_1) = 0.81\,\mu m$, $T(0) = 1.64 \times 10^7$ Pa.

11. $a_{max} = 182g$.

12. $w_2(a + \dfrac{1}{2}A) = \dfrac{1}{EBH^3}\left[6m_o(a + \dfrac{1}{2}A)^2 - \rho ABHg(a + \dfrac{1}{2}A)^3 + \dfrac{1}{2}\rho ABHg(\dfrac{a}{2})^4 - 12D_1(a + \dfrac{1}{2}A) - 12D_2\right]$

where $m_o = \dfrac{\rho ABHg(I_1 A^2 + 6I_1 aA + 6I_2 a^2)}{12(I_2 2a + I_1 A)}$, $D_1 = \left(\dfrac{I_2}{I_1} - 1\right)\left(\dfrac{1}{4}\rho ABHga^2 - m_{.o}a\right)$ and

$D_2 = \left(\dfrac{I_2}{I_1} - 1\right)\left(\dfrac{1}{2}m_{.o}a - \dfrac{1}{6}\rho ABHga^3\right)$.

13. $w(x) = c\left(\dfrac{1}{2}mx^2 - 2M''Lx + 6M'L^2\right)x^2$, where $c = \dfrac{g}{Ebh^3 L}$, $m = \rho bhL$, $M'' = M + m$

and $M' = M + \dfrac{1}{2}m$.

14. $f_1 = 11$ kHz.

15. $f_1 = 70.24$ kHz.

16. (1) Normal vibration, $f_N = 1569$Hz; (2) Lateral vibration, or $f_L = 2.5 f_N = 3.925$ kHz; (3) Angular vibration $f_\varphi = 210$ Hz.

17. (1) $w_{max,Sq} = 0.002\,\mu m$, $w_{max,T} = 0.00248\,\mu m$.
(2) $T_{max,Sq} = 6850$ Pa, $T_{max,T} = 10540$ Pa; (3) $f_{Sq} = 13.85$kHz, $f_T = 12.5$kHz.

18. $f_1 = 4.72$kHz.

19. Eq. (2.3.22).

20. Eq. (2.4.3).

21. Eq. (2.2.36).

22. Eq. (2.2.44).

23. $T = 2.897 \times 10^5$ Pa.

24. $T_b = -1.4 \times 10^7$ Pa.

25. Buckling strain: $\varepsilon_b = 8.22 \times 10^{-5}$; Thermal strain: $\varepsilon_T = 3.15 \times 10^{-3}$. The beam buckles, because ε_T is larger than ε_b.

26. (1) f_d=1538Hz; (2) $A(0)$=0.092μm and $A(\omega_{res})$=0.235μm.

28. Q=1432.

29. Q=6, ζ=0.0836.

30. Stress at edge $T = 5 \times 10^8$ (Pa), rupture pressure $P_R = 6$ atm.

31. Eq. (2.6.16)

32. The stress components at edge center: $T_{xx}(1,0) = 2.55 \times 10^8$ Pa, $T_{yy}(1,0) = \nu T_{xx}(1,0)$ $= 7.65 \times 10^7$ Pa, and $T_{xy}(1,0) = 0$ Pa. The stress components at the point 0.1mm away from the edge center: $T_{xx}(0.9,0)$=1.81×10^8 Pa, $T_{yy}(0.9,0)$=0.501×10^8 Pa and $T_{xy}(0.9,0)$=0 Pa.

33. $f_{00} = 72.3$ kHz.

34. Frequency in air $f_{air} = 83.1$ kHz, frequency in water $f_{water} = 21.4$ kHz.

Problems in Chapter 3

1. $\mu = 8.36 \times 10^{-6}$ Pa·sec.

2. $Q_{water} = 0.0986$ μl/sec and $Q_{air} = 5.55$ μl/sec.

3. Air $R_e = \dfrac{\bar{v}\rho d}{\mu} = 6.3$, water $R_e = \dfrac{\bar{v}\rho d}{\mu} = 1.54$.

4. $v = 1.2$ cm/sec.

5. (1) c=0.0489kg/s; (2) ζ=0.314 and Q=1.68; (3) f_{res}=1.989kHz.

6 (1) $\zeta = 0.493 \dfrac{\mu b^2 L^2}{h^2 d^3 \sqrt{\rho E}}$; (2) ζ=0.0112 and Q=44.6.

7. (1) $\zeta = 3.94 \dfrac{\mu L^2}{bh^2 \sqrt{\rho E}}$; (2) Q=698.

8. d=19.4μm.

9. (1) h=4.48μm; (2) d=19.9μm.

10. c_2:c_1=2.78:1.

11. c_2:c_1=1.014:1.

12. ζ=0.81.

13. $\zeta = 0.42$.

14. Q=703.
15. Q=2012.

Problems in Chapter 4

1. (1) $F(y) = -\dfrac{A\varepsilon\varepsilon_o}{2d^2}V^2$; (2) $F(y) = -7.083 \times 10^{-5}$N.

2. (1) $F(y_1) = \dfrac{nH\varepsilon\varepsilon_o}{d}V^2 \dfrac{y_o^2}{y_1^2}$; (2) 4.722×10^{-9}N.

3. (1) $d_{balance} = V \cdot \sqrt{\dfrac{\varepsilon\varepsilon_o}{8H\rho g}}$; (2) 40.2μm; (3) The balance is not stable, as $\dfrac{\partial F}{\partial d} = \dfrac{A\varepsilon\varepsilon_o}{4d^3} > 0$.

4. The balance displacement is 1.77μm.

5. (1) The equation for displacement Δy is $(1 + \dfrac{\Delta y}{y_o})^2 \dfrac{\Delta y}{y_o} = \dfrac{nH\varepsilon\varepsilon_o}{kd_oy_o}V^2$; (2) $\Delta y = 11$μm.

6 (1) $y_{max} = d/3 = 1.333$μm; (2) $V_{po} = 12.1$ V.

7. $V_{po} = 32.6$V.

8. (1) $y_{max} = 1.421$μm; (2) $V_{po} = 13.3$V; (3) $V_H = 2.06$V.

9. (1) $y_{max} = 1.472$μm; (2) $V_{po} = 9.9$ V; (3) $V_H = 6.82$V.

10. (1) $y_{max} = 1.21$μm; (2) $V_{po} = 7.33$ V; (3) $V_H = 2.96$V.

11. (1) $\varphi_{max} = 0.158°$; (2) $\varphi = 0.0645°$.

12. $V_C = 2.83$ V.

13. $V_C = 3.23$ V.

14. $f_0' = 1526$ Hz.

15. $f_0' = 1516$ Hz.

16. $f_0' = 1314$ Hz.

17. $\omega' = \omega_o \sqrt{1 - \dfrac{C_o}{kd^2}V^2}$, where $\omega_o = \sqrt{\dfrac{k}{M}}$ and $C_o = \dfrac{2A\varepsilon\varepsilon_o}{d}$.

18. $f_0' = 11.24$ kHz.

19. $f_0' = 11.78$ kHz.

Problems in Chapter 5

1. If the electrostatic effect of the bias voltage is neglected, the results are: (1) $S_{open} = 23.7$mV/Pa; (2) $S_F = 14$mV/Pa. If the electrostatic effect of the bias voltage is considered, the results are: (1) $S_{open} = 26.6$mV/Pa; (2) $S_F = 16$mV/Pa.

2. (1) $V_o = 32.8$mV/Pa; (2) $V_{out} = 21.5$mV.

3. $V_1 = 13.4$mV·sinωt.

4. $\omega = \omega_o \sqrt{1 + 4 p_o (\alpha\beta - 1 - \frac{1}{2}\alpha^2)}$, where $p_o = \dfrac{A\varepsilon\varepsilon_o V_o^2}{2kd^3}$.

5. (1) $\Delta f = -1.6$ kHz; (2) $\Delta f = -1.113$ kHz.

6. $V_{eff} = \sqrt{V_o^2 + V_1^2}$.

7. (1) $a_c = 0.67$g; (2) $a_c = 0.45$g.

8. $\Delta t = 0.424 \mu$s.

9. (1) $a_c = 207$g; (2) $a_c = 103.5$g; and (3) $\Delta t = 5.1 \mu$s.

Problems in Chapter 6

1. (1) $E_{x'x'} = \dfrac{1}{\Sigma'_{11}} = \dfrac{10^{11} Pa}{0.764 - 0.175 \sin^2 2\alpha}$ and $\nu_{y'x'} = -\dfrac{\Sigma'_{12}}{\Sigma'_{11}} = \dfrac{0.214 - 0.175 \sin^2 2\alpha}{0.764 - 0.175 \sin^2 2\alpha}$.

(2) For $\alpha = 0$, $E_{x'x'} = 1.31 \times 10^{11}$ Pa, $\nu_{y'x'} = 0.28$; For $\alpha = \pi/4$, $E_{x'x'} = 1.7 \times 10^{11}$ Pa, $\nu_{y'x'} = 0.066$.

2. (1) $\pi_l = \pi_{11} - \frac{1}{2}\pi_o \sin^2 2\alpha$; $\pi_t = \pi_{12} + \frac{1}{2}\pi_o \sin^2 2\alpha$; $\pi_s = -\frac{1}{2}\pi_o \sin 4\alpha$, where $\pi_o = \pi_{11} - \pi_{12} - \pi_{44}$.

(2) For p-Si, we have $\pi_l = \frac{1}{2}\pi_{44} \sin^2 2\alpha$, $\pi_t = -\frac{1}{2}\pi_{44} \sin^2 2\alpha$, $\pi_s = \frac{1}{2}\pi_{44} \sin 4\alpha$.

3. (1) $\pi_l = \pi_{11} - 2\pi_o(\frac{1}{4}\cos^4 \alpha + \cos^2 \alpha \sin^2 \alpha)$; $\pi_t = \pi_{12} + 3\pi_o \cos^2 \alpha \sin^2 \alpha$;

$\pi_s = 2\pi_o \cos\alpha \sin\alpha (\sin^2 \alpha - \frac{1}{2}\cos^2 \alpha)$, where $\pi_o = (\pi_{11} - \pi_{12} - \pi_{44})$.

(2) For p-Si, by $\dfrac{\partial \pi_l}{\partial \alpha} = 0$, we find $\alpha = 35.26^o$. That is the [110] direction.

4. $\pi_l = \pi_{11} - \frac{1}{2}\pi_o$; $\pi_t = \pi_{12} + \frac{1}{6}\pi_o$; $\pi_s = 0$, where $\pi_o = (\pi_{11} - \pi_{12} - \pi_{44})$.

5. $\dfrac{\Delta R}{R} = \frac{1}{2}(1 - \nu)\pi_{44} T_{xx} \cos 2\alpha$, where $T_{xx} = 1.23\dfrac{a^2}{h^2} p$

6. $\dfrac{\Delta V}{V_S} = -\dfrac{w}{L}\dfrac{\pi_{44}}{2}(1 - \nu)T_{xx} \sin 2\alpha$, where $T_{xx} = 1.23\dfrac{a^2}{h^2} p$.

7. $NL = 0.56\%$.

8. $NL = -0.19\%$.

9. For π_{11}, $NL_{11} = 4.3\%$; For π_{12}, $NL_{12} = 0.48\%$.

10. $R_p = 5$kΩ.

Subject Index

Printed and bound by CPI Group (UK) Ltd, Croydon, CR0 4YY

08/05/2025

01864932-0002